THE WORLD OF
Magnolias

THE WORLD OF
Magnolias

DOROTHY J. CALLAWAY

TIMBER PRESS
Portland • London

Copyright © 1994 by Timber Press, Inc.
All rights reserved.

ISBN 978-1-60469-226-6
Printed in the United States of America

TIMBER PRESS, INC.
The Haseltine Building   2 The Quadrant
133 S.W. Second Avenue, Suite 450   135 Salusbury Road
Portland, Oregon 97204-3527   London NW6 6RJ
www.timberpress.com   www.timberpress.co.uk

The Library of Congress has previously cataloged a hardback edition as follows:

Callaway, Dorothy J. (Dorothy Johnson)
    The world of magnolias / Dorothy J. Callaway.
        p.   cm.
    Includes bibliographical references (p.   ) and index.
    ISBN 0-88192-236-6
    1. Magnolia. I. Title.
SB413.M34C34   1994
635.9'33114--dc20
                                                93-2793
                                                CIP

CONTENTS

Color plates follow page 112.

To past and present members of
the Magnolia Society,
all of whom are a part of this book

PREFACE

I consider myself lucky to have grown up in the southeastern U.S., in Georgia, where *Magnolia grandiflora* is used extensively in the landscape. These fifty-foot-tall specimens with their white flowers and wonderful lemony fragrance were the first (and for a long time, the only) magnolias I knew. As I became involved in the Magnolia Society, I discovered the vast array of magnolia hybrids and cultivars available. Many people know only *M. grandiflora* or *M.* × *soulangiana*. My hope is that this book changes that.

Much of my effort in this book involved bringing together scattered information on many topics. The information comes from scientific literature, horticultural publications, personal observations, and conversations with magnolia nursery professionals, breeders, and gardeners. This book is not meant to replace previously published magnolia books, but to complement them with a broad biological, as well as horticultural, discussion of the genus. This is not a taxonomic monograph.

I have not seen all the magnolias discussed in this book. Consequently, I have had to rely on written accounts in the literature and on personal communication for information on some of the plants. In some cases, the available information is voluminous, particularly for the commonly cultivated forms. Yet for others, precious little information is available. Where this is the case, I provide the reader with what scant information is available. This accounts for the brief or incomplete description of some plants (and is one reason I actively encourage registration of hybrids and cultivars). Similarly, I know that as I write this, new cultivars and hybrids are being named which will not be included in this book. This is unavoidable considering the current rate at which selections are being made. I have made every effort to include all that are known to me at the time of this writing. It is not my intention to emphasize North American selections over those offered elsewhere.

This book has been over four years in the making, and during that time I have become indebted to many people for their help. The first of these is my husband, Brett Callaway, whose encouragement and support throughout this process have been unending. My parents, Sydney and Nedra Johnson, deserve similar thanks for their encouragement along the way and for their role in stimulating my interest in plants from an early age.

Members of the Magnolia Society, to whom this book is dedicated, have helped in more ways than I can enumerate. Everyone I have dealt with in the Society has been enthusiastic in their encouragement during the writing of this book, and I extend my gratitude to them all. Certain members who have been especially helpful include Phelan Bright, Bill Dodd, Ted Dudley, Dick Figlar, Frank Galyon, Marj and Roger Gossler, Joe Hickman, Harold Hopkins, August Kehr, Larry Langford, and Phil Savage.

The following people agreed to review parts of the book, and their comments are appreciated: Bob Adams, Julian Beckwith, Brett Callaway, Frank Galyon, Joe Hickman, Sydney Johnson, August Kehr, and Phil Savage. Line drawings in chapter 2 depicting magnolia flowers and fruit were beautifully executed by Bente King and Michael Linkinhoker. I am very grateful for their patience and talent. Slides or photographs were graciously provided by Peter Del Tredici, David Ellis, Dick Figlar, Holly Forbes, Frank Galyon, Martin Grantham, Joe Hickman, Abbie Jury, August Kehr, Lola Koerting, Larry Langford, Jackie Mullen, Barbara Pitschell, and John Tobe. Photos have also been taken from the slide library of the Magnolia Society. Thanks are due to David Dilcher for permission to reprint line drawings of *Archaeanthus* made by artist Megan Rohn.

I would also like to thank the staff of the L. H. Bailey Hortorium at Cornell University for their support, financial and otherwise, during my master's work on magnolias. Bob Dirig and John Freudenstein were also helpful with library work for this book even after I left Cornell.

Finally, several people at Timber Press deserve a great deal of thanks. Richard Abel, later joined by Bob Conklin, guided me through the book-writing process and offered support along the way. Their comments and advice have been invaluable. Micheline Ronningen had the unenviable task of serving as my editor, and I am grateful for her professional advice as well as her patience and support. The entire Timber Press staff has truly been a pleasure to work with.

Dorothy Johnson Callaway

1

INTRODUCTION

"No group of trees and shrubs is more favorably known or more highly appreciated in gardens than magnolias, and no group produces larger or more abundant blossoms." So stated the great plant explorer E. H. Wilson. The vast number of magnolia species and cultivars now available make Wilson's statement even truer today as we near the 21st century. Magnolias have much to offer the gardener and entertain a long and romantic history of introduction and cultivation.

Buddhist monks in China planted *Magnolia denudata* at their temples as early as A.D. 650, the pure white flowers having symbolized purity. *Magnolia denudata* and *M. liliiflora* have been favorite subjects of Chinese art for centuries. Magnolias were first introduced to European gardens in 1688 when John Bannister sent *M. virginiana* to Bishop Henry Compton of London. Introduction of other species, Asian and American, soon followed. The introduction of new magnolias continues today, with over thirty species and numerous cultivars and hybrids in cultivation. Most of the cultivated forms have large flowers borne before the leaves in the spring, producing a dazzling display. Some species are evergreen and produce large white flowers set off by glossy green leaves. The sweet fragrance of species such as *Magnolia grandiflora* perfumes the garden. Magnolia flowers vary in color from purple to pink, yellow, and white and are produced on plants ranging in size from 8-foot (2.4 m) shrubs to 100-foot (30 m) trees. With careful selection of plants, magnolia flowers can grace a garden for nine months of the year. Afterwards, showy red or orange fruits add beauty in the fall, and the smooth gray or scaly brown bark provides winter interest. Many of the shrubby deciduous forms, such as *M. kobus* var. *stellata*, have an artistic shape and architecture that is beautiful even in winter months. The leaves of many magnolias are unusually ornamental. The evergreen *M. grandiflora* has shiny, dark green leaves with brown felted undersurfaces. *Magnolia virginiana* and *M. macrophylla* have silvery white undersurfaces

11

that flash when the wind disturbs the leaves. Magnolias have a well-deserved reputation as "aristocrats" among landscape plants, and their importance in the landscape will be the focus of this book. But magnolias have other uses which are not so widely known, uses which add to the importance and prestige of this group of plants. Many species attract wildlife, especially songbirds, to the garden. Magnolias also offer cut flowers and, in the cases of evergreen species, year-round cut greenery for use in floral displays or decoration; smaller species and cultivars make suitable houseplants. And for centuries magnolias have been used for timber and medicinal purposes.

LANDSCAPING FOR WILDLIFE

The seeds of magnolias are relished by songbirds, which are attracted by the brightly colored seed coat. Having an unusually high fat content, about 40% (Bonner 1974), they provide a source of concentrated energy for migrating songbirds. The southern magnolia (*Magnolia grandiflora*) is especially attractive to birds because of its dense growth form, evergreen foliage, and heavy seed production. Among the birds attracted to its brightly colored seeds are the red-eyed vireo, towhee, summer tanager, kingbird, woodthrush, redstart, bluejay, cardinal, mockingbird, woodpeckers, wild turkey, and bobwhite quail. The seeds are also a favorite food of gray squirrels and certain rodents such as the white-footed mouse. Some magnolia propagators report that birds and mammals may prefer some species over others. Grackles, for example, seem to select *Magnolia acuminata* seeds over other equally available species, while rodents may prefer *M. fraseri* over others. The seeds are occasionally eaten by larger animals such as raccoons and opossums, as well as deer, which also eat the foliage of some species and are often considered garden pests for this reason.

The brightly colored seeds of magnolias, such as this *Magnolia macrophylla*, serve as food for many birds and mammals. Photo from the collections of the Magnolia Society.

CUT FLOWERS AND GREENERY

The large, fragrant flowers of magnolias can be cut and enjoyed in the home as well as in the garden. Cut at an early stage (just before opening) and placed in a cool room, a flower will last for several days. One flower from *Magnolia grandiflora* will easily fill a room with its fragrance. Treseder (1978) suggests that flowers may be forced to open in late winter or early spring by placing the plant or cut stems at 75–80°F (24–27°C) until the flowers open. The fragility, seasonality, and relatively short flower life of magnolias make them unsuitable as a large-scale florist commodity. Their localized use in decoration, however, is more common. *Magnolia grandiflora* is a common decoration in the southern United States, and *M. denudata* is especially prized for its decorative uses in China. In Mexico the flowers of *M. macrophylla* var. *dealbata* are cut from wild plants and used to decorate the local churches during the Easter season, with the blooms sold individually in the marketplace.

Although the flowers of magnolias are not economically important in the commercial cut flower market, the leaves and boughs of *Magnolia grandiflora* have become marketable for use as greenery in cut arrangements and decorations. The dark, glossy green leaves with their brown felted undersides serve as a beautiful backdrop in a floral arrangement. The leaves are especially popular for December holidays, the dark green accenting accompanying holly berries, pine cones, red carnations, and the like. The leaves and boughs are used for decorating mantelpieces or centerpieces, or for making wreaths. They are used year-round as greenery for decoration for weddings. The use of *M. grandiflora* leaves in decorating is not restricted to its native growing region in the southeastern United States; the leaves are shipped to northern markets as well.

Leaves of *Magnolia grandiflora* are often used in holiday decorations. Photo by the author.

HOUSEPLANTS

Several species of *Magnolia* make desirable houseplants. Probably the most suitable species is the slow-growing *M. coco*. It grows 7–13 ft. (2–4 m) and has small, fragrant, white flowers. Its long flowering period provides indoor fragrance and color about nine months of the year. Dwarf evergreen forms of *M. virginiana* can also make desirable indoor potted plants, as do other evergreen *Magnolia*. The Japanese have also grown the deciduous *M. kobus* var. *stellata* in containers for centuries. Desirable traits for indoor plants include slow growth rate, fragrant flowers, profuse blooming, attractive evergreen foliage, and ease of maintenance. Figlar (1986) provides an account of his experience in growing several magnolias as houseplants.

Small forms of some *Magnolia* species make attractive houseplants. Here a dwarf form of *Magnolia virginiana* blooms indoors. Photo by Richard Figlar, Pomona, New York.

TIMBER

Magnolia species, particularly *M. grandiflora*, *M. virginiana*, and *M. acuminata* in North America, and *M. hypoleuca* in Asia, were once more commercially important as timber trees than they are today. Due to the use of other tree species for stronger, more durable wood, and the relative scarcity of large stands of *Magnolia*, magnolias as timber trees have not been of major economic importance for some time. Less than 2% of the hardwood timber harvested in the southeastern United States is from *Magnolia*. *Magnolia grandiflora* and *M. virginiana* are important in Florida, however, where they constitute 10% of hardwood timber harvested. Lumber from magnolias is simply sold as "magnolia," regardless of species, and is sometimes combined in the market with the related tulip poplar (*Liriodendron tulipifera*).

Magnolia wood is moderately light in weight and color. Sapwood is usually of a white or pale yellow color, while the heartwood may have quite unique coloration, varying from green to brown, black, and even dark blue. Magnolia lumber is not very resistant to attack by fungi and insects and is not very strong or durable. It has a fine, straight grain; this, combined with its moderate weight, makes it easily worked. Magnolia lumber was previously used in the production of wooden venetian blinds and is now mostly used for furniture, cabinet work, and some interior finishing.

Perhaps the most widely used *Magnolia* wood has been that of *M. hypoleuca* in Japan. It is used for furniture, utensils, toys, scabbards, and lacquer ware. The wood is also burned to produce a charcoal used to polish metal and lacquer ware. The extensive harvest of this species for wood has led to its scarcity in the wild. Trees are cut as they reach 20–30 ft. (6–9 m) in height; because they do not always reach reproductive maturity before this time, seedlings may not be left to replace the harvested trees.

MEDICINALS

Some of the earliest references to magnolias in the literature involve their purported medicinal properties. These references extend back as far as 1083 B.C. in China, and appear in North American literature in the late 1700s. Early reports from China primarily consider the use of *Magnolia officinalis* in traditional medicine. The bark of *M. officinalis* is used to produce a tonic known as *Hou-phu*. This tonic is generally used to treat neurosis and gastrointestinal disorders. Over the years the extensive cutting of the trees to obtain the high-priced drug has left wild populations sparse.

Watanabe and colleagues undertook a study of the medical properties of *Magnolia officinalis* in 1983. From the bark were extracted two compounds known as *magnolol* and *honokiol*. Magnolol is also found in the bark of *M. henryi*. These compounds were shown to produce sedation and muscle relaxation and act as central depressants. These effects were more gradual, but of longer duration, than the modern drugs with which they were compared. An additional study suggested that tonics of *M. officinalis* bark lessen rigidity and tremor in patients with Parkinson's disease.

The flower buds of *Magnolia officinalis* are reported to be used in a tonic for feminine ailments. While the validity of this use has not been tested, the muscle-relaxing properties of the bark, if also produced in the flower bud, might provide relief from menstrual cramps.

Hsin-i (also called *Shin-i* or *Hsenyi*) is also a widely used drug in many Asian countries. It is made from the flower buds of various species of magnolias, including *Magnolia salicifolia*, *M. liliiflora*, *M. fargesii*, and *M. biondii*. Hsin-i is used to treat headaches, nasal disorders, fever, and allergies. Several studies have been undertaken to determine whether or not medicinal properties have a scientific basis and, if so, what compound is involved. Kimura and colleagues (1983, 1985) extracted alkaloids and neolignans from dried buds of *M. salicifolia* and found that these compounds acted as anti-inflammatory agents, neuromuscular blocking agents, anti-allergy agents, and central nervous system inhibitors. Their 1985 study reports on the potential use of Hsin-i as an anti-cancer drug.

Studies involving *Magnolia fargesii* as a source of Hsin-i show flower bud extracts to be hypotensive, antifungal, and muscle-contracting in nature. Chen and colleagues (1988) report that Hsin-i from *M. fargesii* may serve as an effective treatment for heart disease and hypertension.

Pan and colleagues (1987) isolated several lignans from flower buds of *Magnolia biondii* and provided evidence that Hsin-i may reduce the symptoms involved in various inflammatory, respiratory, and cardiovascular disorders.

American species of Magnolia have not been as widely used for medicinal purposes as the Asian species. Magnolias first appeared in American pharmacopoeias in 1787. The dried bark of *Magnolia virginiana*, *M. acuminata*, and *M. tripetala* was reportedly used as a tonic, astringent, and stomachic, as well as a remedy for respiratory ailments, malaria, rheumatism, and gout (Culbreth 1927). The validity of this use was examined in the late 1800s (Lloyd and Lloyd 1884). Since then more readily available drugs have been found to have the same curative effects as magnolia bark, and instances of magnolia bark use have declined. Species of *Magnolia* have not appeared in North American pharmacopoeias since 1900.

REFERENCES AND ADDITIONAL READING

Bonner, F. T. 1974. Chemical components of some southern fruits and seeds. *U.S. Forest Service Research Note* SO-183.

Chen, C. C., Y. L. Huang, H. T. Chen, Y. P. Chen, and H. Y. Hsu. 1988. On the Ca++ antagonistic principles of the flower buds of *Magnolia fargesii*. *Planta Medica* 54(5):438–440.

Culbreth, David M. R. 1927. *Materia Medica and Pharmacology*, 7th ed. Philadelphia, PA: Lea and Febiger.

Figlar, R. B. 1986. Magnolias as house plants. *Journal of the Magnolia Society* 21(2):1–5.

Howard, Richard A., ed. 1980. Wilson's Magnolias. *Journal of the Magnolia Society* 16(2):3–26.

Kimura, Ikuko, Masayasu Kimura, Masao Yoshizaki, Kazuo Yanada, Shigetoshi Kadota, and Tohru Kikuchi. 1983. Neuromuscular blocking action of alkaloids from a Japanese crude drug "Shin-i" (Flos Magnoliae) in frog skeletal muscle. *Planta Medica* 48:43–47.

Kimura, Masayasu, Jun Suzuki, Tadashi Yamada, Masao Yoshizaki, Tohru Kikuchi, Shigetoshi Kadota, and Satoko Matsuda. 1985. Anti-inflammatory effect of neolignans newly isolated from the crude drug "Shin-i" (Flos Magnoliae). *Planta Medica* 51:291–293.

Lloyd, J. U. and C. G. Lloyd. 1884. *Magnolia*. In *Drugs and Medicines of North America*, vol. 2, Cincinnati, OH: Clarke Publishing. Pp. 21–46.

Pan, Jing-Xing, Otto D. Hensens, Deborah L. Zink, Michael N. Chang, and San-Bao Hwang. 1987. Lignans with platelet activating factor antagonist activity from *Magnolia biondii*. *Phytochemistry* 26(5):1377–1379.

Treseder, Neil G. 1978. *Magnolias*. London: Faber and Faber.

Watanabe, Kazuo, Hiroshi Watanabe, Yoshiake Goto, Masatoshi Yamaguchi, Nobuyuki Yamamoto, and Koji Hagino. 1983. Pharmacological properties of magnolol and honokiol extracted from *Magnolia officinalis*: central depressant effects. *Planta Medica* 49:103–108.

BIOLOGY AND TERMINOLOGY

Magnolia comprises about eighty species of trees and shrubs naturally distributed throughout eastern North America and southeastern Asia (Figure 2.1). *Magnolia* was named by Carl Linnaeus in 1737 in honor of the French botanist Pierre Magnol (1638–1715). It is the largest genus in the family Magnoliaceae, which also contains nine other genera, including *Liriodendron, Michelia, Manglietia,* and *Talauma.* A key to the genera of the Magnoliaceae is given in chapter 6.

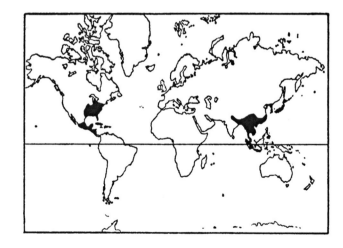

Fig. 2.1. Worldwide distribution of the genus *Magnolia.* From Krüssman 1985.

Magnolias are thought by some botanists to be among the earliest flowering plants (angiosperms) in evolutionary history. Fossilized specimens dating back over 100 million years are thought to be ancestors of existing magnolias. Numerous magnolioid fossils—mostly consisting of complete or partial leaves, flowers, and fruit—have been found in Europe, North America, and Asia. Fossil records suggest that magnolias once occurred throughout western North America, western Asia, and Europe, in addition to their current distribution. Their range became restricted because of climatic changes and glaciation. Literature concerning the fossil record of magnolias is abundant, and will be discussed only briefly here. Readers interested in further information may begin with Tiffney (1977) and Peigler (1989).

The most recent and perhaps the most complete magnolia-like fossil to date was discovered by Dilcher and Crane (1984). The fossilized plant, named *Archaeanthus* ("first flower"), was found in central Kansas, preserved as compressions in the clay soil. *Archaeanthus* has magnolia-like flowers, leaves, and fruits and is estimated to be over 100 million years old (Figure 2.2).

Fig. 2.2. Illustrative reconstruction of flowering (left) and fruiting (right) twigs of *Archaeanthus*. Reprinted with the permission of Annals of the Missouri Botanical Garden, St. Louis, Missouri, and Dr. David Dilcher, University of Florida, Gainesville, Florida. Drawings by Megan Rohn.

An exciting discovery in magnolia paleobotany is the recent extraction of DNA (deoxyribonucleic acid), the genetic blueprint for individual living organisms, from a well-preserved fragment of a magnolia leaf about 17 million years old. Golenberg et al. (1990) extracted DNA from a leaf found in the Clarkia Fossil Beds in northern Idaho. This DNA sample allows scientists to compare the genetic code of the fossil magnolia with that of existing species to determine rates of mutation and relationships between extinct and existing magnolias.

POLLINATION

Magnolias are pollinated by beetles, a fact which may provide additional evidence that they are among the oldest group of plants. Beetles are among the oldest of insects and were abundant at the time of *Archaeanthus*. Bees, wasps, butterflies, and moths appeared on the scene later. Beetle-pollinated flowers are characterized by their large size, white or pink color, lack of nectar, and abundance of pollen. Beetles have strong mandibles and feed on pollen, stigmas, and sugary secretions within the flower. Because beetles can be destructive pollinators, magnolia embryos are protected inside tough carpels that the beetles cannot penetrate.

Beetles are the primary pollinators of magnolias. Here beetles visit an open blossom of *Magnolia virginiana*. Photo by Richard Figlar, Pomona, New York.

Although magnolias lack nectaries (with the possible exception of *Magnolia coco*), some species secrete a nectar-like substance at the base of the tepals and between the stigmas. Beetles are attracted to the flower by fragrances which are emitted from the base of the gynandrophore. Thien et al. (1975) studied the differences in floral odor of North American *Magnolia* as related to their insect pollinators. The authors suggest that certain floral odors attract certain insects and that species with similar floral odors are pollinated by a similar array of beetles. In contrast, Heiser (1962) found *M. grandiflora* and *M. virginiana* to have different pollination faunas, even though they have similar floral odors.

While feeding within the flower, the beetle becomes covered with pollen, which is then transferred to receptive stigmas of the next flower visited. Since stigmas are usually receptive before a flower opens, the insects must also visit buds and unopened flowers in order to be effective pollinators. Smaller beetles may crawl between the closed tepals to get inside the unopened flowers; larger beetles simply chew their way through the tepals.

A vast array of beetles has been collected from magnolia flowers. Heiser (1962), Thien (1974), and Peigler (1989) have produced lists of effective pollinators of North American species of *Magnolia*. Although no one family or species of beetle can be considered most important for all magnolias, the small Nitidulidae are ever-present, as are Mordellidae and Scarabaeidae.

Bees (especially honeybees), moths, and flies may also be frequent visitors of magnolia flowers, yet these insects generally are not effective pollinators. They are attracted by the pollen and the sweet, nectar-like substances produced in the flowers. These insects usually visit only open flowers when the stigmas are no longer receptive; thus pollination is not commonly carried out. Bees or flies are sometimes trapped inside flowers which close for the night, but they cannot gain entry into a flower that has not opened. Occasionally stigmas may be receptive for a short period of time after the flower opens, allowing minimal pollination by bees. Heiser (1962) found flowers of *Magnolia grandiflora* to have receptive stigmas for several hours after opening, and Thien et al. (1975) report that honeybees "undoubtedly effect pollination" in this species. Further study is necessary to determine the extent of honeybee pollination in *Magnolia*. Since all indications are that *Magnolia* flowers are specialized to encourage beetles, it is doubtful that bees account for a significant proportion of successful pollinations.

FLORAL BIOLOGY

The structure of magnolia fruits and flowers is thought to have changed little over the millions of years since the time of *Archaeanthus*. The flower structure (Figure 2.3) of magnolias is not as complex as more highly evolved plant groups. The flower buds are enclosed in fuzzy perules which split and fall as the flower nears opening. These perules cover the spathaceous bract which encloses the flower organs. *Magnolia* flowers have numerous free (not united) carpels and numerous stamens attached in a spiral pattern to a central axis, or gynandrophore. At the base of the gynandrophore the sepals and petals are attached, usually in whorls of three. In magnolias, the sepals and petals are often indistinguishable, and both are commonly referred to as tepals.

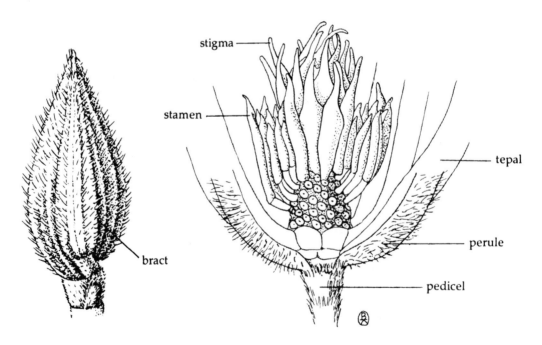

Fig. 2.3. Flower bud enclosed in bracts (left) and diagram of open flower (right) showing floral parts.

In *Magnolia*, stigmas are usually receptive while the flower is still closed. Once the flower opens, the stamens open, releasing the pollen. The stamens may dehisce introrsely, along two sutures on the side of the stamen facing the gynandrophore, or laterally, along sutures on the edges of the anthers (Figure 2.4) When the stamens open, pollen is released to be carried by pollinating insects to receptive flowers.

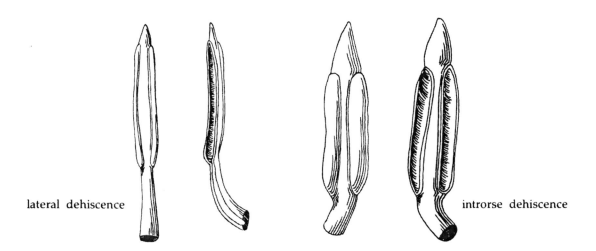

lateral dehiscence   introrse dehiscence

Fig. 2.4. Lateral dehiscence (left) and introrse dehiscence (right).

Once the stamens have shed their pollen, they fall from the central axis of the flower. The tepals eventually fall as well, leaving only the gynandrophore. As the fruits develop, the gynandrophore becomes a woody aggregate which is somewhat unique in its structure (Figure 2.5). Each fruit within the aggregate usually contains 1–2 seeds and, when ripe, the fruit splits along the abaxial (lower) suture or, if there is sufficient room, sometimes along both margins. Generally, individual magnolia fruits are referred to as "follicles," although the strict definition of a follicle refers to multi-seeded fruits splitting along the adaxial (upper) suture only. For lack of a better term, follicle will be used in this book to refer to a single magnolia fruit. The entire aggregate, sometimes called a "cone" in the horticultural trade, is a follicetum, or aggregate of follicles. When the individual fruits are ripe, the seeds are released from the follicle and suspended on a thin thread for a day or so before falling to the ground. The seeds are dispersed by birds, squirrels, and other small animals. *Magnolia* flowers are produced on short "stems" called peduncles. These are often referred to in magnolia literature as pedicels. The term "pedicel," however, denotes the stems of individual flowers within a multi-flowered inflorescence, as in Geranium. Solitary flowers, as in *Magnolia*, have peduncles only. This is a minor distinction in terminology, and the reader may generally consider either term in *Magnolia* literature to mean the stalk on which a flower is borne.

The leaves of magnolias are alternate, sometimes appearing whorled. They are made up of the blade, or the flat portion of the leaf, and the petiole, or leaf "stem." *Magnolia* leaves usually have stipules, or appendages at the base of the leaves. These stipules may be free, not attached to the petiole of the leaf, or adnate, adhering to the petiole (Figure 2.6).

A glossary is provided in Appendix B to enable the reader to get a clearer understanding of botanical terms used throughout this book. Several more comprehensive botanical glossaries can be found in the works listed in the following references.

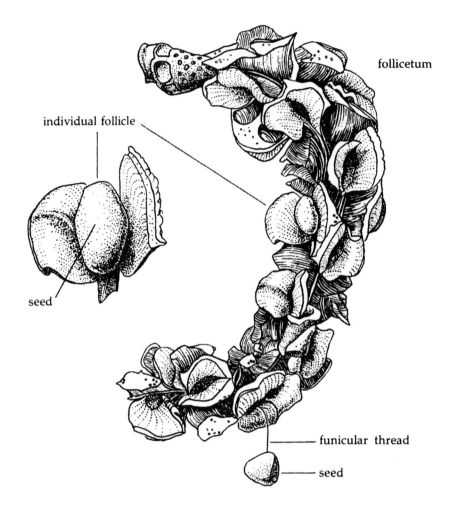

Fig. 2.5. A magnolia fruit aggregate (follicetum) and individual fruit (follicle). Scars on the fruit stem are from fallen stamens and tepals.

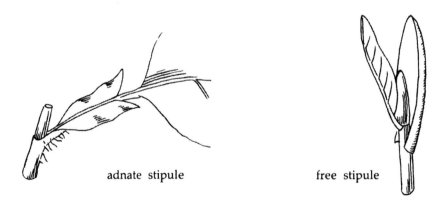

Fig. 2.6. Stipules: adnate (left) and free (right).

REFERENCES AND ADDITIONAL READING

Berry, Edward W. 1923. The magnolia and tulip-tree. In *Tree ancestors*. Baltimore: Williams and Wilkins Co. Pp. 165–180.

Carlson, John D. 1988. Magnolia pollination in Cornwall and Gwent. *Journal of the Magnolia Society* 24(1):18–19.

Dilcher, David L., and Peter R. Crane. 1984. *Archaeanthus*: an early angiosperm from the Cenomanian of the western interior of North America. *Annals of the Missouri Botanical Garden* 71:351–383.

Golenberg, Edward M., David E. Giannasi, Michael T. Clegg, Charles J. Smiley, Mary Durbin, David Henderson, and Gerard Zurawski. 1990. Chloroplast DNA sequence from a Miocene *Magnolia* species. *Nature* 344:656–658.

Good, R. D'O. 1925. The past and present distribution of the Magnolieae. *Annals of Botany* 39:409–430.

Heiser, Charles B. 1962. Some observations on pollination and compatibility in *Magnolia*. *Proceedings of the Indiana Academy of Science* 72:259–266.

Krüssmann, Gerd. 1985. *Magnolia*. In *Manual of Cultivated Broadleaved Trees and Shrubs*, vol. 2. Portland, OR: Timber Press. Pp. 265–277.

Lago, Paul K., and Patricia Ramey Miller. 1986. Notes on Coleoptera associated with the flowers of *Magnolia grandiflora* in northern Mississippi. *Journal of the Mississippi Academy of Sciences* 31:141–146.

Lawrence, George H. M. 1951. *Taxonomy of Vascular Plants*. New York: Macmillan. Includes an excellent glossary of botanical terms.

Leppik, E. E. 1963. Reconstruction of a Cretaceous magnolia flower. *Advancing Frontiers of Plant Sciences* 4:79–94.

——. 1975. Morphogenic stagnation in the evolution of magnolia flowers. *Phytomorphology* 25(4):451–464.

Little, R. John, and C. Eugene Jones. 1980. *A Dictionary of Botany*. New York: Van Nostrand Reinhold.

Niklas, Karl J. 1990. Turning over an old leaf. *Nature* 344:587–588.

Page, Virginia M. 1984. A possible magnolioid floral axis, *Loishoglia bettencourtii*, from the upper Cretaceous of central California. *Journal of the Arnold Arboretum* 65:95–104.

Peigler, Richard S. 1988. A review of pollination of *Magnolia* by beetles, with a collecting survey made in the Carolinas. *Journal of the Magnolia Society* 24(1):1–8.

——. 1989. Fossil Magnoliaceae: a review of literature. *Journal of the Magnolia Society* 25(1):1–11.

Stone, Doris M. 1966. Pollination in the magnolias. *Journal of the Magnolia Society* (3)1:2–3.

Thien, Leonard B. 1974. Floral biology of *Magnolia*. *American Journal of Botany* 61(10):1037–1045.

Thien, Leonard B., W. H. Heimermann, and R. T. Holman. 1975. Floral odors and quantitative taxonomy of *Magnolia* and *Liriodendron*. *Taxon* 24(5/6):557–568.

Tiffney, B. H. 1977. Fruits and seeds of the Brandon Lignite: Magnoliaceae. *Botanical Journal of the Linnean Society* 75:299–323.

Tootill, Elizabeth, ed. 1984. *The Penguin Dictionary of Botany*. New York: Penguin Books.

CULTURE

One reason magnolias are so highly prized as landscape plants is that they are easy to grow and relatively pest-free. Once established, they require a minimum of attention and are forgiving when occasionally neglected. Much information about the culture of magnolias is available, which may make the process of growing them sound more difficult than it really is. This chapter will provide the basic information needed to produce healthy specimens in cultivation.

SITE

One of the most important things to consider when planting any tree or shrub is the ultimate size of the plant. This is especially true for magnolias because of their aversion to being transplanted. The planting site should allow plenty of room overhead and at the base of the plant so that as the tree grows there is no root or shoot competition with nearby plants. A spacing of 20–25 ft. (6.0–7.5 m) should give large plants enough room to mature. Most magnolias do best in full sun or partial shade. They do not suffer adverse effects from the heat of a full-sun location as long as they receive adequate moisture; too much shade will cause plants to become leggy and sparsely flowered.

The planting site should also provide wind protection. Although magnolias are sometimes surprisingly tolerant of wind, they are susceptible to wind damage to branches, leaves, and flowers. Damage is more likely when the wind is gusty rather than steady. Wind protection is especially important for large-leaved species such as *Magnolia macrophylla*. Protection from cold winter winds, ice, and snow is also especially necessary when growing evergreen magnolias. The leaves of these species provide convenient places for ice and snow to accumulate, weighing down the branches and causing them to snap.

Early spring-flowering cultivars should be situated in a location with a northern exposure to reduce the risk of damage to the flower buds by late spring frosts. This delays flowering somewhat, and that delay provides a better chance that the flowers will escape frost damage.

Magnolia hardiness should certainly be taken into consideration when choosing the site for a magnolia planting. By locating your comparable area in Figure 3.1, the United States Department of Agriculture Plant Hardiness Zone Map (USDA 1990), you can determine your hardiness zone. Focus on selecting plants that are suitable for that temperature range. Most nurseries provide such information for the plants they sell. If the selection to be planted is at the northern edge of its hardiness range, it is wise to put it in a protected location in the garden. Variability of environmental conditions within a small area should also be recognized since "cold pockets" or other climatic differences may occur within even a small garden.

SOIL

Magnolias perform best in a well-drained, porous loam soil rich in organic matter. A slightly acidic soil with a pH of 5.5–6.5 is best. Several amendments are available for improvement of poor soils. The most important of these is organic matter. Organic matter is available in many forms including leaf litter, sawdust, bark, and chicken or horse manure.

Lime or alkaline soils provide a special challenge to the magnolia grower. Johnstone (1955) and Treseder (1978) discuss magnolias' aversion to soil with a pH higher than 7. Some species are apparently more tolerant than others to high pH, but Johnstone (1955) reports the success of three hybrids and sixteen species from eight sections cultivated in lime soil in a garden in Ireland. The outward sign of trouble in those species intolerant to lime is chlorosis, or yellowing, of the leaves. This is due to nutrient deficiency. Certain nutrients such as iron become less available to the plant at a pH above 7.0, even if those nutrients are present in the soil. The pH can be lowered by the addition of acidic organic matter, such as peat moss, or by applications of ammonium sulfate. Alternatively, iron deficiency can be treated with applications of iron chelate, a form more readily available for absorption by the plant. Foliar fertilization (discussed later in this chapter) also produces excellent results.

PLANTING

Magnolias should be transplanted during the dormant season, in most cases from November to April. Container plants may be set out at any time of year if the soil is moist and the air is not too dry. A large hole (at least twice the size of the root ball) should be prepared, and the soil amended with organic matter or other additives if necessary. A large planting hole assures that the roots will have plenty of space to develop before reaching the more compact soil outside the prepared area. Magnolias are surface feeders and have shallow roots. They should be planted deeply enough for proper anchorage of the plant, but growth will be set back if they are planted too deeply.

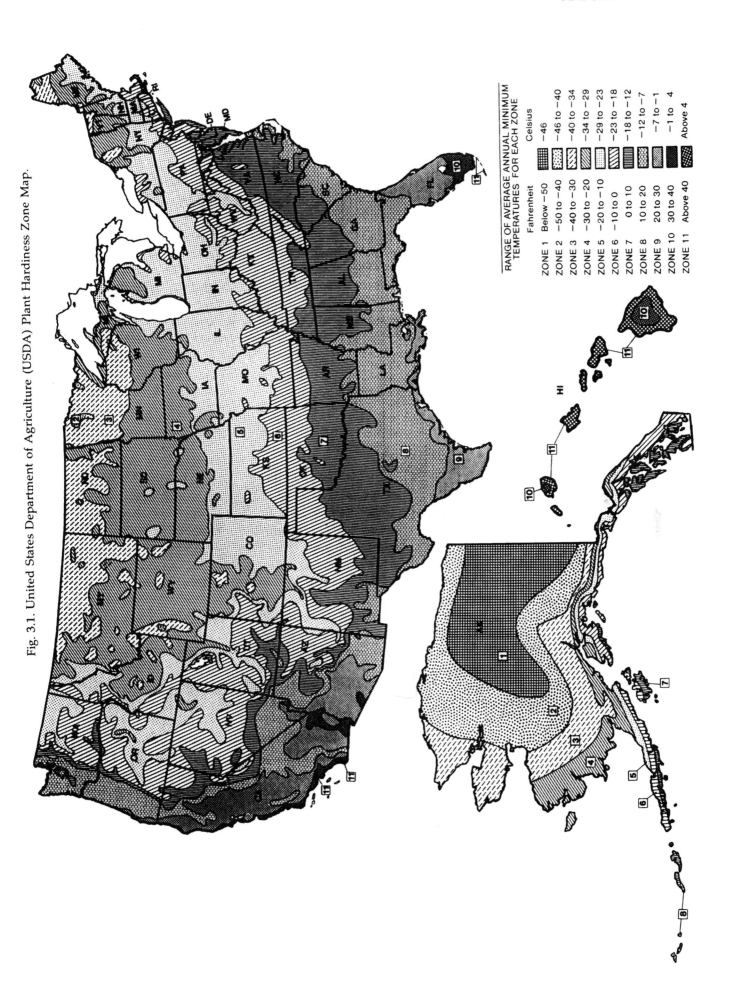

Fig. 3.1. United States Department of Agriculture (USDA) Plant Hardiness Zone Map.

RANGE OF AVERAGE ANNUAL MINIMUM
TEMPERATURES FOR EACH ZONE

	Fahrenheit	Celsius
ZONE 1	Below −50	−46
ZONE 2	−50 to −40	−46 to −40
ZONE 3	−40 to −30	−40 to −34
ZONE 4	−30 to −20	−34 to −29
ZONE 5	−20 to −10	−29 to −23
ZONE 6	−10 to 0	−23 to −18
ZONE 7	0 to 10	−18 to −12
ZONE 8	10 to 20	−12 to −7
ZONE 9	20 to 30	−7 to −1
ZONE 10	30 to 40	−1 to 4
ZONE 11	Above 40	Above 4

A bare-root or balled and burlapped plant should be placed about an inch (2.5 cm) deeper than it was in the container.

Magnolias have soft, fleshy roots that are easily damaged and care must be taken when planting to avoid breakage of the roots. Anchor the plant until it is well established to prevent it from being blown over. The stake should be driven into the soil before the magnolia is planted in order to avoid damaging the roots. Once the magnolia has been properly planted, lightly fertilize and apply mulch to retain moisture and prevent weed growth. The use of both fertilizer, such as an all-purpose N–P–K lawn and garden fertilizer, and mulch at the time of planting was shown by Hensley et al. (1988) to increase plant growth significantly. Proper irrigation during the first growing season is also important for the establishment of the plant.

Some sources suggest cutting back a newly planted magnolia to encourage new growth and allow the root system to become established. This practice may have arisen as a means of stimulating plant growth to avoid "transplanting shock." Plants growing in poor form certainly should be pruned when young to straighten them up. Otherwise, pruning back the top of the plant is not necessary. A similar practice, root pruning, is sometimes carried out to increase the plant's growth rate when transplanting from a container. Gilman and Kane (1990) studied the effects of root pruning on growth and transplantability of *Magnolia grandiflora*. They found that although root-pruned plants initially grew faster after transplanting, there was no difference in size between pruned and unpruned plants one year after transplanting.

CARE

Once established, magnolias need minimal care to become beautiful landscape plants. They should be irrigated as needed, but should not be overwatered. A good layer of mulch around the tree at all times will help keep the soil moist and reduce weeds. Digging or hoeing around magnolias is not advisable since surface roots are easily damaged by hoes and shovels. Chemical weed killers should also be used with caution as the herbicide may easily reach the shallow roots. As with all trees, lawnmower damage to the trunk will allow entry of diseases as well as produce unsightly scars. A mulched area around the trunk of the plant should eliminate the need for intensive weeding or close mowing.

Soil fertility should be maintained by the addition of fertilizers when necessary. Fertilizer formulations containing lime should be avoided to keep from raising the soil pH. Application of fertilizer may continue into the summer if necessary, but should be discontinued after the end of July to allow the plant growth to harden off before winter. Fertilizer should not be placed near the trunk itself. Plant roots are concentrated away from the trunk, at a distance roughly equal to the "drip line" or edge of the branch canopy—it is there that fertilizer should be applied.

Recently some attention has been given to foliar application of fertilizer in magnolias. Smithers (1977) and Adams (1984) reported excellent results with a weekly or biweekly foliar application of macro- and micronutrients throughout the summer. Foliar sprays reportedly decrease heat stress of plants as well as provide nutrients in a readily available form. Continuous vigorous growth was reported throughout the summer and was most noticeable on plants that had also received foliar feeding the previous year. Better root growth was noticed in treated plants after just one season. Foliar spraying is best done in the evenings to avoid leaf burn, and should be discontinued in the fall to allow time for the plants to harden off. Adams (1984) provides information on price and availability of foliar fertilizers as well as practical information on their use.

Pruning is usually not necessary in magnolias, but may be performed as with most woody plants. Certainly misshapen plants can be pruned to create a better form. Suckers from the understock of a grafted plant need to be removed as soon as they are noticed, regardless of season. Also remove unwanted leaders, keeping only those that are appropriate to shape and height requirements. Any weak, dead, or diseased branches can also be pruned out. Precocious species may be pruned in the summer after blooming has ceased. Summer-flowering species are best pruned in late winter.

Root pruning is sometimes advocated as a means of restricting the growth of a large specimen. This may be done during the dormant season by using a shovel to sever the roots about a yard (1 m) from the trunk of the plant. If restricting a plant's growth is necessary, this is a reasonable way to do it; however, it is preferable to avoid the need to do so by carefully siting all plantings. Some gardeners report using this procedure as a method to hasten flowering time. I do not recommend this. The earlier flowering is the plant's reaction to induced stress (the reduced root system) and is not beneficial in the long run.

PESTS

Magnolias have relatively few pests compared to many other landscape plants. Probably the most troublesome magnolia pests are scale insects. Several species of scale in the family Coccoidea are known to attack magnolias. The most commonly found (native to and perhaps still limited to the eastern United States) is known as the magnolia scale, *Neolecanium cornuparvum*. The magnolia leafminer, *Odontopus calceatus*, damages foliage and shoots of magnolias in eastern North America. Common greenhouse insect pests may be a problem for magnolias grown indoors. Mammals such as deer feed on magnolia foliage; a method for repelling them is included in this discussion. The mention and consideration of pesticides or other control methods given here are general recommendations. Be sure to check with local authorities for specific use requirements.

Young plants are particularly susceptible to damage or near defoliation by slugs and snails which feed during the warm season. These may be controlled with methiocarb. For control without the use of chemicals, try setting shallow pans full of beer at ground level near the trees. Slugs are attracted to the beer and literally drown themselves in it. Gardiner (1989) suggests that watering the soil around the trees with a mixture of potassium permanganate and aluminum and copper sulfate will control slugs for several months.

Magnolias of any age are susceptible to various species of scale insects. Scales are probably the most common pests of magnolias in North America, yet they are often overlooked because of their unusual appearance and life cycle. Scale insects may look like dark, leathery or waxy circles or like white fluffy spots, and usually appear first on young twigs. Although life cycles vary a little with the species, their general life cycle is as follows. Eggs hatch into tiny, six-legged "crawlers." These crawlers move to new growth areas on the plant, insert their piercing and sucking mouthparts into the plant tissue, and feed on plant juices. A scaly covering of excrement gradually forms over the insect. Scale species are identified by the color, shape, and texture of this covering. Most species of scale become immobile once they have begun to feed and the covering has begun to form. Adult females lay eggs underneath their scaly covering. The cycle continues when these eggs hatch; young crawlers will move to new growth to feed and reproduce. Male scales pass through a brief winged stage in their life cycle, enabling them to move about to mate.

The damage done to plants by scale insects may be minimal if the plant is healthy and the infestation is light. Plants which are already under stress, or those with extremely heavy scale infestations, can be severely damaged or killed. Additionally some scales excrete honeydew, a sticky, sugary substance that serves as a growing medium for sooty mold fungus. The fungus appears as a gray or black powder on the plant. Sooty mold and scale can both be unsightly, even if the plant itself is relatively undamaged.

Magnolias can be attacked by several species of scale, but the most common is *Neolecanium cornuparvum*, commonly known as the magnolia scale. This species is the largest scale insect in North America, the adult female reaching about half an inch or 13 mm in length. Adults have a yellow, waxy covering resembling chewed gum. Immature magnolia scales are bluish black at first, then covered over with white fluff which weathers to yellow wax at maturity. This species produces large amounts of honeydew. These insects are particularly difficult to control since the waxy scale covering protects the insect from predation and insecticides. Since newly hatched crawlers lack this protection they can be controlled with dormant oil sprays or contact insecticide such as malathion or carbaryl. Adults can be controlled with dormant oil spray or a systemic insecticide. Medium weight, natural petroleum oils are used for dormant sprays. The lighter weight oils are only for summer use. Check locally for appropriate systemic for use on magnolias.

The magnolia leafminer (*Odontopus calceatus*) has recently become a problem on magnolias in eastern North America. Both adult and larval stages are damaging to the plant, causing dieback of terminal shoots and tattered and blistered leaves. The adult weevils feed on leaves and flowers in spring and lay eggs in the leaf apex in May. The white larvae hatch and feed on leaves in early summer, creating "mines" and subsequent browning, while leaves appear blistered. Destroying infested leaves while larvae are active may help control damage. However, foliage damage is almost entirely cosmetic and will seldom affect the tree's overall health. There is no chemical control for the larvae, so the egg-laying adult population must be controlled early in the spring if any chemical treatment is to be effective. Adult leafminers can be controlled with the use of chlorpyrifos.

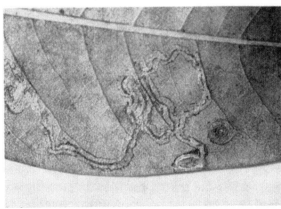

Leafminer on *Magnolia grandiflora.*
Photo by the author.

Many soft-bodied insects can be killed with a mixture of ordinary dishwashing detergent and water, reducing the need for chemical insecticides. Make a light, sudsy mixture of soap and water and spray it onto the infested plants with a mist bottle or sprayer. Test for plant reaction by initially spraying only a small area. The soap acts much like horticultural oil in covering the insect and smothering it.

Larger pests, such as rabbits and deer, are best fenced out or repelled if damage is extreme. Commercial deer repellents are available, but they are only partly effective. In a recent study (Swihart and Conover 1990) bar soap was found to be as effective as two popular commercial repellents. The brand of soap was found to have no influence on effectiveness. Place a bar of soap in an old sock or nylon stocking and hang one in each tree needing protection. Hang the bar soap directly against the trunk about 3 or 4 ft. (1–1.2 m) from ground level. This will protect branches within 3 ft. (1 m) of the soap. While this is not very ornamental during the dormant season, it does deter deer.

Greenhouse pests such as the red spider mite, aphids, mealy bugs, and whiteflies may attack magnolias grown indoors. Spider mites (*Tetranychus telarius*) are probably the worst of these, sucking sap from the leaves and causing yellowing. Once established, they are difficult to control. Spider mites can be discouraged from establishing themselves by spraying plant leaves with water once or twice a day. Warm, dry conditions and dirty or dusty leaves encourage mite infestation, so lower temperatures, increased humidity, and clean leaves may help avoid problems with spider mites. All these greenhouse insects may be controlled with oil sprays or, if well established, with a systemic insecticide.

DISEASES

Magnolias are susceptible to several different foliar diseases, but none of them are necessarily worrisome, life-threatening to the plant, or widespread. Most can be avoided by keeping plants healthy and stress-free. Probably the most common diseases are leaf spots, caused by fungi, bacteria, or algae. Control measures ultimately depend on how much cosmetic damage is allowable. The fungal leaf spots begin as small

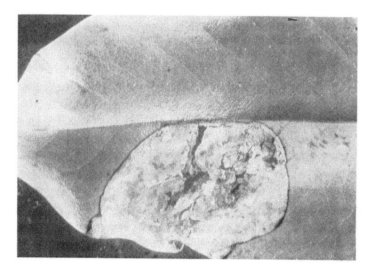

Fungal leaf spot (*Phoma* sp.) on *Magnolia grandiflora*. Photo by Jackie Mullen, Plant Pathologist, Alabama Cooperative Extension Service, Auburn, Alabama.

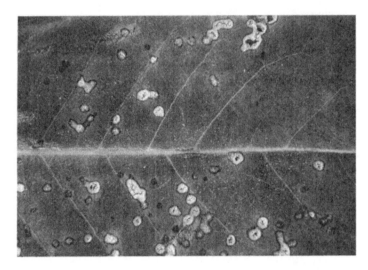

Fungal leaf spot (*Phyllosticta* sp.) on *Magnolia grandiflora*. Photo by Jackie Mullen, Plant Pathologist, Alabama Cooperative Extension Service, Auburn, Alabama.

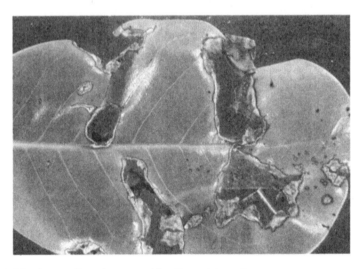

Bacterial leaf spot (*Pseudomonas cichorii*) on *Magnolia grandiflora*. Photo by Jackie Mullen, Plant Pathologist, Alabama Cooperative Extension Service, Auburn, Alabama.

brownish spots which become larger over time. The center of each spot becomes gray and the margin dark. Since these spots include reproductive structures, the spread of the disease can be slowed by removing any infected leaves from the tree and removing leaf litter from the ground around it. Copper fungicides, mancozeb, and dithiocarbamate fungicides may also be used to control the disease.

Bacterial leaf spot in evergreens such as *Magnolia grandiflora* is caused by *Pseudomonas cichorii* and affects mostly the young leaves of the plant. The disease begins as small brown spots which enlarge, become dry, and eventually deteriorate, leaving holes in the leaves. Older leaves are more resistant, and the spots remain small with yellow margins, never deteriorating into "shotholes." In deciduous magnolias, the bacterial leaf spot pathogen is a strain of *Pseudomonas syringae*. Leaf spots in this case are small, dark brown, and interveinal. Copper fungicides are effective in controlling both types of bacterial leaf spots unless rain or high humidity prevails.

Algal leaf spot, caused by *Cephaleuros* species, may occur under damp conditions. The leaf spots are raised and circular with uneven margins. They are first green or brownish in color, becoming whitish. Algal leaf spot can be controlled with the use of copper fungicides. However, with rain and high humidity, control of both bacterial and algal leaf spot can be quite difficult.

Evergreen magnolias may suffer from black mildew (*Meliola magnoliae*) on the leaves. This is easily controlled by spraying the leaves with a wettable sulfur compound. Canker (*Phomopsis* species or *Nectria magnoliae*) in magnolias is rare, yet can occur. Canker is identified as an unnatural swelling developing on the trunk or branch, sometimes accompanied by bark splitting and oozing of sap. The best control is to remove and burn the area involved in order to reduce spread of the disease.

REFERENCES AND ADDITIONAL READING

Adams, Robert W. 1984. New growing aids for magnolias. *Journal of the Magnolia Society* 19(2):15–19.

Gardiner, James M. 1989. *Magnolias.* Chester, CT: Globe Pequot Press.

Gilman, Edward F., and Michael E. Kane. 1990. Growth and transplantability of *Magnolia grandiflora* following root pruning at several stages. *HortScience* 25(1):74–77.

Hensley, David L., Robert E. McNiel, and Richard Sundheim. 1988. Management influences on growth of transplanted *Magnolia grandiflora. Journal of Arboriculture* 14(8):204–207.

Johnstone, George H. 1955. *Asiatic Magnolias in Cultivation.* London: The Royal Horticultural Society.

Krüssmann, Gerd. 1985. *Magnolia.* In *Manual of Cultivated Broadleaved Trees and Shrubs,* vol. 2. Portland, OR: Timber Press. Pp. 265–277.

Mullen, Jackie. 1986. Southern magnolia leaf spot diseases and disease control. *Journal of the Magnolia Society* 22(1):11–13.

Mullen, Jackie M., and G. S. Cobb. 1984. Leaf spot of southern magnolia caused by *Pseudomonas cichorii. Plant Disease* 68:1013–1015.

————. 1986. Control of *Pseudomonas* leaf spot on *Magnolia grandiflora. Journal of the Magnolia Society* 22(1):14–15.

Parrini, C., and U. Mazzucchi. 1985. Angular leaf spot of some deciduous magnolias caused by *Pseudomonas syringae. Phytopathologie Mediterranea* 24(3):245–248.

Pollet, Dale K. 1989. The magnolia leafminer. *Journal of the Magnolia Society* 25(1):24.

Raabe, Robert D. 1962. Diseases and pests of magnolia. *Journal of the California Horticultural Society* 23(1):42–44.

Smithers, Sir Peter. 1977. Foliar feeding of "sulky" magnolias. *Journal of the Magnolia Society* 13(1):19–21.

Suomi, Daniel A. 1990. Scale insects on ornamentals. Pullman, WA: Washington State University Cooperative Extension Service.

Swihart, R. K., and M. R. Conover. 1990. Reducing deer damage to yews and apple trees, testing Big Game Repellent, Ro-Pel, and soap as repellents. *Wildlife Society Bulletin* 18:156–162.

Treseder, Neil G. 1978. *Magnolias.* London: Faber and Faber.

United States Department of Agriculture. 1990. USDA Plant hardiness zone map. *Agricultural Research Service Miscellaneous Publication* No. 1475.

PROPAGATION

Magnolias are usually propagated in two ways: sexually, from seed, and asexually, from vegetative tissue. Each type is more useful to plant propagators in certain situations than the other. Methods of both seed and vegetative propagation of magnolias will be discussed here, together with some of the advantages and disadvantages of each method.

Seeds provide the easiest method of mass propagation of most plants, including magnolias. Since seed results from sexual cross-fertilization, both parents contribute to the genetic identity of the offspring. As a consequence the offspring are genetically diverse and may differ greatly from the parents. In the wild, this genetic diversity is important to the survival of the species. For garden use, however, a different set of plant characteristics may be considered more desirable. The plants producing these desirable characteristics, usually named cultivars, must be vegetatively propagated to avoid seedling variability.

Vegetative, or asexual, propagation involves producing additional plants using only vegetative structures such as roots, stems, and shoots, and not using the sexual structures—flowers, fruits, and seeds. Methods of vegetative propagation most commonly used with magnolias include the rooting of stem cuttings, budding or grafting, layering, and tissue culture. Any one of these methods insures that the offspring produced will be "clones," that is, they will be genetically identical to the parent. These methods must be used when propagating a cultivar or other selected form to retain the desirable character(s) which led to the selection.

The general procedures for seed and vegetative propagation of magnolias provided here have been compiled from the experience of many people, from professional propagators to home gardeners. For further general reading on the methods of propagation, refer to Macdonald (1986), Dirr and Heuser (1987), or Hartmann and Kester (1990). For more specific information on propagation of magnolias, see articles in the list of references at the end of this chapter. The annual *Proceedings of the International Plant Propagators Society (PIPPS)* provides helpful information from horticulturists. Joseph Hickman of Benton, Illinois, and Robert Adams of Shelbyville, Indiana, two experienced *Magnolia* propagators, kindly contributed some of the information in this chapter.

SEED PROPAGATION

Magnolia fruits ripen in late summer and early fall, and seeds are collected when the fruits split to reveal the bright red or orange seed coats. The seeds, when released, are suspended from the fruit on almost invisible funicular threads (Figure 2.5). The seeds will hang in this way for only a day or two before dropping to the ground. Birds and small mammals commonly eat the seeds soon after they have fallen, so it is easier to collect the seed before it is released. Alternatively, a paper bag or nylon stocking may be secured around an entire fruit aggregate so that seeds are contained. If the fruits are picked before they are completely ripe, they must be placed in a warm, dry spot for several days until the follicles open.

Magnolia seed coats are three-layered. The outer layer is a brightly colored, oily, fleshy covering. This outermost seed coat provides an impermeable barrier which prevents the seed from absorbing enough water to germinate prematurely. In addition it acts as a mechanical barrier to gas exchange and embryo expansion and may contain chemicals which also inhibit germination. In nature this coat either degrades after the seed falls to the ground, or is removed when the seed passes through the digestive tract of a bird or small mammal. Beneath the outer covering is a hard, bony, dark- or light-colored inner layer. This coat gives mechanical protection to the embryo and does not inhibit germination. The third layer is a thin membrane. It surrounds the fleshy endosperm, or "food reserve" for the small, undeveloped embryo embedded within. The embryo cannot mature and the seed cannot germinate before the fleshy outer seed coat is broken down and water is absorbed by the seed. After the outer layer has been destroyed, a period of chilling is required to break seed dormancy, after which the embryo develops and expands until germination occurs.

To remove the outer coat of collected seeds, soak them in water for up to three days, then squeeze the seeds or rub them across a screen. Several propagators remove the seed coats by placing seeds into a household blender after replacing the blades with rubber or covering them with plastic tubing, tape, or some other material that will protect the seed from cutting injury. Once most of the coat is removed, the seeds can be washed in water with a mild soap to remove the oily film, then rinsed several times. Since the outer seed coat prevents the seed's water loss as well as absorption, the seeds must be kept moist once the coat is removed. If the seed dries out, viability is reduced or completely lost. Many growers test viability by placing seeds in water, discarding any that float. Although this is not a foolproof test, it does give some indication of viability.

Cleaned seeds may either be planted outdoors in the fall or stored over the winter and planted in the spring. Before planting, wash the seeds again, using dishwashing detergent to remove any remaining oily coating. For fall planting, seeds are sown in a moist but well-drained medium. My sowing medium preference is a peat and perlite mix or soilless potting mix. Magnolia seeds are prone to rot in ordinary garden soil. After sowing, place outside and mulch with leaves or straw to retain moisture. A 2- to 4-month cold period of 33–40°F (0.5–4.4°C) is required for seeds to break dormancy.

If the seeds are to be planted in spring, proper cold treatment and storage must be employed. Cleaned seeds should be rolled up in a "sheet" of damp sphagnum moss, placed in plastic bags, and kept in a refrigerator at 33–40°F (0.5–4.4°C). Browse (1986) suggests a ratio of four parts sphagnum to one part seed, placed in a gas-permeable bag such as an ordinary polyethylene food storage bag. Some people use damp paper towels instead of sphagnum moss, but one experienced propagator called paper towels an "enemy of the gardener, certain to bring mildew and molds." With that warning, it is probably best to avoid their use. The seeds should be rinsed in a fungicide or chlorine bleach solution prior to bagging to protect them from diseases. The bags should be sealed tightly to retain moisture. The duration of cold required to break dormancy depends on species and environmental conditions, and can range from 6 to 17 weeks. In general, 3–4 months of cold treatment are considered optimum. During this time it is important that the seeds be checked every few weeks to spot any obvious fungal growth and to maintain a moist but not soggy soil environment. If seeds cannot be planted immediately after cold treatment, leave them in cold storage at 35–40°F (0.5–4.4°C). Do not allow them to dry out.

In the spring, usually April or May, sow seeds about 0.5 in. (1.3 cm) deep in pots or flats. Potting media include peat moss, sand and peat mixtures, peat and perlite mixtures, sphagnum moss, and other soilless potting mixes. Whatever medium is used, it is important that it be well drained and preferably sterile. It is best to cover the pots or flats with plastic or glass to retain moisture and humidity.

A practice many growers use is to sow seed directly into containers 8–12 in. (20.3–30.5 cm) deep to allow for the eventual, vigorous root system. This practice eliminates the need to transplant seedlings into pots after the first true leaves have formed. Other propagators germinate seeds in plastic bags and plant them in pots or flats after germination.

Seeds germinate at temperatures of about 65–75°F (18–24°C). Germination usually takes place in a few weeks, but as much as several months may be required. After germination, place the seedlings in a sunny location and remove the plastic or glass covering. Transplant seedlings into pots after the first true leaves develop.

In some cases, germination requires 1–3 years after sowing. The vast majority of seeds germinate in the first year, and few commercial propagators keep sowings past one year. The percentage of seeds not germinating within a year that might eventually germinate is generally an unknown.

Protection from birds or small mammals interested in making a meal of the seeds or seedlings, and protection from severe climatic conditions may also be necessary. To protect seedlings from birds or small mammals, put screen material over the pots or flats. Another method is to build a box out of 2 by 4's and screen wire to set the pots or flats in. The screen material will also provide some necessary shading for seedlings in extremely hot weather. The box arrangement provides some wind break under strong wind and desiccating conditions.

VEGETATIVE PROPAGATION

Vegetative propagation of magnolias is more commonly used than seed propagation in commercial nursery production, where uniform offspring are desired. Additionally, vegetatively produced plants usually flower more quickly than seed-grown plants, thus producing a desirable landscape plant more quickly.

Cuttings

The propagation of magnolias from cuttings has become common only recently. Johnstone (1955) reported that it was the least common method of vegetative propagation at that time. Yet with research developments over the past thirty years, techniques and equipment have improved, and we have gained new understanding of the rooting process. Today propagation by cuttings is the most common form of vegetative propagation, though some magnolia species and cultivars, such as *Magnolia denudata*, still remain difficult to root by this means.

Cuttings are made by removing young shoots from the parent plant and inducing the shoot to produce roots. The advantages of using cuttings as compared to grafts are that plants are less expensive to produce on a large scale, the plants produced are more uniform, and graft incompatibilities are avoided since the cutting is on its own roots.

There are, however, disadvantages to cutting propagation. More expensive facilities such as a mist system, virtually mandatory for rooting cuttings, are not necessary with other methods of propagation. Grafting also has the advantage that it can be done over a longer period of the year and usually produces a larger plant more quickly than cuttings.

The following guidelines for propagation of magnolias from cuttings are general, as methods differ in detail from one propagator to another. The information given here can serve as a starting point, while the references listed at the close of the chapter will provide more detailed instructions.

The stock plant from which cuttings are taken must be healthy, free of pests, and not water-stressed. Cuttings from young plants root better than those from old plants, a fact attributed to the increased production of rooting inhibitors with increasing plant age. When plants are young, they favor vegetative growth, while with age, the production of flowers and fruit becomes more important. Cuttings should not be taken from flowering branches, as they are more difficult to root due to changes in hormone levels and energy reserves associated with flowering. Juvenility of stock plants can be maintained by cutting them back every year to encourage a new flush of vegetative growth.

Cuttings of most species are taken from June to August with late June and early July considered optimum. In the more southerly latitudes, cuttings of *Magnolia kobus* var. *stellata* and *M.* × *soulangiana* are sometimes taken as early as April. In general, cuttings from deciduous magnolias are taken when the tissue is soft. Ellis (1988) achieved nearly 100% rooting of various cultivars of deciduous magnolias under mist when the cuttings were "butter soft," meaning at the stage of growth when a knife cuts stem tissue as it might through butter. Alternatively, cuttings from evergreen species, such as *M. grandiflora*, are collected as late as November. The optimal time for evergreen cuttings is after the main flush of growth, when the leaves are mature and the stem is firm. All cuttings are best collected in the morning or evening and must be protected from water loss at all times. They should be collected in plastic bags and placed in an ice chest or in shade to retain moisture. Cuttings from the stock plant are taken at a leaf node and should include three or more leaves. The cuttings may range in length from 3 to 10 in. (8–25 cm), depending on the species. Obviously, small plants such as *M. kobus* var. *stellata* produce shorter cuttings than *M.* × *soulangiana* or *M. grandiflora.* The cuttings should be stored in a refrigerated room or cooler until they are prepared for sticking in the rooting medium.

Several steps are required to prepare the cuttings for the rooting bed. Most propagators remove the terminal buds from the cuttings since there is some evidence that the bud inhibits rooting. A further reason is that the bud tissue is often quite soft, which leads to stem dieback due to fungal or bacterial infection. After removing the terminal bud, strip the lower leaves from each cutting, leaving only the top two or three leaves. On the large-leaved species, such as *M. grandiflora,* these remaining leaves should be cut in half to reduce water loss from excessive leaf surface area.

Wounding the stem before the cutting is placed in the rooting medium has proven beneficial in magnolias as it creates hormonal changes within the cutting which may induce rooting. In addition, wounding allows for increased uptake of water and rooting hormone, if the latter is used. Wounding simply refers to the removal of a strip of tissue from the side of the stem, deep enough to expose the cambium, the thin tissue located between the bark and the wood. The wound should be 0.5–1.0 in. (1–2.5 cm) long, depending upon the size of the cutting. This operation is done with a sharp knife or razor. Large-stemmed cuttings should be wounded on two sides of the stem.

After wounding, the cuttings are usually treated with a hormone to increase rooting percentage and uniformity, speed up the rooting process, and increase root number and quality. The most common rooting hormone used for magnolias is indole-3-butyric acid (IBA). Dirr and Heuser (1987) suggest that napthalene acetic acid (NAA) may be more effective for *Magnolia grandiflora.* While some commercially prepared rooting compounds combine IBA and NAA, most magnolia propagators have had success with IBA in concentrations of 0.3–1.0%. Rates as high as 2.0% have been used for rooting magnolias, but is not generally recommended as high concentrations tend to "burn" the cuttings, causing death of the treated tissue. The most common treatment for magnolias is 0.8% IBA, and this is recommended as a starting point. If burning occurs, the concentration should be reduced. On the other hand, if the cuttings do not respond to 0.8%, a higher level is obviously called for. Hormone mixtures come in powder or liquid formulations which must be adjusted by the propagator, or as ready-to-use commercial preparations. For information regarding the preparation of solutions, and for the identification of commercial hormone preparations and their formulations, see Macdonald (1986) and Dirr and Heuser (1987).

If the hormone preparation does not include a fungicide, most propagators either add one, or first dip the cuttings into a fungicide and then into the hormone. In some cases, the rooting medium may be treated, or the cuttings may be sprayed or drenched with fungicide after placement in the medium. Recommended rooting media for magnolias include sand, sand and perlite, sand and peat, peat and perlite, perlite, or vermiculite. The medium should be loose and well-drained and moistened before the cuttings are inserted. Cuttings may be rooted in a rooting bed, nursery trays, or pots.

Place the cuttings in a greenhouse or other warm, moist environment in a shady location. If the cuttings are covered with a plastic "tent" to retain moisture, a windowsill receiving filtered light provides a satisfactory environment. Water the cuttings frequently enough to keep them consistently moist but not soaked. In commercial operations, an intermittent mist system is used to provide the necessary humidity.

Hickman (1990) has devised an adequate, yet easy and inexpensive, fog system for home propagators. Magnolias respond well to bottom heat, using heating cables or pads placed under the soil to maintain soil temperature at 70–75°F (21–24°C). However, *Magnolia grandiflora* roots better with a higher basal temperature, 75–80°F (24–27°C). When bottom heat for the rooting medium is optimum, the roots emerge a pure white color and elongate rapidly. If the temperature is too high, the roots emerge off-white or orange in color and reduced in vigor. Since these warm, moist conditions are also quite favorable for damping-off and other diseases, a weekly or biweekly fungicide treatment is recommended. *Magnolia* cuttings usually root in 5–8 weeks, but species such as *M. grandiflora* may take twice as long.

Once roots have developed, the propagator is faced with transplanting and over-wintering the cuttings—two traditionally difficult tasks in magnolia propagation. Most authorities emphasize that shoot growth should be induced in deciduous magnolias as soon as possible after rooting to insure successful overwintering. It is therefore best to take cuttings as early in the season as possible to allow plenty of time for rooting and significant shoot growth before winter. Some propagators overwinter cuttings in the rooting containers in the greenhouse and transplant in the spring, yet most transplant in the summer or fall and overwinter the plants in pots. If deciduous magnolias have rooted yet failed to put on new growth, wait to transplant until new growth begins the following spring.

Grafting

Grafting was once the most commonly used method of vegetative propagation of magnolias but has been largely replaced by rooting cuttings. Grafting requires more space and is more labor-intensive and time-consuming than propagation by cuttings. Nurseries have largely come to graft only those species and cultivars which are difficult to root from cuttings. For amateur propagators or home gardeners, however, grafting is an easy and inexpensive method of propagation that can be done nearly year-round. For example, magnolia scions grafted onto a blooming (mature) tree will often bloom in three years, thereby accelerating the first bloom by as much as fifteen years for some cultivars. This is particularly useful when evaluating hybrids.

Grafting involves the uniting of two parts from different plants by inducing them to grow together as a single plant. The part that will become the shoot portion of the grafted plant is called the scion. The scion usually consists of a stem or bud of the desired species or cultivar. The second part used in grafting is the stock (understock or rootstock). This is the plant which will provide the roots of the grafted plant. In magnolias, the rootstock is usually a seedling but occasionally may be a rooted cutting.

In some situations graft incompatibilities arise. For example, a vigorous stock may grow more quickly than the scion, creating a large growth at the graft union. This large growth is usually unsightly and sometimes weak. The only way to know if it is weak is when the union breaks, illustrating a definite graft incompatibility. Or in some instances the graft may appear to grow well for years but the plant eventually dies or the top breaks off at the union. Fortunately, graft incompatibilities are not as common with magnolias as with some other genera, such as maples, so grafted magnolias are usually successful if the process is carried out correctly.

Some amateur and commercial propagators have begun using Schick injector razor blades rather than grafting knives for making cuts. This is a method popularized by the late Joseph C. McDaniel, who was a prominent founder of the Magnolia Society.

Schick blades are sharper than a knife and therefore do less damage to the plant. Another suggestion when grafting is to complete the graft as quickly as possible. Expose the cuts to air for the minimum amount of time possible to avoid cell damage.

Many different methods of grafting exist, based on the type of material used, the part of the plant used, and the particular way in which the two parts are joined. The two methods most commonly used for magnolias are chip-budding, in which a chip with a single bud from the scion is grafted onto the stock, and side-grafting, in which a stem from the scion is grafted onto the side of the stock.

Chip-budding is perhaps the easiest grafting method to learn and perform successfully, so it has become a common grafting technique for magnolias. Charles Tubesing, assistant horticulturist for the Holden Arboretum, Mentor, Ohio, has developed a successful chip-budding method for magnolias. The following discussion is gratefully taken from his work (Tubesing 1983).

Chip-budding may be done in early spring, but most amateur propagators find it easier in the summer when the current year's growth is mature. Collect scion wood while it is dormant; dip in a weak (10%) chlorine bleach or fungicide solution, rinse with water, and keep the scion at about 40°F (4.4°C) in a sealed plastic bag until used. The cut surfaces may be further protected by dipping them in paraffin before storage. Scions should be from shoots produced the previous growing season and should be taken after leaf-fall. If it is necessary to chip-bud graft in the summer, take scion wood from dormant buds on year-old wood and use it as soon as possible, since summer scions do not store well. Any leaves should be removed and the scions stored in a cool, moist place.

Completed chip bud (right) and side graft (left). Photo by Joseph Hickman, Benton, Illinois.

The stock plants should be healthy and vigorous with a straight stem and have a diameter equal to or greater than the scion. One-year old seedlings are commonly used. Stock plants should be actively growing at the time of budding. Clear stock plants of any buds, shoots, or leaves from ground level to several inches above the area to be budded. Then rub the stem with a dry rag and swab it with a fungicide—or use a 10% solution of chlorine bleach—to prevent the spread of disease.

Using a sharp, thin-bladed knife, cut the bud from the scion, making the cuts approximately 0.75 in. (2 cm) above and below the bud itself. The cuts should be smooth and straight. Avoid cutting into the pith, or spongy tissue, in the center of the stem. When the cuts are completed, remove the bud (see Figure 4.1).

The stock plant is cut in a similar way, except that a bud is not included. Make the cut to match the scion as closely as possible. Remove the chip from the stock plant and replace it with the bud chip from the scion, matching the cambium tissue of stock and scion as closely as possible. Once the scion is in place, secure it tightly by wrapping with polyethylene budding tape or a rubber strip, leaving the bud itself uncovered. It is also a good idea to wrap freezer tape around the part of the chip above the bud to secure it and keep water out of the cut. Tubesing (1983) uses 0.5 in. (1.3 cm) clear plastic tape to graft without a rubber strip. Some propagators cover the scion with Parafilm after securing it with a rubber tie.

The graft union should be complete in 2–3 weeks in the greenhouse or 3–6 weeks outdoors. The closing of the union is evident when callus tissue, a light, warty looking tissue, appears on all exposed areas around the bud chip. When the area is healed, remove the tape or rubber strip.

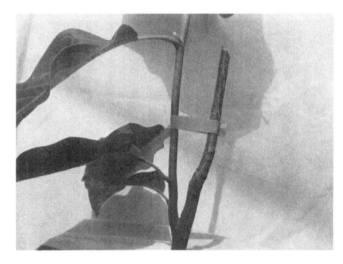

Successful chip bud, with the scion showing good growth. Photo by Charles Tubesing, Assistant Horticulturist, Holden Arboretum, Mentor, Ohio.

Fig 4.1 (opposite). Chip-budding. **A.** The prepared bud stick ready for chip-budding, showing (a) removal of the immature tip. **B.** Front and side views of the prepared scion (bud) chip and rootstock. **C.** Front and side views of the prepared rootstock. **D.** Good matching-up of the scion chip and rootstock. Note (a) the ring of exposed rootstock tissue around the perimeter of the bud chip. **E.** Examples of correct and incorrect matching-up for scion of a different size from the rootstock. (i) Match the scion on one side of the cut only if the scion is smaller in width than the cut. (ii) Incorrect matching-up of a scion smaller in width than the cut. (iii) The scion was cut too long so that it extends above the apex of the cut on the rootstock. **F.** Examples of tying-in. (i) Small and hard buds are covered when tying-in with polyethylene tape. (ii) Soft and prominent buds are left exposed when tying-in with polyethylene tape. Reprinted with permission from Macdonald 1986.

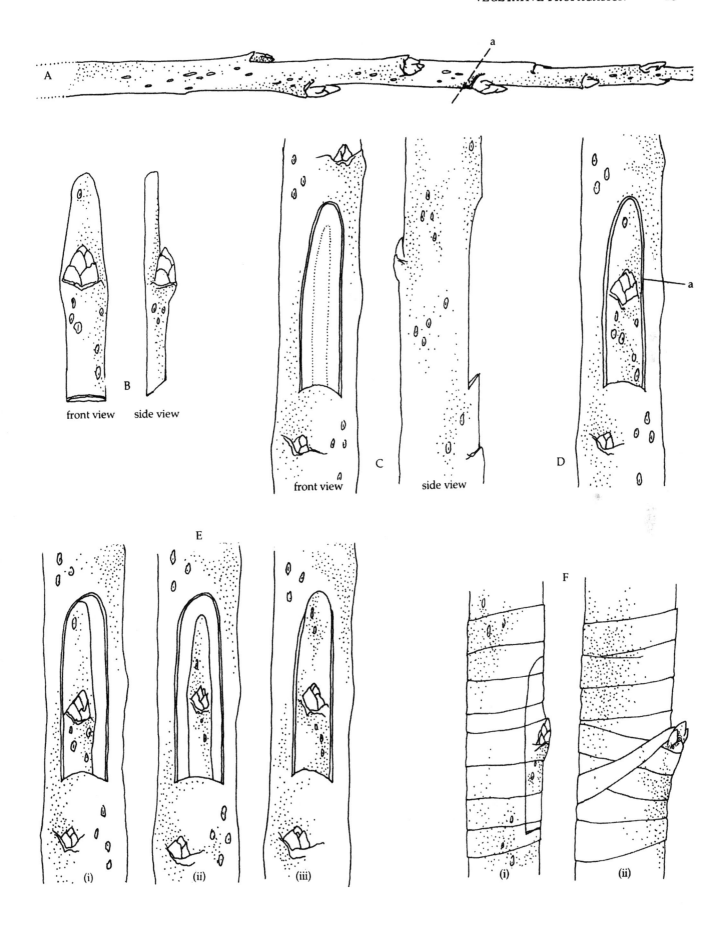

A

B

front view   side view

C

front view   side view

a

D

E

(i)   (ii)   (iii)

F

(i)   (ii)

Close-up of graft union on a successful chip bud. Photo by Charles Tubesing, Assistant Horticulturist, Holden Arboretum, Mentor, Ohio.

A fully completed chip bud; the stub of the rootstock, formerly used for supporting new growth from the scion, has been removed, and the graft union has completely healed. Photo by Charles Tubesing, Assistant Horticulturist, Holden Arboretum, Mentor, Ohio.

Grafts made in spring and early summer can be forced into growth as soon as the union is complete. Grafts made after this time should not be forced until the following spring, to avoid frost damage. Force the new bud into growth by cutting off the top of the stock plant 4–6 in. (10–15 cm) above the graft union. Remove any shoots below the union as well. As growth from the new bud continues, tie it to the remaining stem of the stock to hold the new growth straight. Once the new shoot has become firm, cut the stock just above the graft union and seal the cut with paraffin or grafting wax if desired.

Side-grafting, while not as common as it once was, is a useful grafting technique for magnolias. Side-grafting is best done in the fall, winter, or spring. Stock plants should be at least two years old. The scion wood is taken from stem tips of new growth 8–10 in. (3–4 cm) long from which the foliage is removed. Scions may be stored in a refrigerator at 35–40°F (2–4°C) until used. The scion stems are formed into a thin wedge by two longitudinal cuts on opposite sides of the base, each about 2 in. (5 cm) long (Figure 4.2). A single longitudinal cut of the same length is made on one side of the stock plant, leaving a thick flap. The scion is then inserted into the flap, aligning the cambium layers of scion and stock. The graft is then wrapped with a rubber strip and

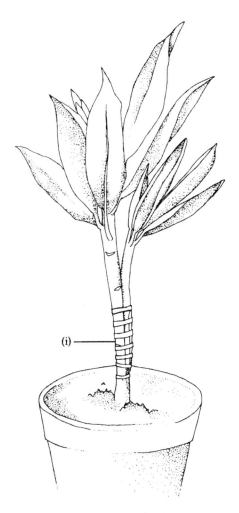

Fig. 4.2. Side-grafting. Completed side wedge graft. The scion was cut on opposite sides to form a wedge which was then matched between a flap of rind and sliver of wood on the rootstock (i). Note that the tying-in was completed below the basal cut on the rootstock. Reprinted with permission from Macdonald 1986.

may be sealed with polyethylene or paraffin, if desired, to protect the graft union from desiccation. The grafted plants are then placed in a sunny location, but protected from hot summer sun by shading. The plants should remain in these conditions until the graft union is complete, usually 3–5 weeks. At the end of the summer, the stem tip of the understock should be removed just above the graft union.

Magnolias may be grafted onto seedling rootstock of the same species or a different species. North American propagators often use *Magnolia acuminata* or *M. kobus* as rootstock for most Asiatic species. *Magnolia sargentiana* var. *robusta* or *M. dawsoniana* are sometimes used as stock for the larger-growing species such as *M. campbellii*. Japanese propagators frequently use *M. grandiflora* as rootstock.

Layering

Layering is the process in which roots are produced on a stem that is still attached to the parent plant (Figure 4.3). It is perhaps the oldest method of vegetative propagation, having been used since at least the time of the ancient Greeks. With the advent of the improved methods of grafting and rooting of cuttings, layering has largely passed out of use except in the case of propagating only one or two plants, or propagating from a very rare form. It is a more expensive method than cutting or seedling production because more space and labor are required. However, layering (as is also true of grafting) produces a larger plant in a shorter period of time than propagation by cuttings. Its advantage over grafting is that plants produced by layering grow on their own roots, thus avoiding graft incompatibility.

Little published information is available on methods and success with layering magnolias. Propagators have reported that *Magnolia grandiflora*, *M.* × *soulangiana*, and *M. liliiflora* are easily layered, but *M. macrophylla*, *M. campbellii*, *M. sargentiana*, and *M.*

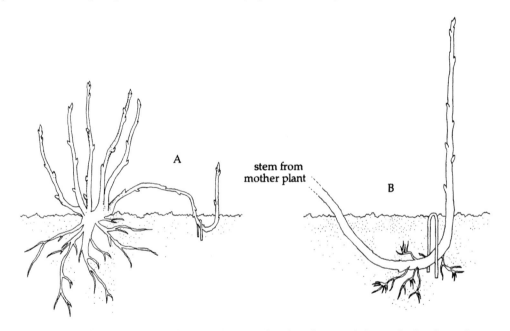

Fig. 4.3. Simple Layering. **A.** The stem is constricted and pegged down during late winter. A cane is sometimes placed adjacent to the shoot and the stem tied to it to keep the shoot vertical. **B.** Sufficient root development has normally occurred by the following fall to allow the stem to be severed from the mother plant. Reprinted with permission from Macdonald 1986.

virginiana present difficulties. Layering is most commonly undertaken by gardeners who want to propagate a particular specimen and only want a few offspring. In such cases, a layering bed is prepared at the place the propagator wishes to root the stem. A bed is usually a wooden box containing a rooting medium such as peat and sand or peat and perlite. It is suggested that layering be done in the spring before new growth begins. A low branch on the tree being propagated is bent to the ground, the stem wounded at the point where it touches the ground, and the wounded area buried in the rooting medium.

Some propagators prefer to layer plants in the summer, in which case the stem should not be wounded. Johnstone (1955) suggests that the best part of the stem to use for layering is the point where new, soft growth emerges from the older, woody growth. Any wounding is done on the older growth. The stem is kept in place by a wooden peg, stake, or rock. The growing tip of the layered branch is then encouraged to grow upwards by propping it up using a forked stick or stake. The layering bed cannot be allowed to dry out. The wounded portion of the stem may be wrapped with sphagnum moss prior to burying in order to protect it against desiccation. Sufficient root development at the wounded area requires 1–3 years. Propagation by layering requires a longer period of time if using older limbs. When well rooted, the new plant is severed from the parent plant and transplanted.

Layering techniques in the nursery industry are somewhat different from those practiced on a smaller scale in the garden. In the nursery, stock or stool plants (the parent plants) are grown in the field specifically for the purpose of layering. These plants may serve as parent plants for many years. Production of numerous, long whiplike shoots (juvenile growth) is encouraged by pruning. These shoots are pegged down to root as described above, or are wounded at the bases and protected by soil mounded around the plants to cover the wounded stems. Rooting typically occurs in less than a year since juvenile tissue roots quickly. Chase (1964), whose discussion of this stooling method is very good, suggests simply wrapping the wounded area with sphagnum moss soaked in a rooting hormone solution to speed the rooting process. This method also has the advantage of allowing the stems to remain upright during rooting.

Tissue Culture

Tissue culture, or micropropagation, involves the regeneration of a whole new plant from a tiny bit of a stock plant. Enormous quantities of plants can be grown from a few cells from a parent plant. This process produces a plant genetically identical to its parent, similar to that from a rooted cutting, but the process is somewhat more complicated. The tiny "cutting," called an explant in tissue culture, can be taken from many different parts of the plant, including shoot or root tips, leaf or flower buds, axillary buds, and leaf pieces. Each plant part contains meristematic tissue which is made up of cells whose final development and function in the plant are not yet determined. These meristematic cells may become leaves, stems, roots, and so forth, depending on environmental conditions as well as signals given by the plant via various hormones.

When the explant is placed in a test tube on a medium containing a rich formulation of nutrients and hormones, the meristematic tissue generates callus tissue which can be repeatedly divided and grown on to produce yet more immature tissue. At some point the callus tissue is placed on a different nutrient and hormone medium formulated to induce the development of shoots. These new shoots are rooted, either by

A small plant of *Magnolia grandiflora* which resulted from tissue culture. After a period of further root growth, this plant can be potted and placed in a greenhouse until large enough to plant outdoors. Photo by John Tobe, Clemson University, Clemson, South Carolina.

further culture or in the greenhouse, and grown to full size in the greenhouse. A simplified outline of the tissue culture process is presented in Figure 4.4. The description presented here is a simplified discussion of a very complex process. The book by Kyte (1983) is an excellent resource for those readers not familiar with micropropagation.

Propagation of magnolias by tissue culture is a difficult, exact, and expensive task. Due to this fact, it is not generally undertaken by the home gardener or small nursery operation. Only two commercial tissue culture operations in North America currently propagate magnolias: Briggs Nursery, Inc., in Olympia, Washington, and Knight Hollow Nursery, Inc., Madison, Wisconsin. Mr. Steve McCulloch, director of the tissue culture laboratory at Briggs, and Dr. Deborah McCown, president of Knight Hollow, were kind enough to provide much of the information included here. Briggs Nursery has been successful in tissue culturing *Magnolia acuminata* var. *subcordata* 'Miss Honeybee'; *M. kobus* var. *loebneri* 'Merrill' and 'Spring Snow'; and *M. grandiflora* 'Victoria', 'St. Mary's', and 'Edith Bogue'. 'Yellow Lantern', 'Elizabeth', 'Ivory Chalis', 'Yellow Garland', 'Sayonara', 'Galaxy', 'Wada's Memory', and *M. kobus* var. *stellata* 'Royal Star' are other successes. McCulloch reports that selections from section *Buergeria* proved most difficult to tissue culture. Knight Hollow focuses their tissue culture efforts on 'Elizabeth', 'Yellow Bird', *M. kobus* var. *loebneri* 'Leonard Messel', and *M. kobus* var. *stellata* 'Royal Star'. Merkle and Wiecko (1990) succeeded in culturing *Magnolia virginiana, M. fraseri*, and *M. acuminata* var. *subcordata* using embryos from immature seeds. This is especially exciting since these North American natives have proven difficult to mass propagate.

As with any propagation method, some species and cultivars are more difficult than others to propagate from tissue culture. Some of this difficulty is attributed to the high phenolic acid content of magnolias, especially yellow-flowered forms. These phenols leak into the tissue culture medium from the explant and inhibit the growth of the cultures, making it necessary to move the explants to fresh media frequently. A second problem, reported by Dr. McCown of Knight Hollow Nursery, is that magnolias have a tendency toward vitreous growth under culture, the shoots appearing crisp and waterlogged. Shoots of this type do not root and seldom revert to normal growth.

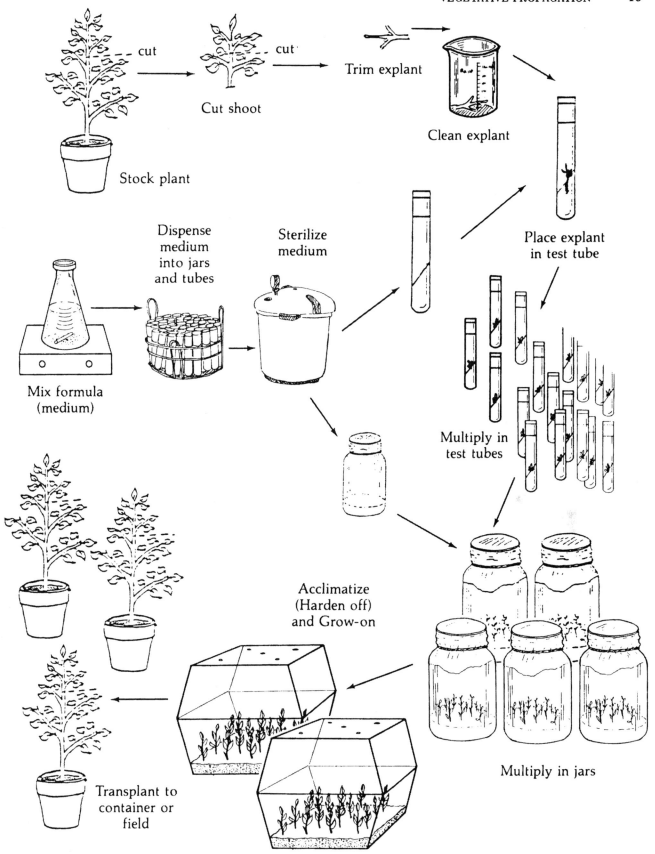

cut

Cut shoot

cut

Trim explant

Clean explant

Place explant
in test tube

Stock plant

Dispense
medium
into jars
and tubes

Sterilize
medium

Mix formula
(medium)

Multiply in
test tubes

Acclimatize
(Harden off)
and Grow-on

Multiply in jars

Transplant to
container or
field

Fig. 4.4. An overview of the tissue culture process showing the sequence of shoot tip micropropagation. Reprinted with permission from Kyte 1983.

Shoot tips are usually used as explants in micropropagation of magnolias. New, actively growing vegetative shoots are most commonly used, yet McCulloch at Briggs Nursery reports success with dormant shoots as well. Biedermann (1987) reports that phenolic acid content is lowest in magnolias after dormancy has broken—in December–March in her study. As a consequence, she suggests taking explants at this time. McCown noted that success rates dropped drastically when explants were taken later in the season.

The tissue culture medium used in magnolia micropropagation varies with the species and cultivar being propagated. Magnolias cannot be cultured on a typical woody plant medium, but prefer a modified solution. McCulloch advises that adjustments both in mineral salts and growth regulators are necessary to tailor the medium to magnolias. Tobe (1990) achieved results on a woody plant medium enriched with benzyladenine. Biedermann (1987) found that a reduced KNO_3 (potassium nitrate) concentration and reduced level of carbohydrates increased success rates in plants with high phenolic compounds.

Magnolias are usually given sixteen hours of light and eight hours of dark and kept at 70°F (21°C) during shoot multiplication and elongation. DeProft et al. (1985) reported best growth with jar or test tube caps that promote gas exchange.

Once the explants have produced sufficient shoots, rooting is induced either in culture or in the greenhouse. This has historically been a difficult task with magnolias as reported by both McCulloch and McCown. Maene and Debergh (1985a, 1985b) addressed this problem in two experiments using *Magnolia × soulangiana*. They found that longer shoots, 1.5 in. (4 cm) in length, were much more likely to produce roots than shorter shoots, 0.75 in. (2 cm) in length. They also found that, although a high-salt culture medium produced more shoots than a low-salt medium, shoots arising from plants cultured in a low-salt medium rooted more quickly than their high-salt counterparts. Both studies showed that rooting was improved by the addition of water to the culture medium for seven days prior to the removal of the plant to the greenhouse for rooting. No difference in rooting success was found using the following rooting media: peat, peat and perlite 3:1, peat and vermiculite 3:1, and peat and polyurethane 3:1. Fungicide applications prior to rooting seem to inhibit rooting, so treatment of the medium after the rooting period seems wise, if such treatment is necessary. Once rooted, the plants are potted and handled in a conventional way.

CHOOSING A PROPAGATION METHOD

Given so many different methods of propagation, which should be used? For commercial growers the answer is usually cuttings, which allow production of many clones from a parent plant and eliminate the offspring variability found in seedlings. If even more clones are required from a smaller amount of parent plant material, a contract for this work with a commercial tissue culture laboratory could be considered. Either method may produce a mature, flowering specimen more quickly than growing a plant from seed.

Gardeners who want a few additional plants should try their hand at grafting or layering. These methods are inexpensive and relatively easy to learn. They also produce clones of the desired specimen, but they do not require the expensive misting equipment often necessary for cutting propagation. Collecting and germinating seeds also offers an inexpensive and easy way for a gardener to increase a magnolia collection, yet since seedlings vary, this method cannot be used to increase plantings of a cultivar or favored specimen. Plant breeders, however, may wish to grow magnolias from seed to exploit their inherent variability in a search for superior forms.

REFERENCES AND ADDITIONAL READINGS

Afanasiev, Michael. 1937. A physiological study of dormancy in seed of *Magnolia acuminata*. Ithaca, NY: *Cornell University Agricultural Experiment Station* Memoir 208.

Biedermann, Ilse E. G. 1987. Factors affecting establishment and development of magnolia hybrids in vitro. *Acta Horticulturae* 212:625–627.

Bojarczuk, K. 1985. Propagation of magnolias from green cuttings using various factors stimulating rooting and growth of plants. *Acta Horticulturae* 167:423–431.

Browse, P. McMillan. 1986. Notes on the propagation of magnolias from seed. *The Plantsman* 8(1):58–63.

Chase, Henry H. 1964. Propagation of oriental magnolias by layering. *Proceedings of the International Plant Propagators Society* 14:67–69. A good article describing one method for commercial production by layering.

DeProft, Maurice P., Ludo J. Maene, and Pierre C. Debergh. 1985. Carbon dioxide and ethylene evolution in the culture atmosphere of *Magnolia* cultured in vitro. *Physiologia Plantarum* 65:375–379.

Dirr, Michael A., and Beth Brinson. 1985. *Magnolia grandiflora*: a propagation guide. *American Nurseryman*, 162(9):38–51. Especially useful for information on rooting cuttings.

Dirr, Michael A., and Charles W. Heuser. 1987. *The Reference Manual of Woody Plant Propagation*. Athens, GA: Varsity Press. A good reference for methods of propagation. Includes a section on propagation of magnolias.

Dodd, Tom. 1953. Propagation of oriental magnolias from soft-wood cuttings. *Proceedings of the International Plant Propagators Society* 3:108–110.

Eisenbeiss, Gene. 1978. Rooting magnolias from cuttings. *Journal of the Magnolia Society* 14(2):17–20.

Ellis, David G. 1988. Propagating new magnolia cultivars. *Proceedings of the International Plant Propagators Society* 38:453–456.

Evans, Clytee R. 1933. Germination behavior of *Magnolia grandiflora*. *Botanical Gazette* 94:729–754.

Hartmann, Hudson T., Dale E. Kester, and Fred T. Davies. 1990. *Plant Propagation: Principles and Practices*, 5th ed. Englewood Cliffs, NJ: Prentice-Hall. A good discussion of principles and techniques.

Heineman, Harry J. 1982. Complete amateur budding from the ground up. *Journal of the Magnolia Society* 18(2):7–10.

Hickman, Joseph W. 1990. Home-made, jerry-rigged, indoor fog propagating set-up. *Journal of the Magnolia Society* 26(1):19–21.

Johnstone, George H. 1955. *Asiatic Magnolias in Cultivation*. London: The Royal Horticultural Society.

Kehr, August E. 1982. A budding magnoliaphile: making game of grafting. *Journal of the Magnolia Society* 18(2):11–13.

Kyte, Lydia. 1983. *Plants from Test Tubes: An Introduction to Micropropagation*. Portland, OR: Timber Press.

Lamb, J. G. D., and J. C. Kelly. 1985. The propagation of deciduous magnolias. *The Plantsman* 7(1):36–39.

Macdonald, Bruce. 1986. *Practical Woody Plant Propagation for Nursery Growers.* Portland, OR: Timber Press. A very complete and thorough reference book for all aspects of plant propagation.

Maene, Ludo, and Pierre C. Debergh. 1985a. Liquid medium additions to established tissue cultures to improve elongation and rooting *in vivo. Plant Cell Tissue Organ Culture* 5:23–33.

———. 1985b. Problems related to *in vivo* rooting of *in vitro* propagated shoots. *Advances in Agricultural Biotechnology* 14:59–72.

McDaniel, Joseph C. 1963. Securing seed production in *Magnolia acuminata* and *M. cordata. Proceedings of the International Plant Propagators Society* 13:120–123.

———. 1973a. Magnolias: try your hand at grafting them. *Journal of the Magnolia Society* 9(1):7–10.

———. 1973b. Chip-budding, a reliable method for propagation of trees. *Journal of the Magnolia Society* 9(1):10–12.

Merkle, Scott A., and A. T. Wiecko. Somatic embryogenesis in three magnolia species. *Journal of the American Society for Horticultural Science* 115(5):858–860.

Perry, Fred B., and H. Max Vines. 1972. Propagation of *Magnolia grandiflora* cuttings as related to age and growth regulators. *Journal of the American Society for Horticultural Science* 97(6):753–756.

Savage, Philip J. 1973. An amateur's method for growing magnolias from seed. *Journal of the Magnolia Society* 19(4):7–15.

Tobe, John David. 1990. *In vitro* growth of *Magnolia grandiflora* cv. Bracken's Brown Beauty. *Journal of the Magnolia Society* 26(1):4–8.

Tubesing, Charles. 1983. Chip-budding magnolias. *Journal of the Magnolia Society* 19(1):17–19 (Part 1) and 19(2):6–14 (Part 2).

———. 1987. Chip-budding of magnolias. *Proceedings of the International Plant Propagators Society* 37:377–379.

5

TAXONOMY

The naming and classification of plants is often a mystery to those who are not involved in the science of plant taxonomy. This chapter is intended to make taxonomy less mysterious and more useful for horticulturists. Its second purpose is to provide a basis for understanding the taxonomic information provided for each species in chapter 7.

Plant taxonomy is the science of identifying, naming, distinguishing, and classifying plants according to their presumed evolutionary relationships. Nomenclature is the naming of plant species, genera, families, and so on, and is strictly governed by a set of rules aimed at maintaining the integrity of the procedures for assigning names and the forms of the names. Nomenclature is an integral part of plant taxonomy, yet it is often seen by horticulturists as an arcane ritual performed by fussy botanic types intent on confusing dedicated gardeners. The business of naming plants correctly is analogous to the work of lexographers who shape and structure language by preparing dictionaries, intending to give words and their meanings some stability. This chapter provides an overview of these disciplines and an explanation of the major principles involved. Of necessity, this will be brief and oriented to users of this book. Readers seeking a more thorough treatment can refer to the publications listed at the end of the chapter.

NAMING PLANTS

The naming of a plant is governed by rules set forth in the *International Code of Botanical Nomenclature*. In the early period of scientific plant taxonomy, such rules had not yet developed. Virtually all early plant scholars and scientists wrote their works in Latin and used a Latin description to assign plant names. Thus, in 1803 *"Magnolia foliis cordatis, subtus subtomentosis: floribus flavis"* was how Michaux identified *Magnolia acuminata* var. *subcordata*. Michaux's description translates as "a magnolia with cordate leaves, subtomentose beneath, flowers yellow." Such descriptions were sometimes several sentences long.

In the mid 1700s Carl Linnaeus, now regarded as the father of taxonomy, proposed the binomial (two-name) system of nomenclature used today. This system involves a genus name (plural: genera), such as *Magnolia*, and a specific epithet, such as *denudata*. The two words together are the scientific name of the species—in this example, *Magnolia denudata*. Linnaeus also wrote the first book of rules governing the naming of plants, the *Critica botanica* published in 1737. It was followed shortly thereafter by several sets of other rules written by Linnaeus's successors. At about the same time, the number of plant names increased very rapidly due to the enormous growth of plant-hunting expeditions associated with and following upon the heels of the European explorations of the wild. With this dramatic growth in the understanding of the enormous plant diversity of the world, it became increasingly obvious that an international set of standard rules was needed to maintain order and organization in botanical nomenclature. At the urging of the Swiss botanist Alphonse de Candolle, an international gathering, the First International Botanical Congress of botanists from North America and Europe, met in Paris in 1867, where they adopted, with modifications, the *Laws of Botanical Nomenclature* proposed by de Candolle. These laws provided that only binomial names, the first of which were provided by Linnaeus, would be considered valid and that the oldest name for a given plant had priority over newer names. Requirements for "valid publication" of new names were also set forth. These rules remain in force today, although they have undergone minor changes over the years and other rules have been added.

After the meeting of the first congress in 1867, numerous meetings of international groups followed. This was due to the fact that many North American botanists disagreed with their European counterparts on proposed additions to the rules. As a result the North American botanists met and agreed in 1907 on a separate American Code, which incorporated the rules for which they had argued. From 1907 until 1930 two separate botanical codes were in use—the American Code and the rules based on de Candolle's *Laws*. Compromise between the two schools was reached at the International Botanical Congress meeting in Cambridge, England, in 1930, from which emerged the first *International Code of Botanical Nomenclature (ICBN)*. These rules have undergone subsequent modification at meetings of the congress in 1936, 1950, 1954, 1959, 1961, 1981, and 1987.

The *ICBN* in its latest form is an extensive and detailed set of rules and recommendations for assigning botanical names. The rules most relevant to the taxonomy and nomenclature of Magnoliaceae are dealt with here. That which follows cannot be viewed as a summary of the *ICBN*, but rather as a presentation of those principles of greatest importance to the reader concerned with the taxonomic treatment of horticulturally important plants.

Ranks of Classification

The *ICBN* provides the basis for the ranks of classification used today. These include in descending order, kingdom, division, class, order, family, genus, species. The *ICBN* also provides for sub-categories of each rank (i.e., subclass, suborder, and so forth) as well as the additional ranks of tribe (between family and genus), section and series (between genus and species), and variety and form (within species). The name of each rank requires a specific Latin ending. The ending-*aceae* is required for family names, such as Magnoliaceae. The term taxon (plural: taxa) is used to denote a taxonomic category of any rank. For example, *Magnolia grandiflora* is a taxon, as is *Magnolia fraseri* var. *pyramidata*.

Type Designation

A type is defined as "the element to which the name of a taxon is permanently attached." In *Magnolia*, this is often a plant specimen collected in the wild by a plant explorer. Thus, when a new name is proposed for a plant species, that name is "attached," or corresponds, to a plant specimen identified by the author. This specimen, which is usually mounted on an herbarium sheet but may be a photograph or figure, represents a permanent, tangible element which is used as a physical example of the plant the author intended to identify with the name given it. The designation of a type specimen and the location of the type is required when a new plant name is proposed.

Valid Publication of Latin Names

The *ICBN* requires that the name proposed for a plant species be written and published in a printed form that can be distributed throughout the botanical community. In order to be validly published, the proposed plant name must be in a format which meets the rules of the *ICBN* and must be accompanied by a diagnosis or description written in Latin. A diagnosis is a short statement of how the plant species being named differs from closely related taxa, by referring to the descriptions of the previously named taxa. Latin remains the language for formal botanical publication because it is still recognized as a universal language that can be written and translated by trained botanists worldwide. Publication in trade catalogs or newspapers after January 1953, or in seed exchange lists after January 1973, does not constitute valid publication of species names.

Priority and Conservation

The correct name of a plant species is that which is the earliest validly published legitimate name, beginning with the publication by Linnaeus in 1753. Names assigned to plants prior to Linnaeus usually were not binomials and so are not considered validly published. The date of publication of the name is assumed to be the date upon which the published account became available to other botanists. There are a few exceptions to the rule of priority; the *ICBN* includes a list of conserved names (*nomina conservanda*). These names are retained, even though they may not be the earliest, due to the fact that they are established and widely used. Conserved names are restricted to plant groups of major economic importance.

Grammar and Orthography

Species names must be in Latin, and the grammar and gender of the genus and specific epithet must be the same. For example, *Tulipastrum americanum* var. *subcordatum* is correct; but when var. *subcordatum* is associated with *Magnolia acuminata*, the ending must be changed to match, becoming *Magnolia acuminata* var. *subcordata*. When a plant is named in honor of a person, the epithet ending depends on whether the person is male or female and whether the person's name ends in a vowel or consonant. There are numerous other rules pertaining to the endings of botanical names; the interested reader is referred to Article 73 of the *ICBN* for particulars. If an author publishes a Latin name with an incorrect ending, however, the name is still considered to be validly published. Richard Howard is credited with valid publication in 1948 of his new species *Magnolia hamori*, though the correct spelling is *hamorii*. Similarly, *Magnolia liliflora* is considered validly published even though the correct spelling is *liliiflora*. These errors in language, referred to as orthographic errors, are simply corrected by the taxonomist discovering the error. The correction is noted in the literature by a brief statement stating the orthographic error and giving the correct spelling. Such errors do not affect the valid publication of the name.

Inappropriate or Disagreeable Epithets

A validly published Latin name may not be rejected simply because the epithet is inappropriate or disagreeable. For example, *Magnolia tripetala* does not have three petals as is suggested by the epithet, but this is the earliest validly published name and therefore must be used.

Author Citation

The *ICBN* requires that the author of a plant name be cited with publication of the name. The author's name is usually represented by standard abbreviations. For example, *Magnolia tripetala* L. was named by Linnaeus and *Magnolia virginiana* var. *australis* Sarg. was named by Charles Sprague Sargent. The citation of authors allows for a clear understanding of which name is being referred to. *Magnolia hypoleuca* was published by Siebold and Zuccarini in 1846 and Diels in 1900 to refer to two different plants. Reference to *Magnolia hypoleuca* Sieb. & Zucc. assures that there will be no mistake as to which authors' name and description are being referred to.

Latin Names of Hybrids

The code considers Latin names of hybrids (e.g., *Magnolia* × *soulangiana*) to fall under the same rules as non-hybrid taxa. Thus, grex (Latin names for hybrids) names are subject to the same rules of priority, grammar, and orthography and must be accompanied by a Latin description or diagnosis and a designated type specimen. Hybrids given cultivar names (e.g., *Magnolia* × 'Galaxy') do not fall under these rules, but are treated as cultivars.

THE *INTERNATIONAL CODE OF NOMENCLATURE FOR CULTIVATED PLANTS*

The *ICBN* requires that plants brought into cultivation from the wild retain the name used to designate the wild plant and also that the Latin names of hybrids, whether a natural cross or occurring in cultivation, comply with the rules governing the naming of non-hybrid taxa. Other than these two rules, the *ICBN* contains no provisions for the nomenclature of cultivated plants. As the breeding, selection, and naming of new forms of cultivated plants increased over the years, it became obvious that the naming of these forms had to be regulated in some way. As a result, the International Botanical Congress Commission for Nomenclature of Cultivated Plants and the International Commission for Horticultural Nomenclature and Registration met in 1952 and drew up the 1953 *International Code of Nomenclature for Cultivated Plants,* the "Cultivated Code." The next year the International Commission for the Nomenclature of Cultivated Plants of the International Union of Biological Sciences was formed, the objective being to produce a Cultivated Code expanded to include plants used in agriculture and forestry in addition to horticulture. This code was published in 1958. The 1961, 1969, and 1980 (current) codes are little changed from that of 1958. At the time this book went to press, a proposal for a new revision was underway. The Cultivated Code does not conflict with the *ICBN* but simply adds provisions for cultivated plants, especially cultivars. The principal rules and recommendations governing the names of cultivated plants are outlined here.

Designation of Cultivars

The term "cultivar" (abbreviated "cv.") is derived from *cultivated variety* and is defined as "an assemblage of plants which is clearly distinguished by any characters and which, when reproduced, retains its distinguishing characters." A cultivar may consist of (1) a single clone or several very similar clones (a clone is a genetically identical group of individuals derived from a single individual) or (2) an assemblage of cross-fertilized individuals which may show genetic differences within that assemblage, but which have one or more characters by which the assemblage may be differentiated from other species or cultivars.

The origination of the character which defines a cultivar is irrelevant insofar as naming the cultivar is concerned. Thus, all dwarf forms of *Magnolia virginiana,* for example, should be included under one cultivar name unless there is another distinct character separating them. A cultivar must be sufficiently distinct from others in one or more characteristics in order to deserve cultivar status. This point cannot be stressed strongly enough. The all-too-frequent naming of numerous questionably distinct or horticulturally unimportant forms only leads to confusion among gardeners and horticulturists.

Valid Publication of Cultivar Names

As is the case with all but a handful of conserved botanical names, the oldest, validly published cultivar name for a particular form is given priority over newer names. An exception to this rule allows the use of a more recent name if it is the cultivar name most commonly used to refer to that form.

Valid publication requires the distribution of a publication including the cultivar name and, for any plant name proposed after January 1959, a description or reference which compares and distinguishes the cultivar from previously described cultivars. Descriptions of cultivars need not be in Latin but should contain the information necessary to distinguish a cultivar from all other forms. It is not, for example, sufficient to write "evergreen with white flowers" in describing a cultivar of *Magnolia virginiana* var. *australis* because that description fits the typical form as well as numerous other cultivars.

A cultivar description should also include information on the origin or parentage of the cultivar and the originator. Reference to a standard color chart such as that provided by the Royal Horticultural Society is suggested. In addition, it is recommended that a photograph be included and a preserved specimen be filed at a public herbarium. The description and photograph should also be sent to the official registrar for inclusion in the register (the address for the *Magnolia* registrar is included at the end of this chapter).

It should be noted here that these rules and recommendations were developed to encourage adequate records of cultivated plants. While some breeders and growers think it unnecessary to go to such lengths to introduce a cultivar, this information will be of use to all those in the future interested in such a group of plants. Like all else subject to human taste, cultivars go in and out of fashion; thus the possession of accurate information, regardless of the current popularity of the cultivar, is of crucial importance.

Selection of Cultivar Names

Prior to 1959 cultivars were commonly given one-word Latin names such as 'Alba' or 'Nana'. These Latin names are subject to the same rules of grammar and orthography as other Latin botanical names. But the Cultivated Code requires that all cultivars named after January 1959 be given "fancy" or non-Latin names such as 'Little Gem' or 'Edith Bogue'. Such names may consist of no more than three words, and preferably one or two. All the words in the name are to begin with capital letters when printed. They may be enclosed in single quotes ('Little Gem') or preceded by the abbreviation "cv." (cv. Little Gem), but never both. Cultivar names are neither italicized nor underlined, but are always printed in roman type.

The Cultivated Code recommends that in selecting a cultivar name, the following should be avoided: abbreviations, especially of proper names; forms of address such as Dr., Mr., Mrs., Mademoiselle; excessively long phrases; names likely to be confused with other cultivars; and names that include the common name of the plant.

Cultivated Hybrids

Under the provisions of the Cultivated Code, hybrids must be given a collective name that includes all progeny of crosses between two parent species. This collective name may be printed as a formula (such as *Magnolia denudata* × *Magnolia liliiflora*), a Latin grex name (for example, *Magnolia* × *soulangiana*), or a name in a vernacular language not to exceed three words (for example, *Magnolia* Freeman Hybrids). When using the formula name, it is customary either to put the parental names in alphabetical order (the practice followed in this book) or to list the female parent first if the direction of the cross is known. The latter is encouraged since it provides more information. The ordering of parental names used in a particular publication must be clearly stated. If the hybrid is given a Latin name, that name is governed by the *ICBN* as is any other Latin name, and all the *ICBN* rules apply.

If a hybrid is introduced into cultivation without a cultivar name, it must be given one if a second form of the same hybrid is introduced.

Cultivar Registration

While the authors of the Cultivated Code recommend that cultivars be registered, registration is not governed by the code. But registration with some central authority is clearly necessary to keep order in naming cultivars. As a consequence, registration authorities, international and national, have been established. Registration may be voluntary or mandatory, depending upon the decision made by each country. In general, acceptance of a cultivar name by the registrar does not imply that the cultivar is distinct or in any way meritorious. However, some registration authorities do test the cultivars for distinctness, which is a highly desirable practice. As noted above, as much information as possible should be given to the registrar when registering a new cultivar. Clearly, registration authorities have become important permanent repositories for horticultural information.

Plant Patents and Trademark Names

Plant patents have become increasingly popular in the United States and other countries as they offer an additional option when introducing a new cultivar. Patenting allows the originator of an exceptional plant form to obtain the protection and rewards, monetary and otherwise, arising from the results of his work. Plants "discovered" in the wild are excluded from patenting; the plant must have been selected from cultivated material or as the result of a breeding program, and must be vegetatively reproduced. A plant patent is obtained through a process similar to patenting an invention. As patenting is a legal process it is not governed by the code.

In the United States an abstract outlining the history, description, and distinctness of the plant, including appropriate photographs, is prepared. A patent application is then submitted to the U. S. Patent and Trademark Office, where reviewers investigate the plant and related forms to determine whether it is distinct enough to be granted a plant patent or not. If a patent is granted, only growers licensed by the patent holder have the authority to propagate and sell the patented cultivar for a seventeen-year period. At this time only eight *Magnolia* cultivars have been patented—*Magnolia grandiflora* 'Bracken's Brown Beauty' (#5520), 'Majestic Beauty' (#2250), 'Russet' (#2617), 'Samuel Sommer' (#2015), and 'San Marino' (#2830); *M.* × *brooklynensis* 'Evamaria' (#2820); *M.* × 'Butterflies' (#7456); and *M.* × 'George Henry Kern' (#820).

A cultivar may also be trademarked. The cultivar is in this case given not only a cultivar name, as usual, but a trademark name as well. The owner of a trademark cannot restrict the use, propagation, or distribution of a cultivar, but only the use of the trademark name.

CLASSIFYING PLANTS

The hierarchical classification scheme—order, family, genus, species, and so forth—outlined in the *ICBN* is the basis for all plant classification. Each category within the hierarchy contains a related group of plants. The closeness of the relativity becomes increasingly narrow further down the hierarchy, thus producing a "family tree." The narrowest categories—that is, the groupings of plants sharing the greatest number of common characteristics—are species, subspecies, and variety.

Once a new species is identified as being related to other plants (such as belonging to the genus *Magnolia*), it must be placed within the existing classification for that genus of plants. If it is thought to deserve the status of species, a species name is given to it. If the plant appears to be very closely related to an existing species, and appears indistinguishable from that species except in minor differences, the new plant may be considered a subspecies or variety of that species. This is not to say that it is exactly the same as that species, as there are recognizable differences. For example, *Magnolia macrophylla* var. *ashei* differs from typical *M. macrophylla* in very small ways even though it is quite difficult to differentiate the two; thus *ashei* is considered a variety. But both *M. macrophylla* and var. *ashei* differ more greatly from *M. tripetala* than from one another and are easily distinguished from the latter. Therefore, *M. tripetala* is viewed as a species separate from *M. macrophylla* and var. *ashei*.

The ranks of subspecies and variety are often used interchangeably, but seldom are both used in classifying a single group of plants. Both are used to denote variation within a species. The *ICBN* leaves the use of these two ranks to the discretion of the author. However, when a variety or subspecies is attributed to a species (for example, var. *ashei* and var. *dealbata* are attributed to *M. macrophylla*), the typical form (*M. macrophylla*) automatically receives a variety name—the same name as the specific epithet. Thus, *M. macrophylla* is composed of three varieties: *M. macrophylla* var. *macrophylla* (typical), var. *ashei*, and var. *dealbata*. The typical variety need not be the most common. The variety named first becomes the typical variety when subsequent varieties are named.

Taxonomists differ in their opinions of what constitutes a difference "great" enough to justify species status for new forms or "minor" enough to require that the plants be recognized as varieties. These disagreements will always exist, for plant classifications are constructed by man and so subject to the limitations of human knowledge. But a few guidelines have come to be recognized as useful in making such distinctions. For the most part these guidelines have grown out of a better understanding of the natural variation of plants.

Some types of variation have been found to be more significant in delimiting species than others. In general, vegetative plant parts (that is, stems and leaves) are easily affected by environmental conditions and often lead to erroneous conclusions if used as distinguishing taxonomic characters. (A character is a plant feature used in comparisons between taxa.) For example, leaf size, thickness, and pubescence may be dependent upon environmental factors, most notably sunlight, water, and temperature. Thus, they are too variable to be used as important diagnostic criteria in plant taxonomy.

Reproductive plant parts (flowers, fruits, and seeds) are less variable and less subject to change by the environment. Reproductive features, such as number and structure of floral parts, are therefore considered to be the most stable characters to use in taxonomic analyses. In addition, the taxonomist must take into account geography, genetics, ecology, biochemistry, and physiology of a plant in order to make the most informed decision as to its taxonomic status.

TAXONOMY OF MAGNOLIAS

Many difficulties are encountered in the classification of the magnolias. Probably the most important of these is the rarity of some of the species and the absence of good information about them in the literature and in herbaria. In addition, the widespread cultivation of magnolias makes it difficult now to come to an understanding of the natural variation and geography of the species. An exhaustive taxonomic treatment of the species of *Magnolia* would take years to complete. I have in this book, therefore, attempted only to synthesize all the published information available on magnolias; I have not undertaken a taxonomic revision or monograph of the genus. Further taxonomic work will be required to determine whether all the species discussed here are truly distinct or not.

Many of the species I consider to be questionable differ from their close relatives largely in degree of pubescence. As noted earlier, vegetative characters such as pubescence are quite variable due to environmental factors which affect the degree of pubescence found on a given plant. Furthermore, pubescence is a function of the maturity of the plant or plant part; the young leaves of magnolias are often densely pubescent, becoming almost glabrous at maturity. This variation makes pubescence a quite questionable taxonomic character unless it is accompanied by more stable characters.

Also noted above, since the ranks of subspecies and variety are used interchangeably by different authors to denote variation within a species, *all taxa below the species level*, with the exception of cultivars, *are treated as varieties in this book*. My decision was based on two criteria: the rank of variety (1) is the one most commonly used in the *Magnolia* literature and (2) is the rank usually used by taxonomists when referring to groups of plants of horticultural importance. Therefore, when faced with the choice of varietal or subspecific rank (for example, in *Magnolia campbellii* [var. or subsp.] *mollicomata*), I have chosen to use varietal rank.

The taxonomic treatment of *Magnolia* in this book is conservative. That is, neither the name nor rank of any taxon is changed if there is doubt that it should be so changed. The inclusion of *ashei* and *dealbata* as varieties of *M. macrophylla* may be considered an example of "lumping" by some, while the separation of *M. sieboldii* and *M. sinensis* may be seen as "splitting." These examples may seem a taxonomic contradiction, but in both cases they reflect my preference for a conservative treatment in which deviations from long-recognized and widely used names are made only if the taxon in question is well understood and there are no doubts as to the wisdom of the change.

It is sometimes confusing to those who are not taxonomists that a plant may be considered by some to be a variety (such as *Magnolia kobus* var. *stellata*) and by others to be a species (*M. stellata*). A nursery professional may wonder which name is correct and which should be given in the nursery catalog. This is usually reduced to a matter of personal preference. Both names have been validly published. The publication of *M. kobus* var. *stellata* by Blackburn in 1955 does not render the previous publication of *M. stellata* invalid. Thus, one who believes the Star Magnolia is a species rather than a variety has every right to call it *M. stellata*, while others may call it *M. kobus* var. *stellata*. Although taxonomists would sometimes like to think their treatment should be the final word on the subject, that is not the case. Undoubtedly I have made some debatable taxonomic decisions in this book. I wish to acknowledge here each reader's right to disagree and to adhere to whatever taxonomic treatment reflects his or her ideas about the plants. A discussion of the taxonomic decisions I have made for each taxon is included in the plant descriptions.

REFERENCES AND ADDITIONAL READING

Bailey, Liberty Hyde. 1933. *How Plants Get Their Names.* New York: Macmillan. Reprint. New York: Dover Publications, 1963. A short, easy to understand discussion of plant nomenclature with a glossary of the meanings of common Latin names.

Brickell, C. D., ed. 1980. *International Code of Nomenclature for Cultivated Plants—1980. Regnum Vegetabile,* vol. 104. The Hague: Dr. W. Junk. Available from the American Horticultural Society, Mt. Vernon, VA 22121, or the Royal Horticultural Society, Vincent Square, London, SW1P 2PE, Great Britain.

Gledhill, D. 1985. *The Names of Plants.* Cambridge University Press. A discussion of nomenclatural principles and an extensive glossary of Latin names.

Greuter, W., chairman, editorial committee. 1988. *International Code of Botanical Nomenclature. Regnum Vegetabile* vol. 118. Germany: Koeltz Scientific Books. The code itself is only 89 pages long. The remainder of the 328-page text is a list of conserved names.

Lawrence, George H. M. 1951. *Taxonomy of Vascular Plants.* New York: Macmillan. One of the classic books on all aspects of plant taxonomy, this 823-page volume contains an excellent glossary.

Moore, James N. and Jules Janick. 1983. Cultivar release and protection. In *Methods in Fruit Breeding.* West Lafayette, IN: Purdue University Press. Pp. 383–397. An excellent source of information on naming and releasing cultivars, as well as good information on plant patents.

Radford, Albert E., William C. Dickison, Jimmy R. Massey, and C. Ritchie Bell. 1974. *Vascular Plant Systematics.* New York: Harper & Row. An excellent reference manual which presents a more advanced treatment than Lawrence.

Stearn, William T. 1983. *Botanical Latin.* Newton Abbot, Devon, U.K.: David and Charles Publishers. The book for anyone who wants to write or understand botanical Latin. Includes a Latin/English and English/Latin dictionary, a glossary of descriptive terminology; and "how-to" instructions for reading and writing Latin descriptions.

International Magnoliaceae Registration Authority:

Dorothy J. Callaway
P.O. Box 3131
Thomasville, GA 31799
U.S.A.

6

KEYS

These keys are included to provide the reader with an understanding of the scope of the genus *Magnolia* and to assist in the identification of the cultivated species. As discussed in the chapter on taxonomy, floral (reproductive) characters are best used in identification; a flowering specimen is often required to completely identify a species. However, leaf and other vegetative characteristics are also provided for use when flowers are not available. References containing additional keys to the magnolia sections and species are listed at the end of this chapter.

KEY TO GENERA OF MAGNOLIACEAE

This key to the genera within the magnolia family (Magnoliaceae) is based on the work of Dandy (1927b). More recent workers including Keng (1978) and Nooteboom (1985, 1987) have combined various genera within Magnoliaceae with the result that Keng recognizes only three genera and Nooteboom only five. I have retained Dandy's concept of the ten genera because it is the most widely accepted and because I find the combination of genera as presented by Keng and Nooteboom unsatisfactory. As described by Dandy, Magnoliaceae is divided into two tribes. The first of these, Liriodendreae, contains only the genus *Liriodendron*, the Tulip Poplar or Tulip Tree. This genus is separated from others in the family based on its very different fruit morphology. The remaining nine genera are placed in the tribe Magnolieae. The genera accompanying *Magnolia* in this tribe are listed in Table 6.1. *Michelia* is composed of forty-five species native to eastern Asia. Several species, the most common of which is *Michelia figo*, are known in cultivation. Referred to as the banana shrubs, these species

Liriodendron tulipifera, a member of the magnolia family, is native to North America.

Michelia doltsopa in cultivation at the Strybing Arboretum, San Francisco, California. Several species of the genus *Michelia* can be found in cultivation. Photo by Richard Figlar, Pomona, New York.

differ from magnolias primarily in having flowers borne in the axils of the leaves rather than at the end of the branches. *Talauma,* a genus of forty species, is native to eastern Asia and tropical America and differs from *Magnolia* in fruit morphology. *Manglietia* is a genus of twenty-five species native to southeastern Asia. These species have four or more ovules per carpel, differing from the two ovules found in *Magnolia. Elmerrillia,* native to the Malay Archipelago, the Philippines, and New Guinea, consists of seven species with axillary flowers. The small genus *Aromadendron,* containing two species, is native to the Malay Peninsula. These two species are different from others in their indehiscent fruit and fused carpels. *Kmeria* contains two species native to southern China and Indochina. These are the only two species in the magnolia family with unisexual flowers; male and female organs are borne in separate flowers. *Pachylarnax*, also containing two species, is native to eastern Asia. These species have only two to three fruits per aggregate, differing from the numerous fruits produced in a single *Magnolia* "cone." The single species of the genus *Alcimandra*, native to the eastern Himalayas, differs from magnolias in the structure of the gynandrophore.

1. Flowers unisexual. . . . . *Kmeria*
1. Flowers bisexual
    2. Flowers borne in the axils of the leaves
        3. Gynoecium borne on a short stalk (stipitate), anthers dehisce laterally
        . . . . *Michelia*
        3. Gynoecium not borne on a short stalk (sessile), anthers dehisce introrsely
        . . . . *Elmerrillia*
    2. Flowers borne at the end of the branches
        4. Fruit an aggregate of samaras. . . . . *Liriodendron*
        4. Fruit not an aggregate of samaras
            5. Carpels 2–3, fruit capsular. . . . . *Pachylarnax*
            5. Carpels numerous, fruit not capsular
                6. Carpels concrescent (growing together as if fused), tepals 18 or
                more . . . . *Aromadendron*
                6. Carpels usually free (not fused), sometimes concrescent, tepals
                fewer than 18
                    7. Carpels free or concrescent, stipules adnate (fused) to the
                    petiole. . . . . *Talauma*
                    7. Carpels free
                        8. Gynoecium stipitate . . . . *Alcimandra*
                        8. Gynoecium sessile or shortly stipitate
                            9. Ovules 4 or more . . . . *Manglietia*
                            9. Ovules 2 . . . . *Magnolia*

Table 6.1. Genera in the tribe Magnolieae, based on Dandy (1927b).

Genus	Approx. number of species	Diagnostic characters
Magnolia	80	Terminal flowers
		Numerous carpels
		1 or 2 ovules/carpel
		Follicular fruit
Michelia	45	Axillary flowers
Talauma	40	Fruit not follicular
Manglietia	25	4+ ovules/carpel
Elmerrillia	7	Axillary flowers
Aromadendron	2	Fruit not follicular
Kmeria	2	Unisexual flowers
Pachylarnax	2	2–3 carpels
Alcimandra	1	Different flower morphology

KEY TO SUBGENERA AND SECTIONS OF *MAGNOLIA*

This key to subgenera and sections of *Magnolia* is based on that of Dandy (in Treseder 1978) and describes the differences setting each of the two subgenera and eleven sections apart. A list of the subgenera, sections, and species is shown in Table 6.2. Descriptions of these are included in chapter 7.

1. Anthers dehiscing introrsely; flowers neither precocious nor with a much reduced (calyx-like) outer whorl of tepals; leaves evergreen or deciduous; fruit various . . . . Subg. *Magnolia*
  2. Stipules adnate to the petiole; leaves evergreen or deciduous
    3. Leaves evergreen; flower buds at first enclosed in one or more deciduous spathaceous bracts which leave as many annular scars on the peduncle; (Asiatic species)
      4. Fruiting carpels short-beaked, the beak not dorsally flattened . . . . Sect. *Gwillimia*
      4. Fruiting carpels long-beaked, the beak forming a dorsally flattened lanceolate coriaceous appendage and finally becoming more or less recurved . . . . Sect. *Lirianthe*
    3. Leaves deciduous (sometimes persistent in section *Magnolia*); flower buds at first enclosed in a single deciduous spathaceous bract which leaves a single annular scar on the peduncle
      5. Leaves usually large or very large; crowded into false whorls at the ends of branchlets (Asiatic and American species) . . . . Sect. *Rhytidospermum*
      5. Leaves not crowded into false whorls at the ends of the branchlets
        6. Anthers with the connective produced into a short, acute appendage; leaves deciduous or sometimes persistent, glaucous on the undersurface; (American species) . . . . Sect. *Magnolia*
        6. Anthers with the connective blunt or retuse and not normally produced into an appendage; leaves deciduous, the undersurface pale green or somewhat glaucescent; (Asiatic species) . . . . Sect. *Oyama*
  2. Stipules free from the petiole; leaves evergreen
    7. Tepals subsimilar in texture; fruit ellipsoid to oblong, sometimes distorted; petiole elongated
      8. Gynoecium sessile; (American species) . . . . Sect. *Theorhodon*
      8. Gynoecium usually short stipitate; plants entirely glabrous; (Asiatic species) . . . . Sect. *Gynopodium*
    7. Tepals of the outer whorl much thinner in texture than those of the inner whorls; fruit more or less cylindric, usually distorted; petiole short; (Asiatic species) . . . . Sect. *Maingola*
1. Anthers dehiscing laterally or sublaterally; flowers precocious and/or with a much reduced (calyx-like) outer whorl of tepals; leaves deciduous; fruit cylindric or oblong, usually more or less distorted . . . . Subg. *Yulania*
  9. Tepals subequal; flowers appearing before the leaves, white to rose or rose-purple; (Asiatic species) . . . . Sect. *Yulania*
  9. Tepals very unequal, those of the outer whorl much shorter, calyx-like
    10. Flowers appearing before the leaves; inner (large) tepals white or pink (Asiatic species) . . . . Sect. *Buergeria*
    10. Flowers appearing with or after the leaves; inner (large) tepals purple or green to yellow; (Asiatic and American species) . . . . . . Sect. *Tulipastrum*

Table 6.2. Listing of *Magnolia* species in each section and subgenus.

Subgenus *Magnolia*

Section *Magnolia*
Magnolia virginiana and var. *australis*

Section *Rhytidospermum*
M. *fraseri* and var. *pyramidata*
M. *hypoleuca*
M. *macrophylla* and vars. *ashei* and *dealbata*
M. *officinalis*
M. *rostrata*
M. *tripetala*

Section *Oyama*
M. *globosa*
M. *sieboldii*
M. *sinensis*
M. *wilsonii*

Section *Theorhodon*
M. *chimantensis*
M. *cubensis*
M. *domingensis*
M. *ekmanii*
M. *emarginata*
M. *grandiflora*
M. *guatemalensis*
M. *hamorii*
M. *hondurensis*
M. *pallescens*
M. *poasana*
M. *portoricensis*
M. *ptaritepuiana*
M. *schiedeana*
M. *sharpii*
M. *sororum*
M. *splendens*
M. *yoroconte*

Section *Gwillimia*
M. *albosericea*
M. *champacifolia*
M. *championii*
M. *clemensiorum*
M. *coco*
M. *craibana*
M. *delavayi*
M. *eriostepta*

M. *fistulosa*
M. *henryi*
M. *nana*
M. *pachyphylla*
M. *paenetalauma*
M. *persuaveolens*
M. *poilanei*
M. *pulgarensis*
M. *talaumoides*
M. *thamnodes*

Section *Gynopodium*
M. *kachirachirai*
M. *lotungensis*
M. *nitida*

Section *Lirianthe*
M. *pterocarpa*

Section *Maingola*
M. *aequinoctialis*
M. *annamensis*
M. *griffithii*
M. *gustavii*
M. *macklottii*
M. *maingayi*
M. *pealiana*

Subgenus *Yulania*

Section *Yulania*
M. *amoena*
M. *campbellii* and vars. *alba* and *mollicomata*
M. *dawsoniana*
M. *denudata*
M. *sargentiana* and var. *robusta*
M. *sprengeri* and var. *elongata*
M. *zenii*

Section *Buergeria*
M. *biondii*
M. *cylindrica*
M. *kobus* and vars. *loebneri* and *stellata*
M. *salicifolia*

Section *Tulipastrum*
M. *acuminata* and var. *subcordata*
M. *liliiflora*

KEY TO CULTIVATED SPECIES OF *MAGNOLIA*

This key is provided to enable the reader to identify the most commonly cultivated magnolias. For keys which include some species not in cultivation, see references at the end of this chapter.

1. Leaves deciduous
    2. Leaves crowded into false whorls at the ends of branches
        3. Leaf base cordate to auriculate
            4. Leaf blades longer than 1 ft. (30 cm), silvery beneath; twigs pubescent; follicetum pubescent, rounded; tepals sometimes having purple tinges at the base
                5. Mature follicles beaked; tepals usually lacking purple tinge at base . . . . *M. macrophylla* var. *dealbata*
                5. Mature follicles not beaked; tepals usually with purple tinge at base
                    6. Follicetum ovoid to globose; flowers 12–18 in. (30–45 cm) across; tree; leaves 2–3 ft. (60–100 cm) long; twigs densely pubescent . . . . *M. macrophylla* var. *macrophylla*
                    6. Follicetum subcylindric; flowers 9–12 in. (20–30 cm) across; shrub; leaves less than 2ft. (60 cm) long; twigs slightly pubescent . . . . *M. macrophylla* var. *ashei*
            4. Leaf blades shorter than 12 in. (30 cm), not silvery beneath; twigs glabrous; follicetum glabrous, elongate; flowers lacking purple tinge at base of tepals
                7. Leaf shape spatulate-obovate 8–15 in. (20–38 cm) long, apex gradually acute; terminal bud purple; tepals rounded at apex; stamens more than 0.2 in. (6 mm) long . . *M. fraseri* var. *fraseri*
                7. Leaf shape rhombic, 6–9 in. (15–23 cm) long, apex abruptly acute; terminal bud greenish brown; tepals pointed at apex; stamens 0.1–0.2 in. (3–6 mm) long . . *M. fraseri* var. *pyramidata*
        3. Leaf base rounded or cuneate
            8. Leaf base narrowly cuneate; flowers with unpleasant odor; follicetum 2–4 in. (5–10 cm) long . . . . *M. tripetala*
            8. Leaf base rounded to broadly cuneate; flowers with pleasant fragrance; follicetum greater than 4 in. (10 cm) long
                9. Mature follicles with curved beaks over 0.2 in. (6 mm) long; leaves with reddish hairs on lower midrib and veins, petioles light green . . . . *M. rostrata*
                9. Mature follicles with short beaks up to 0.1 in. (3 mm) long; leaves with pale hairs on lower midrib and veins
                    10. Young branchlets and petioles purple; pedicel glabrous; follicetum 5–8 in. (12–20 cm) long . . . . *M. hypoleuca*
                    10. Young branchlets yellowish, petioles yellow-green; pedicel pubescent; follicetum 4–5 in. (10–13 cm) long; leaf apex sometimes emarginate . . . . *M. officinalis*
    2. Leaves alternate, not crowded into false whorls at ends of branches
        11. Flowers precocious or with calyx-like outer whorl of tepals, tepals white to pink, yellow, or purple; fruit usually distorted (except *M. cylindrica*)
            12. Outer whorl of tepals calyx-like; flowers precocious or appearing with the leaves

13. Flowers precocious; tepals white to pink (occasionally purple)
    14. Leaves lanceolate, broadest at or below the middle; petaloid tepals 6–9(12)
        15. Flowers with erect tepals, held erect on branches, not drooping; branchlets lightly pubescent; fruit cylindrical, not distorted.................*M. cylindrica*
        15. Flowers with drooping tepals, held horizontally on glabrous branches; fruit usually distorted
            16. Leaves anise-scented; pedicels glabrous; stamen filaments much shorter than anther..........
............................*M. salicifolia*
            16. Leaves not anise-scented; pedicels pubescent; stamen filaments equal in length to anther ...
...............................*M. biondii*
    14. Leaves elliptic to obovate, broadest above the middle
        17. Petaloid tepals 5–9
            18. Number of petaloid tepals 9; flowers vase-shaped, erect; shrub or small compact tree ...
..........................*M.* × *soulangiana*
            18. Number of petaloid tepals 5–7; flowers opening almost flat, sometimes drooping; large tree with open habit ....*M. kobus* var. *kobus*
        17. Petaloid tepals 11 or more; shrub or small tree with compact branching habit
            19. Petaloid tepals 12–16, spatulate, obovate, or oblanceolate 1–2 in. (2.5–5 cm) wide, sepaloid tepals persistent at anthesis................
.....................*M. kobus* var. *loebneri*
            19. Petaloid tepals (12)15–33, narrowly oblong or strap-shaped, 0.2–0.5 in. (.5–1.5 cm) wide, sepaloid tepals often falling before anthesis...
.......................*M. kobus* var. *stellata*
13. Flowers not precocious; tepals yellow-green or purple
    20. Flowers yellow or yellow-green, occurring after the leaves are produced; leaves acuminate; large trees or shrubs
        21. Flowers yellow-green, twigs glabrous...*M. acuminata* var. *acuminata*
        21. Flowers yellow, twigs pubescent...*M. acuminata* var. *subcordata*
    20. Flowers purple, precocious, continuing after the leaves are produced; leaves usually ± cuspidate; small shrub.....
....*M. liliiflora*
12. Outer whorl of tepals petaloid; flowers precocious
    22. Flowers erect; leaves abruptly acuminate
        23. Tepals 9, white..........................*M. denudata*
        23. Tepals 12–16, pink or white
            24. Innermost whorl of tepals held erect, enclosing the gynandrophore, outer whorls held horizontally, forming a "cup-and-saucer" shape; leaves elliptic or narrowly ovate, rounded at the base (rarely cuneate)
                25. Tepals white ....*M. campbellii* var. *alba*
                25. Tepals pink

26. Pedicels glabrous . . . .
. . . . *M. campbellii* var. *campbellii*
26. Pedicels with yellow pubescence . . . .
. . . . *M. campbellii* var. *mollicomata*
    24. Innermost whorl of tepals not enclosing the
        gynandrophore, all tepals held ± horizontally;
        leaves obovate . . . .*M. sprengeri*
22. Flowers horizontal or nodding; leaves rounded at the apex
  (occasionally emarginate)
    26. Shrubs; tepals 9–12, all roughly equal in length . . . .
    . . . .*M. dawsoniana*
    26. Trees; tepals 10–14, the inner tepals longer than the
        outer . . . . *M. sargentiana* var. *robusta*
11. Flowers not precocious; outer whorl of tepals petaloid; tepals white;
  follicetum cylindric, rarely distorted, often pendulous
    27. Pedicels usually less than 5 in. (2 cm) long; stamens creamy white to
        yellowish; flowers erect; leaves deciduous to persistent on current
        year's growth, glaucous beneath
        28. Twigs and pedicels glabrous; leaves deciduous; shrub to 30 ft.
            (10 m) tall . . . .*M. virginiana* var. *virginiana*
        28. Twigs and pedicels pubescent; leaves persistent; tree to 100 ft.
            (30 m) tall . . . .*M. virginiana* var. *australis*
    27. Pedicels usually 1–2 in. (2.5–5 cm) long; stamens red or crimson;
        flowers nodding or pendent; leaves deciduous, pale green to
        glaucescent or with yellowish or rufous pubescence beneath
        29. Twigs yellowish tan . . . . *M. sieboldii*
        29. Twigs brown, becoming dark brown or purplish
            30. Leaves with lower surface rufous pubescent, leaves
                usually obovate (occasionally elliptic) . . . .*M. globosa*
            30. Leaves with lower surface silver-gray pubescent, leaves
                elliptic to lanceolate . . . .*M. wilsonii*
1. Leaves persistent
  31. Stipules adnate to the petiole . . . . *M. delavayi*
    32. Leaves with silvery gray pubescence beneath . . *M. virginiana* var. *australis*
    32. Leaves pale green beneath, glabrous or pubescent
        33. All plant parts entirely glabrous; leaves no longer than 8 in. (20
            cm) . . . .*M. coco*
        33. Twigs, leaves, and pedicels pubescent; leaves up to 12 in. (30 cm)
            long . . . .*M. delavayi*
  31. Stipules free from the petiole
    34. Gynoecium stipitate; all plant parts completely glabrous . . . . . . *M. nitida*
    34. Gynoecium sessile; lower surface of leaves pubescent . . . . *M. grandiflora*

REFERENCES AND ADDITIONAL READING

Bailey, Liberty Hyde. 1949. *Manual of Cultivated Plants.* New York: MacMillan. A useful reference, slightly out-of-date.

Dandy, James E. 1927a. Key to the species of *Magnolia. Journal of the Royal Horticultural Society* 52:260–264. Includes cultivated species as well as some non-cultivated.

_____. 1927b. The genera of Magnolieae. *Kew Bulletin* 7:257–264. Descriptions and keys to genera in the Magnoliaceae, excluding *Liriodendron.*

_____. 1950. A survey of the genus *Magnolia* together with *Manglietia* and *Michelia. Camellias and Magnolias: Report of the the Royal Horticultural Society,* 4–5 April 1950. Includes a key and descriptions of the sections of *Magnolia.*

Johnstone, George H. 1955. *Asiatic Magnolias in Cultivation.* London: Royal Horticultural Society. Includes keys to Asiatic species within each section.

Keng, Hsuan. 1978. The delimitation of the genus *Magnolia* (Magnoliaceae). *Garden Bulletin, Singapore* 31(2):127–131.

Nooteboom, H. P. 1985. Notes on Magnoliaceae. *Blumea* 31:65–121.

_____. 1987. Notes on Magnoliaceae II. *Blumea* 32:343–382.

Rehder, Alfred. 1940. *Manual of Cultivated Trees and Shrubs Hardy in North America,* 2nd ed. New York: Macmillan. Includes a key and descriptions for cultivated species.

Seidl, William J. 1983. Genus *Magnolia:* a taxonomic and hybridizing diagram. *Journal of the Magnolia Society* 19(1):11–15.

Seitner, Philip G. 1968. A taxonomic diagram of the genus *Magnolia. Journal of the Magnolia Society* 5(1):3–5.

Spongberg, Stephen A. 1974. A tentative key to the cultivated magnolias. *Arnoldia* 34(1):1–11. A preliminary report of Spongberg's 1976 key.

_____. 1976. Magnoliaceae hardy in temperate North America. *Journal of the Arnold Arboretum* 57(3):250–312. A good, up-to-date key to cultivated species and some hybrids.

Treseder, Neil G. 1978. *Magnolias.* London: Faber and Faber. Includes a key to sub-genera and sections by J. E. Dandy.

Vazquez-Garcia, José Antonio. 1990. Taxonomy of the genus *Magnolia* (Magnoliaceae) in Mexico and Central America. Unpublished Master's Thesis, University of Wisconsin, Madison. Includes keys to some of the Mexican and Central American magnolias.

7

SPECIES OF MAGNOLIA

Over eighty species and varieties of magnolias are distributed throughout eastern North America and southeastern Asia (see Figure 2.1). About thirty-five of these are cultivated in North American and European gardens, leaving approximately forty-five species and varieties yet to be introduced. Many of these are tropical and lack the hardiness necessary for extensive cultivation, and many are inferior in ornamental value to what is already available. In this chapter, I attempt to provide as much information, for as many species, as possible. Historical information about some cultivated species is discussed in the books by Treseder (1978), Johnstone (1955), and Millais (1927), and those discussions are not reproduced here.

The discussions of species in this chapter generally include the following information in roughly this order:

- The species name and author (that is, who named the plant). This is followed by any synonym and/or rejected names in brackets. These are given in alphabetical order.

- When and where the name was first published.

- Common name(s).

- The type specimen and where that specimen is located.

- A reference to color plates in the present book, if any.

- A description of the taxon, including information available from the literature.

- Native range.

- Historical information regarding discovery and naming of the plant, and what the common and scientific names mean.

- Ethnobotanical uses of the plant.

- Date, location, and means of introduction of the plant into cultivation.

- Garden merits, disadvantages, and cultural requirements of the plants if different from the general coverage of those topics in chapters 1–4.

- Estimated flowering and fruiting dates based on literature and experience. These dates will obviously vary with climate.

- Hardiness zone rating based on the U.S. Department of Agriculture hardiness map provided in chapter 3.

- List of cultivars of the species along with a description of how each cultivar differs from the typical form, who selected it, and when it was introduced. Some descriptions are more complete than others, based on available information.

- List of the hybrids involving the species. Crosses which have not led to introductions are also included for those interested in magnolia breeding. In all crosses the parent species are listed in alphabetical order rather than with the female parent first, since direction of the cross is not always known.

I have not grown all the magnolias listed in this chapter; out of necessity I have relied on written accounts and the experience of past and present magnolia gardeners for much of the information included. As previously mentioned, this is not meant to be a monographic treatment of the genus.

Species are arranged in this chapter by sections of the genus. Table 6.2 outlines the sections and the species which they include. Section descriptions are given at the beginning of each section. The alphabetical index of plant names at the back of the book will help the reader locate a particular species or cultivar.

MAGNOLIA subgenus *MAGNOLIA* L.
Species Plantarum 1:535. 1753.

Species in this subgenus are evergreen or deciduous trees or shrubs. In the deciduous species, the flowers appear with the leaves (not precocious). The outer whorl of tepals is petaloid, not calyx-like. The anther sacs open toward the inside of the flower (dehisce introrsely).

This subgenus comprises eight sections.

Magnolia contains only the American species *Magnolia virginiana*.

Rhytidospermum, the umbrella magnolias, includes three American and three Asian species.

Oyama includes four Asian species with white flowers and maroon stamens.

Theorhodon contains eighteen American species, of which only *Magnolia grandiflora* is cultivated.

Gwillimia contains eighteen Asian species including *Magnolia coco.*

Gynopodium contains three Asiatic species, including *Magnolia nitida.*

Lirianthe contains only the Asian *Magnolia pterocarpa.*

Maingola includes seven Asian species, of which none are cultivated to any extent.

Magnolia subgenus *Magnolia* section *Magnolia* DC.

[*Magnolia* section *Magnoliastrum* DC.] *Regni Vegetabilis Systema Naturale* 1:450. 1818.
TYPE: *Magnolia virginiana*

The one species now contained in this section is an evergreen or deciduous tree or shrub. The leaves are glaucous beneath with stipules adnate to the petiole. The flowers are cream to white and fragrant. The species is diploid with a chromosome number of $2n = 38$.

This section was named *Magnoliastrum* by Augustin P. de Candolle (DC.) in 1824 and originally included all the American species of *Magnolia*, while all the Asian species were placed in section *Gwillimia*. Nomenclatural rules established in 1954 require the use of *Magnolia* for this section because it contains *Magnolia virginiana*, the type species of the genus. Of the nine species originally included in this section by de Candolle, seven are now placed in other sections of *Magnolia*, while one species has been referred to *Talauma*, a change de Candolle himself considered. The only species remaining in the section is the type species, *M. virginiana*, which has no close allies.

Magnolia virginiana L.

[*M. fragrans* Salisb., *M. glauca* L., *M. palustris* Hort. ex Pampan.] *Species Plantarum* 1:535. 1753. Sweet Bay, Swamp Bay, Swamp Laurel, Beaver Tree. Type specimen: *Linnaeus, Hortus Cliffortianus 222* (1737). Clifford Herbarium, British Museum, London. **Plates 77–80.**

Magnolia virginiana is a deciduous or evergreen tree reaching 30 ft. (9.1 m) in height, often multi-stemmed and shrubby. It has smooth, gray-brown bark; twigs are yellow-green or sometimes almost black. The twigs may be glabrous or pubescent. The oblong to elliptic leaves are 2–8 in. (5–20 cm) long and 1–3 in. (2.5–7.6 cm) wide. They are a glossy medium green above and whitish or silvery beneath, with an acute apex and a cuneate, or occasionally rounded, leaf base. The green or brown petioles are about 0.5 in. (1.3 cm) long and glabrous or pubescent. *Magnolia virginiana* has lemon-scented flowers 2.5–4.0 in. (6–10 cm) across and white or cream in color. The flowers have 8–12 tepals which are 1.5–2.5 in. (4–6 cm) long and 0.5–1.0 in. (1.0–2.5 cm) wide. The stamens are creamy white. The glabrous follicetum is ovoid, 1–2 in. (2.5–5.1 cm) long, and 0.5–1.0 in. (1–2.5 cm) wide. The outer seed coats are bright red to orange. Chromosome number: $2n = 38$.

Magnolia virginiana is native to swampy areas of Georgia and Florida, west to eastern Texas, Oklahoma, and Arkansas, and north to Pennsylvania, New York, and Massachusetts. The most northerly occurrence is a small population in Essex County,

Massachusetts, which has been discussed in the literature for many years. Like *M. grandiflora*, *M. virginiana* is a coastal plain species and is seldom found very far inland. It is, however, occasionally found in the piedmont of Tennessee and Georgia and in Arkansas and Oklahoma. The species is cited in written records as early as 1584, when Arthur Barlow, in a journey up the Pamlico Sound to Roanoke Island, North Carolina, described the plant now known as *M. virginiana*. This name was given by Linnaeus in 1753.

Magnolia virginiana is commonly known as the Sweet Bay, Swamp Bay, Swamp Laurel, and Beaver Tree. "Bay" and "Laurel" are common names often given to broad-leaved, evergreen trees with laurel-shaped leaves. The common name Beaver Tree came about as a result of the tendency for beavers to eat the fleshy roots of this species; early settlers in Pennsylvania are thought to have baited beaver traps with pieces of the root.

The bark of the young twigs of the Sweet Bay was used by the Indians of the southern United States and by early settlers as a medicinal treatment for rheumatism and fever, or as a general tonic or stimulant. The seeds were used to prepare a decoction for treatment of diarrhea and stomach disorders. Many studies on the possible medicinal qualities of *Magnolia virginiana*, as well as several other North American *Magnolia* species, were carried out throughout the 1800s (Lloyd and Lloyd 1884). These studies failed to support views of *Magnolia* as an important medicinal plant. Its use in herbal medicine had declined considerably by the 1880s due to the appearance of more effective alternatives.

Magnolia virginiana was the first magnolia to be introduced into cultivation. In 1688 it was sent by the missionary John Banister to Bishop Henry Compton of London. The plant is now grown throughout the United States. It is not commonly cultivated in Europe, and when it is grown there it does not set seed regularly (see book references Millais 1927, Treseder 1978). In cultivation it is usually a shrub or small tree seldom more than 15 ft. (4.7 m) tall. The Sweet Bay often forms root suckers, creating a colony of plants. Because of its shrubby habit and small leaves and flowers, it is more suitable for smaller plantings than some other *Magnolia* species. The Sweet Bay blooms at an early age, about five years from seed, although in cool climates it may mature more slowly . The flowers are borne over a period of several weeks. Although it is not as spectacular as some other magnolias, it is a valuable addition to any landscape. The delicate, cream-colored flowers produce an intense lemony fragrance, and the red-orange fruit provide ornamental interest in the fall. The silvery undersides of the leaves are interesting year-round and can especially be of advantage in evergreen forms (see *M. virginiana* var. *australis*). Planted in a row, shrubby forms of this species make a nice screen. If an evergreen screen is desired, care must be taken in selection and pruning of the plants, since evergreen forms (*M. virginiana* var. *australis*) are usually more treelike in habit. The small stature and fragrant flowers make *M. virginiana* an excellent selection for placement near a porch or patio, and larger forms can be used as specimen trees. The leaves of this species are small; the amount of leaf litter produced is less than of most other magnolias and does not present a problem.

While its natural habitat is somewhat swampy, plants in cultivation cannot tolerate saturated soil. Treseder (1978) suggests a sandy loam mixed with peat, but the Sweet Bay tolerates a wide range of soil. It appears that any moist, well-drained medium is sufficient. Light shade is preferred, yet plants can be grown in full sun if they are not subjected to drought. Hardiness varies within the species, but several selections are hardy to Zone 6. Flowers May–June; fruits August–October. Zones (5)6–9.

CULTIVARS

'Dwarf Form'   Selected at the U.S. National Arboretum in 1970 for dwarf habit.

'Everblooming'   Flowers over a long period of time. Introduced by Louisiana Nursery, Opelousas, Louisiana.

'Grandview'   Leaves and flowers larger than typical. Introduced by Louisiana Nursery, Opelousas, Louisiana.

'Havener'   Flowers up to 4.5 in. (11 cm) across, creamy with a pink tinge, double, reproduces true from seed. Selected in 1970 by J. C. McDaniel from a seedling in Mt. Pulaski, Illinois.

'Mayer'   Multi-stemmed shrub; flowers when young, sometimes when less than 12 in. (30.5 cm) tall. Selected in 1970 by J. C. McDaniel from a seedling in Champaign, Illinois.

'Opelousas'   Tall, pyramidal deciduous tree with broad leaves, flowers 4 in. (10.2 cm) across. Selected by Ken Durio of Louisiana Nursery, Opelousas, Louisiana, ca. 1980.

'Wada's Clone'   Broad-leaved form; flowers when young. Listed by Treseder's Nurseries of Cornwall, England, ca. 1973.

HYBRIDS

Magnolia virginiana has been experimentally crossed with *M. fraseri* var. *pyramidata* and *M. acuminata*, but plants have not been released into cultivation. August Kehr, Hendersonville, North Carolina, has crossed *M. virginiana* (pollen) with *M. macrophylla* var. *ashei*, producing a hybrid which is more dwarf than the hybrids 'Birgitta Flinck' and 'Karl Flinck'. Kehr has also crossed *M. virginiana* with the tetraploid *M. sieboldii* 'Genesis'. Neither of Kehr's hybrids has been named. The following cultivated hybrids have been named.

  M. globosa × *M. virginiana* = 'Porcelain Dove'
  M. grandiflora × *M. virginiana* = 'Freeman', 'Griffin', 'Maryland'
  M. hypoleuca × *M. virginiana* = 'Nimbus'
  M. macrophylla × *M. virginiana* = 'Birgitta Flinck', 'Karl Flinck'
  M. tripetala × *M. virginiana* = *M.* × *thompsoniana*

Magnolia virginiana var. *australis* Sarg.

[*M. australis* Ashe, *M. australis* var. *parva* Ashe, *M. virginiana* subsp. *australis* (Sarg.) E. Murray, *M. virginiana* var. *parva* Ashe] *Botanical Gazette* 67:231–232. 1919. Evergreen Sweet Bay, Southern Sweet Bay. **Plate 81.**

Sargent (1919) described a southern variety of *Magnolia virginiana* which he named var. *australis*. Found from the Carolinas south to Florida and west to eastern Texas, plants of this variety reach 90 ft. (27 m) tall, the twigs and pedicels are covered with thick, silky white pubescence, and the leaves persist through the winter. Typical *M. virginiana* lacks the pubescence on the twigs and has deciduous leaves. Over the years there has been some disagreement as to the distinctiveness of *M. virginiana* var. *australis*. Dandy, in 1950, recognized the southern var. *australis*, but in 1965 he cited the considerable intergradation and extensive geographic overlap of var. *virginiana* (the northern, typical variety) and var. *australis* from North Carolina to Florida as his main reasons for not recognizing the two varieties at that time. Before his death, however, Dandy again reverted to recognizing the existence of the southern variety, noting that "Sargent's account of the variety . . . leaves no doubt about its identity" (Dandy, unpublished). Spongberg (1976) did not recognize var. *australis*, noting that, although there are differences in northern and southern populations, these differences represent the normal variation that can appear in a species with such a wide range. On the other hand, Ashe (1931) recognized var. *australis* as a distinct species, *M. australis*, based on his observation that no intermediate forms connected the glabrous northern variety with the pubescent southern variety, even though they overlap in South Carolina and eastern Georgia. McDaniel (1966b, 1967), who has made extensive observations of the variation in *M. virginiana*, also supported the recognition of distinct northern and southern forms, either as varieties or separate species. In addition to Sargent's (1919) original set of characters by which var. *australis* can be recognized, McDaniel (1966b) added several more: in var. *australis*, pollen is of a paler color, flowers are more lemon-scented, growth begins about three weeks later in the spring, flowers open and close two to five hours later, fruits take about twenty more days to mature, and trunks are fewer with a smaller crown spread than var. *virginiana* (the typical, or northern, variety). McDaniel (1967) also pointed out that the two varieties are able to hybridize, creating intermediate forms.

The field experience of Ashe and McDaniel, and McDaniel's insights gained from his hybridizing work with these two varieties are important contributions to the understanding of these taxa. While further study is needed to thoroughly understand the variation in the species, a study of the literature and personal observations in herbaria and in the field have led me to conclude that the two varieties of *Magnolia virginiana* which have been recognized for some time are real and should be sustained.

A population of plants in the Florida everglades has the confusing combination of a shrubby habit and evergreen leaves. Similar characteristics seem to occur in a population found in Louisiana and eastern Texas. McDaniel (1966b, 1967) opined that there might be various races and forms within each variety of *Magnolia virginiana*. Another possible explanation of the combination of these characters is the maintenance of the shrubby habit due to some localized environmental condition. This hypothesis could be tested by planting specimens from these populations in different locations and observing the growth habit over time.

Magnolia virginiana var. *parva* and var. *pumila* are described as being dwarf, low-growing forms with small leaves and flowers, found in Florida. These varieties are not recognized here, as the small form and shrubby habit are probably maintained by frequent fires. If planted under more favorable conditions, these forms grow like typical *M. virginiana* var. *australis*.

Variety *australis* is not as commonly cultivated as typical *Magnolia virginiana*. This is mostly because, being a southerly form of the species, it is not as hardy. When var. *australis* is cultivated, it is selected for its evergreen character and more treelike habit, both of which characteristics distinguish it from typical *M. virginiana*. In these respects, *M. virginiana* var. *australis* is sometimes more desirable in the landscape than the typical variety when a larger or evergreen plant is needed. Otherwise, landscape use for this variety is the same as for typical *M. virginiana*. Flowers May–June (July); fruits August–September. Zones (5)6–9.

CULTIVARS

'Crofts'   Exceptionally glossy upper surfaces of the leaves. Selected by J. C. McDaniel from Polk County, Tennessee, ca. 1976.

'Henry Hicks'   Hardy to Zone 5 with dense, persistent foliage and pyramidal outline. Flowers April–June. Selected by J. C. McDaniel from a seedling at Swarthmore College, Swarthmore, Pennsylvania, and registered in 1967.

'Milton'   Evergreen selection from Milton, Massachusetts, where its leaves are persistent even at −10°F (−23°C). Selected by Peter Del Tredici of the Arnold Arboretum, Cambridge, Massachusetts, in 1981.

'Satellite'   Single-stemmed form which is evergreen as far north as Washington, D.C. Fragrant, cream-colored flowers. Selected by Frank Santamour of the U.S. National Arboretum ca. 1980 from seed collected in Tennessee.

HYBRID

M. grandiflora 'Samuel Sommer' × *M. virginiana* var. *australis* = 'Sweet Summer'

Magnolia subgenus *Magnolia* section *Rhytidospermum* Spach.

Histoire Naturelle des Végétaux, Phanérogames 7:474. 1839. TYPE: *Magnolia tripetala*.

Plants of this section are characterized as deciduous trees or shrubs with smooth, light gray bark. The leaves are alternate and crowded into false whorls at the ends of the branches, not clustered on new shoots, and apparently lack axillary buds. The flowers are white to creamy yellow and are produced in the center of a false whorl of leaves. The flowers have 9–12 (rarely 15) tepals, the outer 3 often reflexed. The tepals are occasionally tinged with purple at the base. The species are diploid with a chromosome number of $2n = 38$.

Section *Rhytidospermum* contains six species and several varieties native to North America and eastern Asia. It is one of only two sections in the genus which includes both American and Asian species. The most striking characteristic of this section is the "whorled" appearance of the very large leaves. This false whorl resembles an umbrella, giving this group, and particularly the type species, *Magnolia tripetala*, the common name Umbrella Tree. The section name *Rhytidospermum* means "wrinkled seed," referring to the presence of striations on the inner seed coat. These striations are especially prominent in *Magnolia tripetala*.

The three American species in this section include the type species, *Magnolia tripetala*, which is native to the mountains of the southeastern United States. *Magnolia fraseri* has a similar mountainous distribution but is also represented in the coastal plain by var. *pyramidata*. *Magnolia macrophylla* is the most widespread of the American species. It is found from Ohio to Kentucky south to Georgia and west to Arkansas and Louisiana. It is represented by var. *ashei* in southeastern Alabama and the Florida panhandle, and by var. *dealbata* in the cloud forests of southern and eastern Mexico.

The three Asian species include *Magnolia hypoleuca*, widespread throughout Japan, *M. officinalis* of Hubei and Sichuan Provinces, China, and *M. rostrata* of Yunnan

Province, China, Upper Burma, and northeastern India. These Asian species are considered quite closely related and are not as easily distinguished as the American species.

None of the species in this section is widely cultivated. In general these species are large and coarse in texture, requiring careful placement in a garden. They especially require a spot which provides protection from strong winds. These species are also difficult to propagate vegetatively, yet seeds generally germinate readily. Phil Savage of Bloomfield Hills, Michigan, has made numerous crosses of species in this section. His efforts are discussed by McDaniel (1975) and Savage (1976).

Magnolia tripetala L.

[*M. frondosa* Salisb., *M. umbellata* Steud., *M. umbrella* Desr., *M. virginiana* var. *tripetala* L.] *Systema Naturae* ed. 10, 2:1082. 1759. Umbrella Tree. Type: Catesby in *Natural History of Carolina* (1743). **Plates 75–76.**

Magnolia tripetala is a deciduous tree reaching 30–50 ft. (9–15 m) in height. It has an open, coarse texture and a rounded habit. The bark is a smooth, light gray; the twigs and terminal buds are glabrous and purple, often with a waxy appearance. The leaves are oblong-obovate, 12–24 in. (30–60 cm) long and 6–8 in. (15–20 cm) wide. They are medium green and glabrous above, gray-green and pubescent beneath while young, becoming almost glabrous at maturity. The leaves have an acute apex and a cuneate base. The petioles are 1.0–1.5 in. (2.5–4.0 cm) long and yellow-green in color. The white, upright flowers are 6–10 in. (15–25 cm) across and have an unpleasant fragrance. The flowers have 6–9(12) tepals, each 4–5 in. (10–13 cm) long and 1–2 in. (2.5–5 cm) wide. The 3 outer tepals are reflexed, light green, and shorter than the inner tepals. The stamens are 0.5–1.0 in. (1.3–2.5 cm) long with purple filaments. The pedicels are 2–3 in. (5–7 cm) long, slender, and waxy. The follicetum is oblong-ovoid, 2–4 in. (5–10 cm) long, glabrous, and rose-pink. The outer seed coats are scarlet. Chromosome number: $2n = 38$.

This species grows in deep, moist soil in woods along mountain streams or swamp margins from Pennsylvania to Georgia, west to Arkansas and Mississippi. A small population was reported from northern Florida; however, it is doubtful that the species exists there. *Magnolia tripetala* is widely distributed throughout the southern Appalachians, yet nowhere is it common. The species was first described by Mark Catesby, the author and publisher of *Natural History of Carolina*, published in 1743 in London. In 1753 Linnaeus named the plant *M. virginiana* var. *tripetala* based on Catesby's description and plate. Linnaeus later (in 1759) raised var. *tripetala* to specific rank, *M. tripetala*. This is a misnomer, as the Umbrella Tree has six or more petals (inner tepals). The name *M. tripetala* refers to the three petaloid sepals (outer tepals) which are reflexed. Why Linnaeus gave it this name is not known, as both the text and the plate published by Catesby show it "composed of ten or eleven petals, the three outermost of which are a pale green." Recognizing that this name was misleading, in 1791 Desrousseaux substituted the more appropriate name, *M. umbrella*, for *M. tripetala*. However, even though Linnaeus's name is botanically inaccurate, it was the first validly published name for the species, and current nomenclatural rules require its use. *Magnolia umbrella*, therefore, is only a later synonym for *M. tripetala*.

This species was grown in European gardens as early as 1752. It is not now commonly grown as an ornamental in the United States or Europe. There are several possible reasons for this. The tree is not as graceful as some other magnolias, and as is the case with all species in this section, the large leaves make for a rather coarse texture and require a large viewing area in order to be effective. The flowers are not pleasantly fragrant, as in magnolias of most other sections. Savage described the odor as resembling that of a billy goat. While the fragrance is not strong and one must be close to the plant to smell it, some people choose to avoid the plant altogether. *Magnolia tripetala* has its advantages, however, as it is a fast grower and blooms at 7–8 years from seed. It is quite cold hardy and is exceptional in its production of fertile fruit. I consider the large purplish maroon fruit to be this tree's greatest asset. This quality is emphasized in the cultivar 'Woodlawn'. *Magnolia tripetala* provides ornamental interest for large gardens but can be overwhelming in small areas. It is at its best if grown as a single-stemmed tree where any suckers are pruned out. Light shade and rich, moist soil provide an environment most like the native habitat of the species, but it can be grown in full sun provided it is given enough moisture. This species is sometimes confused with *M. macrophylla* or *M. hypoleuca* in the nursery trade. Flowers April–May; fruits July–October. Zones 5–8.

CULTIVARS

'Bloomfield'   Exceptionally long and thick leaves, flowers slightly larger than species, often with extra tepals; fruit pale pink to almost white. Selected from a Pennsylvania seed source and registered by Phil Savage, Bloomfield Hills, Michigan, in 1974.

'Koeler'   A vigorous pyramidal form offered by Louisiana Nursery, Opelousas, Louisiana.

'Variegata'   A golden-variegated form introduced from Harrisburg, Pennsylvania, in 1896. Probably no longer grown.

'Woodlawn'   Fruit exceptionally large, flowers somewhat larger than species. Selected by J. C. McDaniel, Urbana, Illinois, from the Woodlawn Cemetery in Urbana. Registered in 1974.

HYBRIDS

Magnolia tripetala has been crossed with *M. fraseri, M. officinalis,* and *M. macrophylla.* In the 1970s, Phil Savage of Bloomfield Hills, Michigan, crossed *M. tripetala* pollen on the flowers of the bilobed form of *M. officinalis.* This cross produced hybrids with unlobed leaves. *Magnolia macrophylla* pollen on *M. tripetala* produced plants with little vigor. Pollen of *M. fraseri* on *M. tripetala* produced hybrids with auriculate leaf bases. August Kehr, Hendersonville, North Carolina, used *M. tripetala* as the seed parent in crosses with *M. hypoleuca* and *M. macrophylla.* Kehr also used *M. tripetala* as the pollen parent with *M. officinalis, M. sieboldii,* and *M. sieboldii* 'Genesis'. None of these hybrids has been named. The following named hybrids involve *M. tripetala:*

M. *hypoleuca* × M. *tripetala* = 'Silver Parasol'
M. *sieboldii* × M. *tripetala* = 'Charles Coates'
M. *tripetala* × M. *virginiana* = M. × *thompsoniana*

Magnolia fraseri Walt.

[M. *auricularis* Salisb., M. *auriculata* Desr., M. *auriculata* Bartr.] *Flora Caroliniana* 159. 1788. Fraser's Magnolia, Ear-leaved Magnolia, Mountain Magnolia. Type specimen: *Walter 70.* Walter Herbarium, British Museum, London. **Plates 21–22.**

Magnolia fraseri is a coarse-textured, openly branching tree 30–60 ft. (9–18 m) in height, often multi-stemmed. It has smooth, dark brown to gray bark and glabrous, red-brown twigs. The terminal buds are 1–2 in. (2.5–5.0 cm) long and 0.5–1.0 in. (1–2 cm) wide, glabrous, and purple. The leaves are obovate to rhombic (kite-shaped), 6–12 in. (15–30 cm) long, and 3–7 in. (7.6–17.8 cm) wide. They are thin, medium green and glabrous above, pale green beneath, with an obtuse to acute apex and auriculate base. Young, vigorous plants may produce leaves almost twice as large as this, and leaves may have a reddish purple tint. The petioles are 2–4 in. (5–10 cm) long, slender and glabrous. The fragrant flowers are vase-shaped, becoming saucer-shaped as they open. They are creamy white to pale yellow and 4–12 in. (10–30 cm) across. Of the 9 tepals, the inner 6 are 3–6 in. (7.6–15.2 cm) long and 1–2 in. (2–5 cm) wide. The outer 3 tepals are shorter. The stamens are 0.5–1.0 in. (1.3–2.5 cm) long and white. The pedicels are 1–3 in. (2.5–7.6 cm) long and glabrous. The follicetum is ovoid, 2–5 in. (5–13 cm) long and 1–2 in. (2.5–5.0 cm) in diameter. It is glabrous and rose-red but becoming brown or purplish when mature. The outer seed coats are red. Chromosome number: $2n = 38$.

Magnolia fraseri inhabits woods in valleys and coves and along streams and moist slopes at about 2000–3500 ft. (600–1050 m) elevation. It is native to the southern Appalachian Mountains from Virginia and Kentucky to northern Georgia. It is abundant in cool, moist localities and grows to 60 ft. (18 m) in height.

Magnolia fraseri was described by Walter in 1788 and named by him in honor of his friend and co-worker John Fraser. In 1776, William Bartram described plants of *M. fraseri* found in his travels through the southern Appalachian Mountains. In 1791, after Walter's *M. fraseri* had been published, Bartram published the name *M. auriculata* for this species, which he found in Georgia, thinking it a new species. In the same year, Desrousseaux independently published *M. auriculata*, unnecessarily rejecting *M. fraseri*, as he had done with *M. tripetala*. Both Bartram's and Desrousseaux's names were published after Walter's *M. fraseri*, and were intended to denote the same species. Their name *M. auriculata* was preceded by the correct name *M. fraseri*.

Magnolia fraseri was introduced into cultivation in England independently by Fraser and Bartram in 1786 and into France by Michaux in 1789. It flowers in 5–6 years from seed, maturing earlier than many other members of the genus. It is probably the most graceful of the umbrella magnolias with its thinner leaves and finer texture. The flowers open nearly flat and have an almost plastic appearance because of the thick texture of the tepals. The fragrance is milder than that of *M. virginiana* or *M. grandiflora*, yet it is pleasant. This is one of the few magnolias having autumn coloration. Although it will not compete with maples in this sense, it does have a reddish brown or yellow fall color. In warm climates, partial shade is best for this species. It grows well in full sun in cooler areas. Rich, moist soil provides the best growing conditions. *Magnolia fraseri* is a fast grower and is easily propagated from seed; it is also often used as an understock for grafting. Surprisingly, this species is not widely cultivated. Flowers April–May; fruits July–August. Zone (5)6–8(9). No cultivars of this species have been selected.

HYBRIDS

Phil Savage, Bloomfield Hills, Michigan, has crossed *Magnolia fraseri* with *M. virginiana*, *M. tripetala*, and *M. hypoleuca*. He has found the auriculate leaf base to be dominant in all the crosses he has made using *M. fraseri* pollen. Crosses with *M. hypoleuca* produced seedlings very much like *M. fraseri*, including the purple coloring on the new leaves. Hybrids with *M. virginiana* also showed the purple young leaves with small auriculate bases. None of these hybrids has been introduced into cultivation.

Magnolia fraseri var. *pyramidata* (Bartram) Pampanini.

> [*M. auriculata* var. *pyramidata* (Bartr.) Nutt., *M. fraseri* subsp. *pyramidata* (Bartr.) E. Murray, *M. pyramidata* Bartr.] *Bulletino della Societa Toscana di Orticultura* 40:230. 1915. Pyramid Magnolia. **Plate 23.**

Plants of *Magnolia fraseri* found in southern Georgia, Alabama, and Florida west to southern Texas belong to *M. fraseri* var. *pyramidata*. This differs from var. *fraseri* in being a smaller tree, and in having more rhombic or kite-shaped leaves, smaller flowers, smaller gynoecia, shorter stamens, and shorter folliceta. *Magnolia fraseri* var. *pyramidata* is found in the coastal plain, whereas typical var. *fraseri* grows exclusively in the mountains. It is local and somewhat rare, found especially on slopes, bluffs, and uplands along the Ochlochnee, Apalachicola, and Escambia rivers of the Florida panhandle and southwestern Alabama. It is not as common elsewhere in its range.

Bartram first described this plant as a new species, *Magnolia pyramidata*, from southwestern Alabama. In 1800 Willdenow included the species in typical *M. fraseri*. Recently this taxon has been treated either as a species or a variety. It is my opinion that, in view of the slight differences between this taxon and typical *M. fraseri* noted above, the coastal plain plants are best recognized as *M. fraseri* var. *pyramidata*.

This variety was introduced into cultivation in England by Bartram in 1806. It is now seldom cultivated, due in part to the difficulty of propagation. There is little advantage to growing this variety over typical *M. fraseri* other than the smaller stature. Flowers April–May; fruits July–August. Zones 7–9. No cultivars have been selected from nor hybrids made with this species.

Magnolia macrophylla Michaux.

> *Flora Boreali-Americana.* 1:327. 1803. Bigleaf Magnolia, Bigleaf Cucumber Tree. Type specimen: *Michaux* specimen, without number, collected in the Cumberland Mountains of eastern Tennessee in 1795. Paris Herbarium. **Plates 55–58.**

Magnolia macrophylla is a coarse-textured, openly branching tree or shrub reaching 30–50 ft. (9–15 m) in height and 16–20 in. (40–50 cm) in diameter at maturity. It has smooth, light gray bark and twigs which are yellow-green and tomentose when young, later becoming glabrous and brown. The terminal buds are 1.5–3 in. (3.8–7.6 cm) long and covered with white hairs. The leaves are oblong-obovate to spatulate, 10–40 in. (25–100 cm) long, and 4–18 in. (10–45 cm) wide. They are green and glabrous above and silvery gray and pubescent beneath, especially along the midrib, with a rounded or acute apex and cordate or auriculate base. The petioles are 2–4 in. (5–10 cm) long and pubescent to tomentose. The fragrant, white flowers are cup-shaped, 12–18 in. (30–45 cm) across, and 3–5 in. (7.6–12.7 cm) high. They have 9 reflexed tepals. The inner 6 are fleshy, usually of irregular shape, 3–8 in. (7.6–20.3 cm) long and 1–4 in. (2.5–10 cm) wide, often with purple blotches at the base. The outer 3 are membranaceous and greenish, ovate or oblong, rounded at the apex, and smaller than the inner 6. The white stamens are 0.5–1.0 in. (1.3–2.5 cm) long. The pedicels are 1–2 in. (2.5–5.0 cm) long and tomentose, becoming glabrous with age. The follicetum is ovoid or globose, 2–4 in. (5–10 cm) long and densely pubescent, becoming glabrous at maturity. It is rose-pink in color, becoming brown with age. The outer seed coats are bright rose-red to dark brown. Chromosome number: $2n = 38$.

Magnolia macrophylla is native to woods of Ohio and Kentucky south to Georgia, west to Arkansas and Louisiana, often growing in moist, shady areas along gorges. It is probably most abundant in south-central Mississippi, yet it is not common at any locality. Most populations are small and isolated.

Magnolia macrophylla was named by Michaux in 1803 following his 1795 journey through the Cumberland Mountains of northeastern Tennessee. The name *macrophylla* refers to the giant leaves ("macro" = large, "phylla" = leaf)—the largest of the genus, sometimes reaching 3 ft. (0.9 m) in length.

Magnolia macrophylla was introduced into European gardens in 1800 but is now seldom cultivated there. In the United States it is probably the most commonly cultivated species in this section, however it still is not grown as widely as magnolias of other sections. The huge flowers, also the largest of the genus, are breathtaking when examined closely. However, they are generally hidden from view by the equally large leaves. Overall, this tree has a tropical appearance. The huge, rippled leaves show the silvery undersurfaces when the wind blows them. The fruits are also quite attractive, being reddish orange, produced in a large follicetum. The plant is best grown in a large garden, for the large leaves and coarse texture seem overwhelming in a small area. The leaves of this species must be protected from wind and direct sunlight, and a rich, moist soil is ideal for growth. Flowers (April) May–June; fruits (July) August–September. Zones (5)6–8.

CULTIVARS

'Holy Grail'   Flowers chalice-shaped; selected from a tree in Santa Cruz, California, in 1963 by D. Todd Gresham of Santa Cruz, California.

'Julian Hill'   Blooms at earlier age than the species; very fertile, hardy to Zone 6; selected in 1982 by Polly Hill of Vineyard Haven, Massachusetts.

'Mississippi Clone'   Leaves larger than the species, flowers pure white. Introduced by Louisiana Nursery, Opelousas, Louisiana.

'Palmberg'   Flowers larger than the species. Introduced by Louisiana Nursery, Opelousas, Louisiana.

'Sara Gladney'   Flowers without purple at base of tepals, long-flowering; registered in 1974 by J. C. McDaniel of Urbana, Illinois; parent plant in Gloster, Mississippi.

'Whopper'   Exceptionally large flowers with extra tepals and especially prominent purple bases; selected from seedlings of unknown source by J. C. McDaniel, Urbana, Illinois, in 1974.

HYBRIDS

Phil Savage, Bloomfield Hills, Michigan, has crossed *Magnolia macrophylla* (pollen) with *M. tripetala* and *M. hypoleuca*, producing hybrids of poor vigor. Hybrids from crosses with *M.* × *wieseneri* are more vigorous. August Kehr, Hendersonville, North Carolina, has used *M. macrophylla* pollen in crosses with *M. tripetala* and *M. sieboldii* 'Genesis'. None of the above has been introduced into cultivation. The following is the only *M. macrophylla* hybrid which has been named:

M. macrophylla × *M. virginiana* = 'Birgitta Flinck', 'Karl Flinck'

Magnolia macrophylla var. *ashei* (Weatherby) D. Johnson.

[*M. ashei* Weatherby, *M. macrophylla* subsp. *ashei* (Weatherby) Spongberg.] *Baileya* 23(1):55–56. 1989. Ashe Magnolia. **Plate 59.**

Plants of *Magnolia macrophylla* endemic to deciduous forests along the bluffs and steepheads of the Florida panhandle, especially the Knox Hill region, and southeastern Alabama belong to *M. macrophylla* var. *ashei*. The most reliable characters used to distinguish this taxon from var. *macrophylla* are habit and fruit shape and size. Variety *ashei* is shrub-like in habit and smaller in all ways than var. *macrophylla*. The fruit aggregates are shorter (1–2 in. [2.5–5 cm] long) and neither as round nor as broad (1–2 in. [2.5–5.0 cm] in diameter). Variety *ashei* is less densely pubescent on the undersides of the leaves. Miller (1975) described var. *ashei* as an obligate understory plant often growing almost horizontally toward light gaps, whereas var. *macrophylla* is co-dominant in the forest canopy. Plants of var. *ashei* flower when only a few years old, while plants of var. *macrophylla* require about fifteen years from seed to flower.

Magnolia *macrophylla* var. *ashei* is very rare and in danger of extinction in the wild. It grows in the sandy soil along the banks of the Apalachicola River in the Florida panhandle in association with *M. grandiflora*. This taxon along with var. *macrophylla* is reported from Texas by Vines (1960). Although his description appears to be that of var. *ashei*, many taxonomists believe the plants to which he referred are naturalized specimens of var. *macrophylla*, which is cultivated in that state.

Many authors have included *Magnolia macrophylla* var. *ashei* in typical *M. macrophylla*, while others consider it to be a distinct species. Spongberg (1976) was the first to combine this form with *M. macrophylla*, recognizing the taxa as subspecies. I agree with Spongberg that the "extremely close relationships" of these taxa should be recognized; however, in order to maintain consistency within the genus, they should be treated as varieties rather than subspecies.

Magnolia *macrophylla* var. *ashei* has never been widely cultivated. Small (1933) suggested its use as a border plant since it flowers when quite young, occasionally when less than 1 ft. (30.5 cm) tall. The smaller, shrubby habit as well as early flowering age makes this variety suitable for garden use, yet it is not common in the trade. Flowers May–June; fruits August–September. Zones 6–9.

CULTIVAR

'Weatherby'   Selected at Louisiana Nursery, Opelousas, Louisiana, for exceptional flower quality.

HYBRIDS

August Kehr, Hendersonville, North Carolina, has crossed *Magnolia macrophylla* var. *ashei* with *M. virginiana*, producing a hybrid which is more dwarf than 'Birgitta Flinck' and 'Karl Flinck'. Kehr has also crossed var. *ashei* with *M. sieboldii*. These crosses have not yet been named.

Magnolia *macrophylla* var. *dealbata* (Zuccarini) D. Johnson.

[*M. dealbata* Zucc.] *Baileya* 23(1):56. 1989. Mexican Cow-Cumber, Eloxochitl. Type specimen: *Karwinski*, collection without number. Munich Herbarium. **Plates 60–61.**

Magnolia *macrophylla* var. *dealbata*, native to the cloud forests of eastern and southern Mexico, is the only deciduous *Magnolia* found in the tropics. It has been considered a separate species, though most acknowledge its close relationship with *M. macrophylla*. Variety *dealbata* usually lacks the purple coloration at the base of the tepals and has "beaked" follicles. The purple tepal coloration is not a good distinguishing characteristic because it varies somewhat in both varieties. The presence or absence of follicle beaks, however, appears to consistently distinguish between the two. In addition, var. *dealbata* is usually more pubescent on the undersurface of the leaves and bears larger leaves than typical *M. macrophylla*. Flowering age also differs between these two taxa: var. *dealbata* flowers when the tree is seven to ten years old, while var. *macrophylla* requires about fifteen years.

This variety has been reported to occur in association with oaks (*Quercus* spp.) and sweetgum (*Liquidambar styraciflua*) at 4000–5000 ft. (1200–1500 m) above sea level in the states of Hidalgo, San Luis Potosi, Oaxaca, and Vera Cruz, Mexico. It is rare, however, and is considered endangered because trees are felled to harvest flowers which are sold for decoration in the home or to decorate churches at Easter. Because of this constant harvest of flowers, the trees never become very large, and most of those reported from the wild appear to be root suckers. The leaves, often browsed by cattle, are thought to be of medicinal value and are harvested for use as a poultice.

This variety was first described and pictured by Francisco Hernandez, a Spanish physician and explorer. Hernandez spent seven years in Mexico, by order of Philip II, studying the natural history of the country. His work, including the description of this magnolia, was published posthumously in 1651. Zuccharini named the plant *Magnolia dealbata* in 1837. The common name, Eloxochitl, is derived from the Mayan *elotl* (green ear of corn with husk) and *xochitl* (flower), probably referring to the unopened flower bud surrounded by the spathaceous bract.

Due to its presumed marginal hardiness, this variety is seldom cultivated. However, it is actually quite hardy and has survived −5°F (−20°C) with no damage. Trees at the U.S. National Arboretum in Washington, D.C., have done well. Other than earlier flowering age, there is little advantage to growing this variety over typical *Magnolia macrophylla*, and the two fill the same niche in landscaping. Zone (7)8–9. No cultivars have been selected from nor hybrids made with this variety.

Magnolia hypoleuca Siebold & Zuccarini.

[*M. obovata* Thunb.] *Abhandlungen der Mathematisch-Physikalischen Klasse der Königlich Bayerischen Akademie der Wissenschaften* 4(2):187. 1846. Japanese White-bark Magnolia, Hono-ki. Type specimen: *Siebold* collection. Leiden Herbarium, the Netherlands. **Plates 38–40.**

Magnolia hypoleuca reaches 80 ft. (24 m) in height and 20–25 in. (50.8–63.5 cm) diameter, with spreading branches and smooth, light brown bark. The twigs are glabrous and green becoming purple-brown with age. The buds are glabrous and purple-green. The obovate leaves are 6–18 in. (15–45 cm) long and 4–8 in. (10–20 cm) wide with an acute apex and cuneate base. The upper surface is light green and waxy, the lower surface is glaucous blue-green with scattered white hairs. The petioles are 1–2 in. (2.5–5.0 cm) long, usually glabrous, with purple shading. The creamy white, cup-shaped flowers are heavily scented and reach 7–9 in. (18–23 cm) across with (6)9–12 tepals. The tepals are 4–5 in. (10–12 cm) long and 1.5–2.0 in. (4–5 cm) wide. The outer 3 tepals are shorter than the inner and are pale green, tinged pinkish. The styles and stamen bases are crimson; the gynoecium is light green. The follicetum is cylindric to oblong, 4–7 in. (10–18 cm) long and bright red. The outer seed coats are bright red. Chromosome number: $2n = 38$.

Magnolia hypoleuca is native to damp, rich forests of Japan from the Kurile Islands and Hokkaido south to Kyushu and Ryukyu Islands at altitudes of 2000–5600 ft. (600–1680 m). Records of *M. hypoleuca* from China probably refer to cultivated specimens, or to the closely related *M. officinalis,* which is native to that country.

Magnolia hypoleuca was described by Thunberg in 1794 as *M. obovata*. At this time Thunberg mistakenly attributed to this species certain references made by Kaempfer which should have been attributed to *M. denudata* and *M. liliiflora.* The name *M. obovata* was then attached to *M. liliiflora* rather than to the species Thunberg had intended. In

1846 Siebold and Zuccarini published the name *M. hypoleuca* for the species, referring to the pale undersides of the leaves. This name is correct in view of Thunberg's errors in describing *M. obovata.* These errors are explained in detail by Dandy (1973) and Ueda (1986a), who also conclude that *M. hypoleuca* is the correct name.

Magnolia hypoleuca is often confused with *M. officinalis.* The two are very closely related and may actually be two varieties of the same species. *Magnolia hypoleuca* is native to Japan, whereas *M. officinalis* is native to China. The two species have been cultivated together for timber and medicinal purposes, and it is likely that hybrids have resulted. In general, leaves of *M. hypoleuca* are more obovate, and the petioles have purple shading, which is absent in *M. officinalis.* The gynandrophore of *M. hypoleuca* is longer than that of *M. officinalis,* as is the follicetum. The pedicels of *M. hypoleuca* are glabrous, whereas those of *M. officinalis* are pubescent.

The timber of *Magnolia hypoleuca* is highly prized in Japan for making utensils, tools, and toys. Trees in central Japan are often cut for timber as soon as they have reached a significant size, thus endangering the wild population in that area.

Magnolia hypoleuca has been cultivated in the United States since 1865. It was introduced into Germany in 1877 and was grown in England at least as early as 1883. The species requires from fifteen to twenty years to flower from seed. It is among the hardiest of the Asiatic magnolias, and, since the flowers open in the summer, they escape damage by late frost. It is also less brittle than many magnolias but still requires protection from wind. Johnstone (1955) reports a golden yellow fall color. The gray-green undersurfaces of the leaves and the red fruit aggregates add to the ornamental value of this species. As with other members of this section, *M. hypoleuca* prefers a rich, moist soil. It, too, can be most appreciated in large gardens. Flowers June–July; fruits August–September. Zones (5)6–9.

CULTIVARS

'Caerhays Clone' Form selected ca. 1973 by Treseder's Nurseries of Cornwall, England. No description available.

'Lydia' Tree with narrow, upright, oval habit, flowers 6–7 in. (15–18 cm) across with pink flush on outside of outer tepals. Very fertile form with ornamental, maroon fruit aggregate. Selected by Polly Hill, Vineyard Haven, Massachusetts, ca. 1985 from seedlings sent from Japan.

'Millais Clone' Form selected ca. 1973 by Treseder's Nurseries of Cornwall, England. No description available.

'Pink Flush' Flowers white with a slight pink flush at the base of the tepals. Offered in 1989 by Otto Eisenhut, Switzerland.

HYBRIDS

In the 1970s Phil Savage, Bloomfield Hills, Michigan, crossed *Magnolia hypoleuca* with pollen from *M. macrophylla,* producing weak, slow-growing hybrids. Crosses with pollen of *M. fraseri* yield hybrids almost indistinguishable from the pollen parent. A natural hybrid between *M. hypoleuca* and *M. tripetala* is said to grow in a park in Pruhonia, near Prague, Czechoslovakia (see Vasak 1973). Tor Nitzelius of Sweden successfully crossed *M. hypoleuca* and *M. wilsonii* in the mid 1970s in an attempt to combine the beauty of *M. wilsonii* with the hardiness of *M. hypoleuca.* August Kehr, Hendersonville, North Carolina, has used *M. hypoleuca* pollen in crosses with *M. tripetala* and *M.*

officinalis. None of the above plants has been named. The following named hybrids are available.

> *M. hypoleuca* × *M. sieboldii* = *M.* × *wieseneri*
>
> *M. hypoleuca* × *M. tripetala* = 'Silver Parasol'
>
> *M. hypoleuca* × *M. virginiana* = 'Nimbus'

Magnolia officinalis Rehder & Wilson.

[*M. biloba* (Rehder & Wilson) Cheng, *M. hypoleuca* Diels (not Siebold and Zuccarini), *M. officinalis* var. *biloba* Rehder & Wilson.] In Sargent, *Plantae Wilsonianae* 1:391. 1913. Medicinal Magnolia, Hou-Phu. Type specimen: *Wilson 652* of the Arnold Arboretum Expedition of 1907–1909. Arnold Arboretum Herbarium, Harvard University, Cambridge, Massachusetts. **Plate 62.**

Magnolia officinalis reaches 60–70 ft. (18–21 m) in height and has smooth, rusty-ash bark and yellowish twigs. The elliptic-obovate leaves are 12–16 in. (30.5–40.7 cm) long and 6–7 in (15–18 cm) wide. Leaves are green and glabrous above, greenish gray below with hairs on the midribs and veins. The leaves have a papery texture, rippled margins, an acute apex, and a cuneate base. The petioles are 1–2 in. (2.5–5.0 cm) long, pubescent, and pale green-yellow. The creamy white flowers are often somewhat irregular in shape, 6–8 in. (15–20 cm) in diameter. There are 9–15 tepals with incurved margins and wrinkled surfaces. The outer 3 tepals are oblong-spatulate, 4 in. (10.2 cm) long, 2 in. (5.1 cm) wide, and pale green, tinged with pink. The inner tepals are fleshy, obovate to spatulate, 3–4 in. (7.6–10.2 cm) long, and 1.5–2.0 in. (3.8–4.0 cm) wide. They are concave, forming a dome over the gynoecium. The stamens are reddish, but less so than those of *M. hypoleuca.* The pedicel is pubescent. The follicetum is oblong and 4–5 in. (10.2–12.7 cm) long, and the outer seed coats are bright red. Chromosome number: $2n = 38$.

This species is native to woodlands in the mountains of the Chinese provinces of Hubei and Sichuan at altitudes of 2000–5500 ft. (600–1650 m). It is also cultivated extensively throughout China. It is very closely related to the Japanese *Magnolia hypoleuca* and was included within that species until 1913, when Rehder and Wilson described *M. officinalis.* The name *officinalis* refers to the use of the plant as a medicinal. At that time the species was only known from cultivation. There is now some evidence that wild specimens may still exist in the eastern provinces of China.

Magnolia officinalis was mentioned in herbals as long ago as 200 B.C. The bark is highly valued for its medicinal qualities, and the species has almost been destroyed in its native habitat due to harvest of the bark. Bark and flower buds are exported at high prices from China to other parts of Asia. The bark is used in a tonic for coughs and colds, the flower buds are used for feminine ailments, and the wood is used for timber. Pink-flowered forms are reported to exist in the wild, but none has been introduced into cultivation.

Magnolia officinalis was introduced into cultivation by E. H. Wilson in 1900. It is seldom cultivated in the United States and is represented in Europe primarily by the var. *biloba*, which is here considered to be part of typical *M. officinalis.* The flowers are quite fragrant, but perhaps not as pleasant as those of *M. grandiflora.* It has cultural requirements similar to those of other species in this section. A moist soil, partial shade, and shelter from high winds are preferred. This species is not as hardy as *M. hypoleuca.* Flowers May–June; fruits July–August. Zones 6–9.

Plants of *Magnolia officinalis* with bilobed leaves (that is, emarginate apices) have previously been recognized as *M. officinalis* var. *biloba.* The variety was first described by

Rehder and Wilson from specimens cultivated in eastern China. Later, additional specimens were collected which seemed to indicate a geographical variety with bilobed leaves. Further evidence, however, has shown that this is not the case. Plants from western China occasionally have bilobed leaves, and plants from the eastern part of the range often have typical leaves. In addition, some individuals produce both types of leaves. Dandy (unpublished) notes that a typical *M. officinalis* grown from seed collected by Rehder and Wilson in western Hupeh Province shows a variety of leaf apices ranging from bilobed to rounded to slightly pointed. In addition, both *M. hypoleuca* and *M. tripetala* have occasional emarginate apices, though less frequently than *M. officinalis*. Spongberg (1976) noted that leaf apex is an inconsistent character for defining *M. officinalis* var. *biloba*, yet he did recognize the variety based on the presence of pubescence on the stipular scales, absent in typical *M. officinalis*. I have found that this is not a consistent character for defining the variety. The existence of a true var. *biloba* has been questioned for over thirty years. Based on the information presented, plants with bilobed leaves cannot be recognized as a botanical variety but may be designated as the cultivar 'Biloba'.

CULTIVARS

'Biloba'   Forms with a great number of emarginate leaves. These plants were formerly known as var. *biloba*, but the bilobed leaf characteristic is not consistent enough to warrant botanical recognition of these plants as a variety. **Plate 63.**

HYBRIDS

August Kehr, Hendersonville, North Carolina, has crossed *Magnolia officinalis* with pollen of *M. hypoleuca*, *M. tripetala*, *M.* × *wieseneri*, and *M. sieboldii*. Phil Savage, Bloomfield Hills, Michigan, crossed *M. officinalis* 'Biloba' with *M. tripetala* pollen, producing a plant with unlobed leaves. No hybrids resulting from these crosses have been named.

Magnolia rostrata W. W. Smith.

Notes from the Royal Botanic Garden, Edinburgh 12:213. 1920. Beaked Magnolia. Type specimen: *Forrest 15052*. Herbarium of the Royal Botanic Garden, Edinburgh.

Magnolia rostrata is a deciduous tree reaching 50–100 ft. (15–30 m) tall, with an open habit and ash-gray bark. The obovate-oblong leaves are 14–20 in. (35.6–50.8 cm) long and 8–12 in. (20.3–30.5 cm) wide. They are bright glossy green above and glaucous beneath, with the midribs and veins covered with reddish brown tomentum until fully expanded. The leaf apex is broadly rounded, and the base is rounded to subcordate. The petioles are 2.5–3.0 in. (6.4–7.6 cm) long, glabrous, and light green to bronze. The melon-scented flowers have 11 tepals. The outer 3 tepals are oblong, reflexed, pinkish above, and green below. The inner tepals are white and fleshy, forming a dome over the gynoecium before maturity. The base of the stamens is red-purple, and the gynoecium is tinted pink with cream stigmas. The follicetum is oblong, 4–5 in. (10.2–12.7 cm) long, and 1.5–2.5 in. (3.8–6.4 cm) wide, containing beaked follicles. The follicle beaks are curved, 0.2–0.4 in. (0.6–1.2 cm) long. The outer seed coats are reddish pink. Chromosome number: $2n = 38$.

Magnolia rostrata is native to the eastern edge of the Himalayas to northwestern Yunnan Province in China, and northeastern Upper Burma at 5000–13,000 ft. (1500–3900 m) above sea level, as well as northeastern India associated with *M. campbellii* and *M. globosa* at 6500–10,000 ft. (1950–3000 m) above sea level.

Magnolia rostrata was named by Wright Smith based on three collections made by Forrest (*15052, 17301, 16403*) in northwestern Yunnan. It is distinguished from *M. hypoleuca* and *M. officinalis* in having new leaves covered in reddish brown indumentum. In addition, the curved "beaks" which appear on the follicles of *M. rostrata* do not occur in either of the other species.

This species has not been cultivated as widely as *Magnolia hypoleuca* and *M. officinalis* due to its lack of hardiness, and it is seldom grown. It is more susceptible to breakage by wind and must be in a quite protected location to survive. *Magnolia rostrata* flowers in approximately fifteen years from seed. Pink-flowered forms have occasionally been reported, but none has been named. No cultivars have been selected nor hybrids made with this species. Flowers May–June; fruits June–October. Zones 7–9.

Magnolia subgenus *Magnolia* section *Oyama* Nakai.

[*Magnolia* section *Cophanthera* Dandy.] *Flora Sylvatica Koreana* 20:117. 1933. TYPE: *Magnolia sieboldii.*

The species in section *Oyama* are deciduous trees or shrubs bearing white flowers. The flowers have maroon stamens and are nodding or pendent on the branches. The anthers are blunt or rounded at the apex. The follicetum is pendent. These species are diploid with a chromosome number of $2n = 38$.

The sectional name derives from *oyama-renge,* the Japanese name for *Magnolia sieboldii.* Dandy's 1936 synonym *Cophanthera* refers to the blunt apex of the anthers, a character which distinguishes this section from all others of *Magnolia.*

This section is composed of four species native to eastern Asia. The type, *Magnolia sieboldii,* is the easternmost species and inhabits Japan, Korea, Manchuria, and eastern China. A geographical gap occurs between the distribution of this species and two others in this section, *M. sinensis* and *M. wilsonii* in western China. *Magnolia globosa* has the westernmost distribution, occurring from eastern Nepal along the Himalayas to northwestern Yunnan.

These species form a homogeneous group easily distinguished from other sections of the genus. The individual species, however, can be difficult to differentiate. The main characters used to identify them include color and density of pubescence on leaves and twigs, leaf shape, and posture of flowers. As discussed in chapter 5, these characters are quite variable and may be influenced by a number of environmental factors. Specimens collected from the wild show more of this variability than cultivated specimens, since cultivated plants come from a limited number of sources. The difficulty of defining the species is increased by the small number of specimens (living or dried) available for study. A look at the synonymy of these species shows the extent to which the plants are confused. *Magnolia sinensis,* for example, has been considered by various authors as a variety of *M. sieboldii* or *M. globosa,* while some think it is most closely related to *M. wilsonii.* As more knowledge is gained, it seems likely that this section will undergo some taxonomic changes. But rather than make hasty decisions based on a poor sampling of material, I have presented the plants here as four species which have been recognized for some time.

While all four of these species are found in cultivation, they have never received the attention they deserve. They are small trees or shrubs of a good size for the home landscape. They are less frequently damaged by wind than other magnolias and are easily propagated. The flowers are produced with the leaves in the summer over a period of up to three months. The summer blooming time allows the flowers to escape frost damage.

Magnolia sieboldii Koch.

[*M. oyama* Kort., *M. parviflora* Sieb. & Zucc. (not Blume), *M. verecunda* Koidz.] *Hortus Dendrologicus.* 4(11). 1853. Siebold's Magnolia, Oyama Magnolia. Type specimen: unknown. **Plates 71–72.**

Magnolia sieboldii is a deciduous tree or shrub reaching 20 ft. (6 m) in height, with light gray bark. The twigs are yellowish or grayish tan and slightly pubescent, becoming glabrous with age. Terminal buds are slightly pubescent. The obovate or elliptic leaves are 2–6 in. (5–15 cm) long and 2–4 in. (5–10 cm) wide with an acute apex and cuneate or rounded base. The upper surface is dark green, and the lower surface is light green with yellowish pubescence. The petioles are 0.5–2.0 in. (1.3–5.1 cm) long and grayish brown. The white, fragrant flowers are saucer-shaped and nodding or pendent with 9–12 obovate tepals 1–3 in. (2.5–7.6 cm) long by 1–2 in. (2.5–5 cm) wide. The showy magenta stamens are 0.2–0.4 in. (0.6–1.2 cm) long. The flower is borne on a pedicel 1–3 in. (2.5–7.6 cm) long covered with yellowish pubescence. The follicetum is cylindric, pendent, and 1–3 in. (2.5–7.6 cm) long. The outer seed coats are bright red. Chromosome number: $2n = 38$.

Magnolia sieboldii is native to southern Manchuria, Korea, southern Japan (Honshu, Shikoku, and Kyushu), and southern China (Anhui). It is the easternmost species in this section. The most western population of *M. sieboldii* occurs in the Hwang-shan mountains of southern Anhui Province of China, 800–950 mi. (1300–1500 km) from the other species in the section. The Hwang-shan population is well removed from the main area of distribution in Japan, Korea, and Manchuria.

Siebold and Zuccarini first described this species from Japan in 1846, naming it *Magnolia parviflora*. This name was used until the 1930s, when the homonym rule was adopted by the 1930 International Botanical Congress. This rule rendered Siebold and Zuccarini's name invalid since *M. parviflora* had been used by Blume in 1825 to denote another member of the magnolia family, the plant now known as *Michelia figo*. The correct name for the plant described by Siebold and Zuccarini then became *M. sieboldii*, published by Koch in 1853 but based on the original description of Siebold and Zuccarini. The common name Oyama Magnolia, or *oyama-renge*, refers to the mountain of that name in Nara prefecture, Japan, where the species is native. The plant apparently has been commonly used in Japanese tea cottage gardens, where the guests, seated on the floor for tea, may enjoy the nodding flowers from beneath.

In a recent study Ueda (1980) divided *Magnolia sieboldii* into two subspecies. Plants native to Korea and Manchuria were included in subsp. *sieboldii*, while those native to Japan and southern China were designated as subsp. *japonica*. The latter plants differ from typical *M. sieboldii* in having stamens of a lighter color (light rose to pink or even yellow) and a more procumbent or "creeping" habit. These pink- and yellow-stamened forms have been reported elsewhere, yet it is not clear that the distinction between these two subspecies is warranted. Pending further study, Ueda's subspecies are not recognized here.

Magnolia sieboldii differs from *M. globosa* in having yellow or silvery pubescence rather than rust-colored pubescence, and from *M. wilsonii* in having more obovate leaves in contrast to the elliptic leaves of *M. wilsonii*. It differs from both species in having yellowish tan twigs rather than dark brown or purple twigs. *Magnolia sieboldii* differs from *M. sinensis* in having straight hairs on the undersurfaces of the leaves (in contrast to the undulate hairs of *M. sinensis*), and in having glabrous twigs and a rounded leaf apex. This species also has shorter internodes just below the flower than other species in this section.

Magnolia sieboldii has long been cultivated in Europe and, to a lesser extent, North America. Although the exact date and means of introduction into cultivation is unknown, this species is thought to have been introduced about 1865 and has been prized for its white flowers with red to maroon stamens and its reddish pink fruit ever since. It is probably the most commonly cultivated species in this section, yet surprisingly, it is little used in North America. Although *M. sieboldii* can reach 25 ft. (8 m) in height, it is usually grown as a small plant of 10 ft. (3 m) or less and is suitable for use in most any size garden. It flowers at an early age, and a plant may bloom for up to six weeks. The nodding flowers have a pleasant fragrance. The combination of magenta stamens and pure white tepals make these flowers quite beautiful when set against the green leaves. The small pink fruits also have ornamental value. *Magnolia sieboldii* requires partial shade and a moist but well-drained soil. According to Gardiner (1989), it is quite hardy, surviving −38°F (−39°C) and frozen ground with no lasting adverse effects. This species is also the most wind tolerant of the species in this section. It is easily propagated by layering. Flowers May–June (August); fruits August–September. Zones 6–8(9).

CULTIVARS

'Genesis'   A colchicine-induced tetraploid form with flowers of typical size yet with larger stamens, leaves, and stems. The flowers are of heavier texture than typical. Produces tetraploid seedlings with flowers much larger than typical. Produced by August Kehr of Hendersonville, North Carolina, and registered by Kehr in 1985.

'Kwanso'   Double-flowered form registered in 1961 by D. Todd Gresham, Santa Cruz, California. This is the long-established Japanese name for double-flowered plants. Double forms are sometimes sold under the names 'Semi-Plena', 'Michiko Renge', and others, but, as discussed in chapter 5, unless these cultivars are significantly different from 'Kwanso', they actually should not carry these new names. Michiko-renge is the Japanese common name for this form.

'Minor'   Smaller than typical in all ways. Introduced from Japan in 1888 and published by Nicholson in 1901.

'Variegata'   A white-variegated leaf form introduced from Mt. Chiisan, Korea, and published by Nakai in 1933.

HYBRIDS

August Kehr, Hendersonville, North Carolina, produced hybrids between *Magnolia sieboldii* and *M. macrophylla* var. *ashei*, *M. tripetala*, and *M. officinalis*, but the plants have not been named. He has also produced unnamed hybrids between the tetraploid *M. sieboldii* 'Genesis' and *M. tripetala*, *M. virginiana*, and *M. macrophylla*. The following named hybrids are available:

M. hypoleuca × M. sieboldii = M. × wieseneri
M. sieboldii × M. tripetala = 'Charles Coates'

Magnolia globosa Hook f. & Thoms.

> [*M. tsarongensis* W. W. Smith & Forrest, *Yulania japonica* var. *globosa* (Hook f. & Thoms.) P. Parment.] *Flora Indica* 1:77. 1855. Globe-flowered Magnolia. Type specimen: none designated. Lectotype: a specimen collected by Hooker in 1849. Herbarium of the Royal Botanic Gardens, Kew.

Magnolia globosa is a deciduous tree or shrub reaching 15–40 ft. (4.6–12.4 m) tall, with smooth, gray bark and dark brown to reddish brown twigs which are first pubescent, later becoming glabrous. The terminal buds are elongate and covered with brown pubescence or may be nearly glabrous. The leaves are elliptic or usually ovate or obovate, 4–8(10) in. (10.2–20.3 [25.4] cm) long and 2–4(6) in. (5.1–10.2 [15.2] cm) wide. They are dark green above and light green with rusty or yellowish pubescence below. On young leaves this chestnut-colored tomentum can be quite noticeable and ornamental. The leaf apex is acute or acuminate, and the base is usually rounded, rarely more or less cordate. The petioles are 1–2 in. (2.5–5 cm) long and covered with yellowish or rusty pubescence, or nearly glabrous. The white flowers are fragrant, cup-shaped, nodding or pendent, with 9–12 obovate tepals, each 2–3 in. (5–7.6 cm) long and 1–2 in. (2.5–5 cm) wide. The flower buds are egg-shaped. The stamens are 0.5–1.0 in. (1.3–2.5 cm) long and magenta. The pedicels are 1.5–2 in. (4–5 cm) long and covered with reddish pubescence. The follicetum is cylindric, 2–3 in. (5–7.6 cm) long, and reddish brown. The outer seed coats are red. Chromosome number: $2n = 38$.

This species is native to forests and thickets at altitudes of 8000–11000 ft. (2400–3300 m) in the eastern Himalayas from eastern Nepal through southeastern Tibet to northwestern Yunnan Province, China. It was first named and described in 1855 by Hooker and Thomson, and collected by them in Sikkim at an altitude of 9000–10,000 ft. (2700–3000 m).

In 1920 George Forrest collected the plant from the opposite extreme of its natural range, the Tsarong region of southeastern Tibet. Forrest published it as a new species, *Magnolia tsarongensis*. Forrest's plant was included in *M. globosa* by E. H. Wilson in 1926, and it is now considered only a variant within the species. Treseder (1978; see book references), Johnstone (1955), and Cowan (1938) present extensive accounts of the ways in which the "Indian" form (typical *M. globosa* collected by Hooker and Thomson from Sikkim) and the "Chinese" form (collected from Tibet and named *M. tsarongensis* by Forrest) differ. All these authors have grown both forms and note that they can generally be differentiated in the following ways: the Chinese form begins growth earlier in the season than the Indian form; chestnut pubescence is more noticeable on the Chinese form; the undersides of the leaves are silvery gray in the Chinese form and reddish gold in the Indian form; and the Chinese form is less hardy and is bushier than the Indian form. Johnstone (1955) and Treseder (1978) note, however, that these characters are not consistent, while Cowan (1938) found it difficult to set criteria by which the two can be distinguished. All three authors maintain that the plants can be distinguished if seen growing together.

The epithet *globosa* probably refers to the almost globose (rounded) flower buds. The buds, however, are usually more egg-shaped than round, and the Nepalese call this species the Hen Magnolia. *Magnolia globosa* differs from other members of this section in producing rufous tomentum on the branchlets and undersurfaces of the leaves, and in having more globose flowers.

This species was introduced into cultivation in Britain in 1919 by George Forrest from his collections in Tibet, the "Chinese" form. It is likely that most of the cultivated plants of *Magnolia globosa* derive from this introduction. The "Indian" form appears to have been introduced around 1930. The species is cultivated for its fragrant, egg-shaped flowers, large leaves, and chestnut brown tomentum on the branchlets and petioles. Seedlings flower at about 8–10 years of age. Johnstone (1955) and Treseder (1978) report that the flowers often turn brown if dampened. This species differs little from *M. sieboldii* in landscape value, and the two can be used interchangeably where they both are hardy. *Magnolia sieboldii* and *M. wilsonii* are probably better choices for cooler climates since *M. globosa* initiates growth earlier in the spring and is therefore susceptible to frost damage. *Magnolia globosa* may be preferred by some gardeners for the egg-shaped flowers which do not open very wide. Cultural requirements are quite similar to those of other species in this section: rich, moist soil and a shady location. Flowers May–June; fruits August–September. Zones 5–8(9). No cultivars of this species have been selected.

HYBRIDS

Todd Gresham, Santa Cruz, California, crossed this species with *Magnolia wilsonii*, producing a hybrid combining desirable qualities of the two species. This hybrid was described by Gresham (1966) yet never named. The following named hybrid is available:

M. globosa × *M. virginiana* = 'Porcelain Dove'

Magnolia wilsonii (Finet & Gagnep.) Rehd.

[*M.* × *highdownensis* Dandy, *M. liliifera* var. *taliensis* (W. W. Smith) Pampan., *M. nicholsoniana* Rehder & Wilson, *M. parviflora* var. *wilsoni* Finet & Gagnep., *M. taliensis* W. W. Smith, *M. wilsonii* f. *nicholsoniana* Rehder & Wilson. *M. wilsonii* f. *taliensis* (W. W. Smith) Rehder.] In Sargent, *Plantae Wilsonianae* 1:395. 1913. Wilson's Magnolia. Type specimen: *Wilson 3137*. Musée National d'Histoire Naturelle, Paris.

Plants of *Magnolia wilsonii* are deciduous trees or shrubs to 30 ft. (10 m) tall with brown bark and dark brown to grayish purple twigs which are pubescent, becoming glabrous with age. The terminal buds are brown and pubescent. The elliptic to oblong leaves are 2–6(8) in. (5–15.2 [20.3] cm) long and 1–3(4) in. (2.5–7.6 [10.2] cm) wide. They are dark green above, lighter green and pubescent beneath, with an acuminate or acute apex and rounded base. The petioles are 0.5–2 in. (1.3–5 cm) long and pubescent. The ovoid flower buds open to fragrant flowers which are saucer-shaped and nodding or pendent with 9(12) spatulate tepals. Each tepal is 2–3 in. (5–7.6 cm) long and 1–2 in. (2.5–5 cm) wide. The red or magenta stamens are 0.2–0.4 in. (0.6–1.2 cm) long. The pedicels are 0.5–2.0 in. (1.3–5 cm) long, brown, and often more or less pubescent. The follicetum is cylindric, 2–4 in. (5–10 cm) long, and reddish brown. The outer seed coats are red. Chromosome number: $2n = 38$.

This species is native to woods, thickets, and open fields in eastern Sikang, northern Yunnan, and western Sichuan provinces of China at altitudes of 4000–10,000 ft. (2000–3400 m). Wilson discovered the species in 1904 in eastern Sikang Province during his Veitch expedition. In 1906 he published its description under *Magnolia globosa*, although he noted that there were differences between the plant he discovered and that species. The same year Finet and Gagnepain designated the plant as *M. sieboldii* var. *wilsonii*. In 1913 Rehder raised the variety to specific rank, creating *M. wilsonii*, the currently accepted name for this species. In the same article Rehder and Wilson published a new species, *M. nicholsoniana*, which differed from *M. wilsonii* in having leaves nearly glabrous except on the primary nerves of the lower surface. Before his death in

1930, Wilson wrote that "my mature judgement now, however, is that it is doubtful if it should be distinguished even as a form. Confusion has been caused in gardens by an error on my part" (see Howard 1980). Thus, *M. nicholsoniana* is reduced to a synonym of *M. wilsonii*. This species is sometimes spelled *wilsoni*, but the correct spelling is *wilsonii* pursuant to *ICBN* article 73.10 and Recommendation 73C.1(b).

Magnolia wilsonii differs from *M. sieboldii* and *M. sinensis* in having narrower, more elliptic leaves and dark twigs. It differs from *M. globosa* in having yellowish rather than rufous pubescence on branchlets and leaves.

The name *Magnolia* × *highdownensis* has been applied to purported hybrids between *M. sinensis* and *M. wilsonii*. The original plant was an unlabeled seedling from Caerhays given to Col. F. C. Stern and grown in his garden at Highdown, West Sussex. As the plant grew it showed characteristics intermediate between those of *M. sinensis* and *M. wilsonii*. Dandy (1950b) published the name *M.* × *highdownensis* for the assumed hybrid. He reported that in habit and size of leaf and flower, this plant resembled *M. sinensis*. The leaf shape, however, approaches that of *M. wilsonii*, especially in its more pointed apex. Since Dandy's publication of *M.* × *highdownensis*, many authors have been of the opinion that the purported hybrids are actually only variant plants of *M. wilsonii*. Similar plants have been grown from *M. wilsonii* seed without possibility of pollination from *M. sinensis*. Treseder (1978) reports that Dandy himself was never quite content to call the Highdown plants hybrids. Probably the most important reason for not considering the Highdown plant to be a hybrid is that it consistently reproduces true from seed. Based on all the above information, *M.* × *highdownensis* is here considered a synonym of *M. wilsonii*.

Magnolia wilsonii was introduced into cultivation in 1904 by E. H. Wilson and is grown for its large, pendent, white flowers with crimson stamens, its dark purple twigs, its treelike habit, and its hardiness. This small, multistemmed tree is the hardiest, and perhaps the most distinct, species in this section, yet it is still little grown. The dark brown, almost black, branches add interest to the winter landscape. *Magnolia wilsonii* prefers at least partial shade and moist soil. It is easily propagated from seed and seedlings flower at 5–7 years of age. Flowers May–June; fruits August–September. Zones 6–8(9).

CULTIVAR

'Bovee'   A hardy form reaching 25 ft. (7.8 m) tall with abundant pendent flowers and
    large ornamental fruit.

HYBRIDS

D. Todd Gresham, Santa Cruz, California, crossed *Magnolia globosa* with *M. wilsonii*, producing a hybrid combining the desirable characteristics of each parent. The hybrids are vigorous trees with the tan wood of *M. globosa*. The fragrant flowers nod, but are not pendent as in *M. wilsonii*. See Gresham (1966) for a complete description of this unnamed hybrid. Tor Nitzelius of Sweden successfully hybridized this species with *Magnolia hypoleuca* in the mid 1970s in an attempt to combine the beauty of *M. wilsonii* with the hardiness of *M. hypoleuca*. None of these plants has been named.

M. sinensis × *M. wilsonii* = 'Jersey Belle'

Magnolia sinensis (Rehder & Wilson) Stapf.

[*M. globosa* var. *sinensis* Rehder & Wilson, *M. nicholsoniana* Hort. ex Millais (not Rehder & Wilson), *M. sieboldii* subsp. *sinensis* (Rehder & Wilson) Spongberg.] *Curtis's Botanical Magazine.* 149:9004. 1924. Chinese Oyama Magnolia. Type specimen: *Wilson 1422* of the Arnold Arboretum Expedition of 1907–09. Arnold Arboretum Herbarium, Harvard University, Cambridge, Massachusetts.

Plants of *Magnolia sinensis* are deciduous trees or shrubs of spreading habit reaching 20 ft. (6 m) tall, with light gray bark and grayish brown twigs with brown pubescence, twigs becoming glabrous. The leaves are leathery, broadly ovate, 3–4(8) in. (7.6–10.2 [20.3] cm) long, and 2–5(7) in. (5–12.7 [17.8] cm) wide, with a rounded or acuminate apex and cuneate base. The upper surface is bright green and slightly pubescent to glabrous; the lower surface is covered with light gray or tan pubescence. The petioles are 1–3 in. (2.5–7.6 cm) long and covered with gray pubescence. The fragrant flowers are white, cup-shaped, and nodding, about 3 in. (7.6 cm) across, with 9–12 obovate tepals. The tepals are 2–3 in. (5–7.6 cm) long and 1–2 in. (2.5–5 cm) wide. The magenta stamens are 0.4–0.6 in. (1.2–1.6 cm) long. The pedicels are 1–2 in. (2.5–5 cm) long. The follicetum is pendulous, oblong, 2–3 in. (5–7.6 cm) long, and pink. The outer seed coats are scarlet. Chromosome number: $2n = 38$.

Magnolia sinensis is found only in woods and thickets at 6500–8500 ft. (1950–2550 m) altitude in the Sichuan Province of China near Wenchuan-hsien. Wilson discovered the plant and collected the type material in 1908 during his Arnold Arboretum Expedition. In 1913 Rehder and Wilson published it as a variety of *M. globosa*. In 1926 Wilson suggested that the variety should not be recognized but that this plant should be included within *M. globosa*. In the meantime, however, Stapf (1923) had raised var. *sinensis* to specific rank, creating *M. sinensis*, a name which has been recognized ever since. In 1976, Spongberg published a new combination, treating *M. sinensis* as a variety of *M. sieboldii*. This change has not been widely accepted, due to the absence of convincing evidence that it is warranted. I presented evidence (Johnson [Callaway] 1989a) that *M. sinensis* should be included within *M. sieboldii*; however Ueda (1980) presented evidence that the two should remain separate. In both cases, the authors reported that the limited number of specimens examined left the matter unresolved. For this reason, due to the conflicting data presented, and until a more thorough study of wild and cultivated specimens of all the species in this section can be made, I consider them as separate species, as they have been considered for many years.

Magnolia sinensis differs from *M. wilsonii* in the rounded apex of the leaf, more pubescent leaf undersurfaces, lighter color of twigs, and more spreading habit. This species differs from *M. globosa* and *M. sieboldii* in having larger, more pendulous flowers and a rounded leaf apex. *Magnolia sinensis* differs from *M. sieboldii* in having more pubescent twigs.

Magnolia sinensis was introduced into cultivation in North America in 1908 by E. H. Wilson. It was distributed by the French nurseryman Chenault under the name of *M. nicholsoniana* (a synonym for *M. wilsonii* as well as *M. sinensis*), and many plants cultivated today still carry that name. Its flowers are larger than those of the other species in the section, yet it is not widely grown as an ornamental. Although it does not seem to be particular about its environment, *M. sinensis* does best in partial shade. The spreading habit of this species in nature can be adapted by pruning to produce a neat, upright tree. *Magnolia sinensis* has the largest flowers of the oyama magnolias and has large pink fruit. It is perhaps the species of this section which is most tolerant to hot, dry conditions. It is easily propagated from seed; seedlings flower at 5–7 years of age; cuttings and layers also root easily and bloom quickly. Flowers May–June; fruits August–September. Zones (5)6–8(9).

CULTIVAR

'Floreplena'   Double-flowered form blooming three weeks earlier than typical. Offered by Otto Eisenhut, Switzerland, in 1990.

HYBRID

M. sinensis × *M. wilsonii* = 'Jersey Belle'

Magnolia subgenus *Magnolia* section *Theorhodon* Spach.

Histoire Naturelle des Végétaux, Phanérogames 7:470. 1839. TYPE: *Magnolia grandiflora*.

Species in section *Theorhodon* are large evergreen trees with glabrous to densely tomentose twigs and leaves and free stipules. The flowers are creamy white and fragrant with 9–12 tepals. The gynoecium is sessile. Chromosome numbers vary among species.

Little is known about most species in this section. Chromosome counts have not been made for all species, but *Magnolia grandiflora* and *M. schiedeana* are hexaploid ($2n = 114$) and the few tropical species which have been examined are diploid ($2n = 38$). Most species have a dark brown heartwood, which is used in cabinetmaking. The section name is comprised of the Greek words *theo*, "god," and *rhodo*, "red." Spach published the name without reference as to why it was so named.

As treated here, section *Theorhodon* contains eighteen species native to the American tropics, with the exception of *Magnolia grandiflora* from the southeastern United States. Dandy (unpublished) divided *Theorhodon* into two subsections: *Grandiflora* is distributed throughout continental America from the southeastern United States south to Venezuela; subsection *Splendentes* is found throughout the West Indies.

The continental American group consists of the following ten species: *Magnolia grandiflora* from the southeastern United States, *M. schiedeana* and *M. sharpii* from Mexico, *M. guatemalensis* from Guatemala, *M. yoroconte* and *M. hondurensis* from Honduras, *M. poasana* from Costa Rica, *M. sororum* from Panama, and *M. ptaritepuiana* and *M. chimantensis* from Venezuela. This group is characterized by having the bract inserted immediately below the flower, and by lacking the bristle tip to the anthers, found in the West Indian species of this section.

The West Indian group consists of the following eight species: *Magnolia cubensis* from eastern Cuba; *M. domingensis*, *M. emarginata*, and *M. ekmanii* from Haiti; *M. pallescens* and *M. hamorii* from the Dominican Republic; *M. portoricensis* from western Puerto Rico; and *M. splendens* from eastern Puerto Rico. The West Indian species are distinguished from the continental American species in having an unusual means of releasing pollen. The stamens of these species have at their apex a bristle-like tip, usually about the same length as the anther itself. When the flower is still in the bud stage, these tips are held tightly against the gynoecium. As the gynoecium forms, these stamen tips become embedded in the tissue of the gynoecium or are caught between the carpels. As the flower opens, the stamens dehisce at the base but are still held by the tips embedded in the gynoecium so that it appears that they are growing out of the gynoecium. By this time the stigmas are no longer receptive, and the stamens release pollen. This unusual method of pollination is not known in any magnolias other than these West Indian species (Howard 1948).

The delimitation of species within section *Theorhodon* is not an easy task based on the available information. Howard (1948) provided a thorough account of the West Indian species, and his work is the source of much of the information presented here. Recently, Vazquez (1990) undertook a much-needed study of the continental species, proposing four new species and three new subspecies. These include *Magnolia tamaulipana, M. iltisiana, M. panamensis, M. pacifica* (with its subspecies *pugana* and *tarahumara*), and *M. sororum* subsp. *lutea*. At the time of publication of this book, the new species and subspecies proposed by Vazquez had not yet been validly published. I believe it is best to await further information and validation of these species before discussing them here.

Magnolia grandiflora L.

[*M. angustifolia* Millais, *M. elliptica* Link, *M. exoniensis* Don, *M. ferruginea* Collins ex Raf., *M. foetida* (L.) Sarg., *M. hartwegii* Hort. ex Gard., *M. lacunosa* Raf., *M. lanceolata* Link, *M. longifolia* Ser., *M. maxima* Ser., *M. microphylla* Ser., *M. obovata* Ait. ex Link (not Thunb.), *M. obtusifolia* Ser., *M. praecox* Ser., *M. pravertiana* Millais, *M. stricta* Ser., *M. tardiflora* Ser., *M. tomentosa* Ser., *M. virginiana* var. *foetida* L., *Talauma plumieri* Stahl.] *Systema Naturae* 10, 2:1082. 1759. Southern Magnolia, Bull Bay. Type specimen: none designated. Lectotype: Miller, *Figures of Plants* 2:115 (1760). **Plates 24–28, 36.**

Magnolia grandiflora is an evergreen tree reaching 100 ft. (30 m) tall and 3 ft. (1 m) in diameter. Trees are usually rounded to pyramidal in shape with a straight trunk and light brown to gray bark. The twigs and terminal buds are yellowish green to brown and pubescent, becoming dark brown to black and glabrous with age. The elliptic or ovate to obovate leaves are 2–12 in. (5–30.5 cm) long and 1–5 in. (2.5–12.7 cm) wide. Leaves are dark green and glossy above, light green to brown and pubescent to tomentose beneath, with an acute apex and cuneate base. The petioles are 0.5–2 in. (1–5 cm) long, brown, and pubescent. The intensely fragrant flowers are 6–12 in. (15–30 cm) across and white or cream white with 9–15 spatulate to ovate tepals 2–6 in. (5–15 cm) long and 1–4 in. (2.5–10 cm) wide and usually concave. The 3 outer tepals are sepaloid. The stamens are 0.5–1.0 in. (1.3–2.5 cm) long and cream-colored with crimson bases. The gynoecium is crimson and pubescent. The pedicels are 0.5–2.0 in. (1.3–5 cm) long, stout, and covered with chestnut brown tomentum. The follicetum is ovoid, 2–5 in. (5–12.7 cm) long, yellowish or reddish brown, and densely pubescent. The outer seed coat is red, orange, or rarely yellow. Chromosome number: $2n = 114$.

Magnolia grandiflora is native to moist woods in the coastal plain from central Florida, north to North Carolina, and west to Texas. It is rarely found more than about 150 miles (93 km) inland but does occur in southern Arkansas. In the wild, trees may reach 100–120 ft. (30–36 m) tall and become quite large, lacking branches for most of their height and producing them only in the top of the tree as part of the forest canopy. In parts of its range, *M. grandiflora* is the dominant species in the forest.

Magnolia grandiflora is highly variable in habit, leaf and flower size, color and amount of indumentum on the lower surface of the leaves, fruit characteristics, hardiness, ease of propagation, and age of maturation. This variability is due, in part, to its broad natural range. Some authors contend that it is also due to long-continued introgressive hybridization with *M. virginiana* where the ranges overlap. These authors cite as evidence the resemblance that some forms such as 'Exmouth' and 'Griffin' bear to the Freeman Hybrids, known hybrids between *M. virginiana* and *M. grandiflora*.

This species was first mentioned in the literature by Plukenet in 1705 and then by Phillip Miller in his *Gardener's Dictionary* (1731). Miller gave a rather brief discussion of it as "another species" under *Magnolia virginiana*. In a later edition (1759) Miller described the species more thoroughly, and a reference to a plate accompanied the description, as did a reference to Catesby's (1743) description. In 1759, citing the descriptions of Miller and Catesby, Linnaeus named the plant *M. grandiflora*. He left no type specimen, and his brief description was taken from Miller. The epithet *grandiflora* refers to flower size ("grand" = large, "flora" = flower).

The common names for *Magnolia grandiflora* include Southern Magnolia, Bull Bay, and Great Laurel Magnolia. Southern Magnolia is the most common name and simply refers to its distribution in the southern United States. *Magnolia grandiflora*, together with *M. virginiana* and many other broadleaved evergreen trees, is referred to as Bay or Laurel because of the resemblance of the leaves to those of the Red Bay or Laurel, *Persea borbonia*. Bull Bay is used to distinguish *M. grandiflora* as the largest of the Bay Magnolias, or possibly because cattle reportedly eat the leaves of this species.

This species is grown for timber in the southeastern United States, although its use for this purpose is declining. The wood was once used for venetian blinds but now more commonly appears in furniture, veneer, cabinets, boxes, crates, baskets, and as an interior finish for houses. It is also cut for fuel. Ornamental qualities of the plant are utilized as well: the glossy evergreen leaves are used commercially for greenery in floral arrangements, wreaths, centerpieces, and mantle decorations, especially during the December holidays. Once cut, the foliage keeps well, so it is commonly shipped to cities in the northern United States for sale for decorative purposes.

Magnolia grandiflora is such a strikingly beautiful and romantic plant that it has become a symbol of the American South. The flowers were one of the symbols of the Confederate Army during the Civil War. It is the state flower of Mississippi and Louisiana, and the state tree of Mississippi as well. To many southerners, this is the *only* magnolia. Perhaps this sense is derived from its widespread occurrence throughout the southern states, for although other *Magnolia* species are native, none is as widespread. The universities in the southern United States plant the species widely on campuses, and in the summer the trees fill the air with their lemony fragrance. Most open-grown trees are left unpruned, allowing the lower branches to touch the ground; but some specimens are "limbed up," allowing the smooth, dark gray trunk to show. This species, as well as most other magnolias, attracts birds and small mammals that eat the seeds.

The exact date of the introduction of *Magnolia grandiflora* into European cultivation is not known, but it is generally thought to have been in the late 1720s (see Treseder 1978). The trees certainly had been grown for ornament in the United States prior to that date, and some early landscape plants may have been sent to Europe. *Magnolia grandiflora* does not grow as well in England as in the United States, and plantings there will seldom set fruit. In addition it is generally used as a "wall plant," cultivated near buildings or espaliered, to obtain maximum protection for the plant from wind and cold. The species is also cultivated throughout Central and South America and parts of Asia.

In the United States the plant may be grown as far north as Washington, D.C., and even further if hardy cultivars are grown and protection is provided. On the West Coast, some cultivars of the species may be grown with care as far north as British Columbia. Wherever it is grown, *Magnolia grandiflora* is prized for its glossy, evergreen leaves and large, white, fragrant flowers. It requires enough space to spread out; remembering its ultimate size when first planting the tree is better than having to transplant it later. *Magnolia grandiflora* does not tolerate wet soil. It requires less sun than most evergreens and does best in partial shade. It is easily propagated from seed but has proven somewhat difficult to propagate by cuttings. Flowers (April) May–September; fruits September–November. Zones (6)7–9.

Perhaps the only disadvantage of this magnolia as a garden plant is the leaf litter as new leaves replace the old throughout autumn and spring. The thick, brown, leathery leaves which fall to the ground do not decay quickly and can become unsightly. Considering the beauty of the species, however, this is probably only a minor nuisance. If the lower branches are not pruned out, but are allowed to reach the ground, the leaf litter will remain hidden beneath the tree.

McCracken (1985) observed a decline and, in some cases, death of Southern Magnolia in urban areas of Mississippi and some adjacent states. This problem does not appear to be widespread, and its cause has not been established.

As discussed above, there is great variability within *Magnolia grandiflora*, especially in characters such as age of flowering, habit, and flower size. Seed-grown plants are, therefore, risky, so it is best to buy a known cultivar. Well over 100 cultivars have been selected to take advantage of the variability of the species. Many of the older cultivars are no longer grown, and new forms are replacing them. Some selections were originally named as varieties, a rank of which they are not worthy; over the years, they have been reduced to cultivar status.

CULTIVARS

'Alabama Everblooming'   Flowers more fragrant than those of the species. Bloom period lasts into September. Selected by J. C. McDaniel from a seedling in Cullman, Alabama, in 1930.

'Anne Pickard'   Sport of 'St. George'. Identical to that cultivar but with variegated leaves. Hardy to −15°F (−26° C). Registered by A. A. Pickard of Magnolia Gardens, Canterbury, Kent, England, in 1968.

'Angustifolia'   Narrow-leaved form with typical flowers. In cultivation since 1817.

'Baby Doll'   Dwarf form cultivated in Tampa, Florida, and registered by J. C. McDaniel of Urbana, Illinois, ca. 1969. **Plate 27.**

'Baldwin'   Form with good foliage characteristics, including dark tomentum on underside. Upright growth habit, late blooming, flowers slightly larger than typical. Collected in Baldwin County, Alabama, by Tom Dodd, Jr., and registered ca. 1963.

'Blackwell'   Glossy leaves with undulating margins. Blooms when quite young and listed as "exceptionally hardy." Introduced by Louisiana Nursery, Opelousas, Louisiana, ca. 1986.

Blanchard™   Broadly elliptic leaves, dark green and glossy above with copper indumentum on lower surfaces. Compact, pyramidal growth form. The original tree of this form was found in the garden of D. D. Blanchard in Wallace, North Carolina, in the early 1960s by Robbins Nursery of Willard, North Carolina. The name is a registered trademark of Robbins Nursery.

'Bracken's Brown Beauty'   Compact form with dense pyramidal habit; foliage with undulating margins and dark, rust-colored indumentum. Leaves 5 in. (12 cm) long and 2 in. (5 cm) wide. Flowers about half typical size, averaging 5–6 in. (12.7–15.2 cm) across. Withstood temperatures of −20°F (−29°C) with only slight damage. Selected in 1968 by Ray Bracken at his nursery in Easely, South Carolina. Registered in 1987. Plant Patent #5520. **Plate 30.**

'Bronze Beauty'   New leaves bronze color, remaining so for 2–3 weeks. Selected from a tree on the University of Florida campus in Gainesville and registered by Harold Hume of Gainesville, Florida, in 1961.

'Cairo'   Leaves extremely glossy; columnar habit; hardy to at least 0°F (−18°C). Selected in 1966 by J. C. McDaniel of Urbana, Illinois, from the original tree in Cairo, Illinois.

'Celestial'   Flowers larger than species; tree blooms heavily. Registered by D. Todd Gresham of Santa Cruz, California, in 1963.

'Charles Dickens'   Follicetum brighter red, larger, and more ovoid than that of species. A tetraploid form, possibly a hybrid between *Magnolia grandiflora* and *M. macrophylla*, although it is highly fertile. Discovered by Jewel Templeton in the garden of Charles Dickens in Franklin County, Tennessee. Registered in 1965. **Plate 29.**

'Claudia Wannamaker'   Leaves about 6 in. (15 cm) long by 2 in. (5 cm) wide. Flowers 3–4 in. (7.6–10 cm) in diameter, blooming over a long season. Both flowers and leaves smaller than typical, produced on a broadly pyramidal tree. A good selection for a screen or hedge. Selected in the late 1960s by Johnny Brailsford of Shady Grove Plantation, Orangeburg, South Carolina. Registered in 1988.

'Conger'   Flowers up to 14 in. (35.6 cm) across; possibly the largest flowers of all *Magnolia grandiflora* cultivars. Registered by Inez B. Conger, Arcadia, Florida, ca. 1980.

'Edith Bogue'   Very cold-hardy and tolerates snow load. A vigorous, bushy tree reaching 35 ft. (10.7 m) high and wide with narrow leaves and light indumentum. Selected from a seedling sent from Florida and cultivated in the Montclair, New Jersey, garden of Miss Edith A. Bogue. Tolerates −13°F (−25°C) with minor marginal leaf burn. Registered in 1961.

'Eldorado'   Large columnar tree. Selected from a specimen in Eldorado, Arkansas. Listed in 1989 by Louisiana Nursery, Opelousas, Louisiana.

'Emory'   Dense, columnar growth habit. Leaves with dark indumentum on undersurface. Registered ca. 1980 by Richard Stadtherr of the Horticulture Department at Louisiana State University, Baton Rouge, Louisiana.

'Empire State'   Selection from a tree cultivated on Long Island, New York. Registered by Harold Hume of Gainesville, Florida. Year of registration unknown.

'Exmouth' (also 'Exoniensis', 'Exionensis', 'Lanceolata')   Growth form is somewhat fastigiate; leaves narrow; flowers when young. Introduced by John Colliton prior to 1737 and still common in cultivation. Hardy to Zones (5)6.

'Exionensis'   See 'Exmouth'

'Exoniensis'   See 'Exmouth'

'Fairhope'   Leaves almost round with a blunt tip. Introduced by Magnolia Nursery, Chunchula, Alabama, ca. 1988.

'Ferruginea'   Undersides of leaves with dark indumentum; rounded tree. Cultivated since 1817.

'Florida Giant'   Leaves and flowers larger than typical. Offered by Louisiana Nursery, Opelousas, Louisiana.

'Floridiana'   Flowers heavily, habit broad, pyramidal; leaves whitish beneath with black spots. Selected from a wild population in Florida. Published by Henry Nehrling, Gotha, Florida, in 1944.

'Gallisonniere'   Hardy, pyramidal form with large flowers. Imported from France in 1745 by Baron Galisonnière.

'Gloriosa'   Flowers at a young age; flowers large with more tepals than typical. Cultivated since 1860.

'Goliath'   Flowers larger than species; late-blooming. Leaves short, broad, undulate or bullate. Compact growth habit. Selected in 1910 by Caledonia Nursery, Guernsey. **Plate 27.**

'Griffin'   This plant is sometimes listed as a cultivar of *Magnolia grandiflora*, but is probably a cross between *M. grandiflora* and *M. virginiana*. A description is given in chapter 10.

'Harold Poole'   Compact, shrubby form with narrow leaves. Registered by Ken Durio of Louisiana Nursery, Opelousas, Louisiana, ca. 1980. **Plate 27.**

'Harwell'   About 20% of the leaves have fused margins, forming a hollow tube. Selected as a seedling by L. H. Harwell of Van Buren, Arkansas. Registered by Bon Hartline of Anna, Illinois, in 1983.

'Hasse'   Upright form with small, glossy, dark green leaves. Introduced by Shady Grove Nursery, Orangeburg, South Carolina, ca. 1986. **Plate 31.**

'Josephine'   Upright form with numerous, slightly smaller flowers, continuing late into the season. Registered by Malcolm Manners, Florida Southern College, Lakeland, Florida, ca. 1985.

'Lanceolata'   See 'Exmouth'

'Laurifolia'   Leaves "laurel-like"; dense growth, pyramidal form. Published by Henry Nehrling, Gotha, Florida, in 1933.

'Little Gem'   Growth compact, habit narrow columnar. Leaves and flowers smaller than those of species; flowers when quite young. Relatively easy to propagate by cuttings. Possibly the least hardy cultivar of this species, yet one of the most commonly grown. Selected by Warren Steed of Steed's Nursery from a tree in Candor, North Carolina, in 1952 and registered in 1966. **Plates 27, 32.**

'Louisiana'   Rounded form with very large flowers. Published by Henry Nehrling, Gotha, Florida, 1933.

'Mainstreet'   Fastigiate form selected by Cedar Lane Farms Nursery in Madison, Georgia, ca. 1986.

'Madison'   Compact habit, Late-blooming form selected in the early 1950s in Madison, Alabama, and registered by J. C. McDaniel of Urbana, Illinois, in 1968.

'Majestic Beauty'   Leaves and flowers larger than typical; leaves thick and dark green. Habit pyramidal. Registered by Monrovia Nursery, Azusa, California, in 1963. Approximately 40,000 plants of this cultivar are produced annually. Plant Patent #2250. Hardy to Zones (7)8. **Plates 27, 33–34.**

'Majestica'   Conical form. Leaves and flowers larger than those of species; flowers sometimes double. Published by Henry Nehrling, Gotha, Florida, in 1933.

'Margaret Davis'   Flowers larger than species, reaching 8–10 in. (20.3–25.4 cm) in diameter; leaves narrow. Introduced by Shady Grove Nursery, Orangeburg, South Carolina, ca. 1986, registered 1988.

'Margarita'   Rounded, compact habit; large, lustrous leaves with prominent veins. Registered by the Saratoga Horticultural Foundation, Saratoga, California, in 1958.

'Meyers'   Propagated from one of the largest specimens in the country. Original tree in Tifton, Georgia. Offered in 1989 by Louisiana Nursery, Opelousas, Louisiana.

'Milton's Wavy'   Leaf margins undulate. Offered by Louisiana Nursery, Opelousas, Louisiana.

'Monland' (Timeless Beauty™)   Mature tree forms a broad oval. Profuse flowering continues into fall. Flowers 10–12 in. (25.4–30.5 cm) across with 12 tepals. Leaves narrow with long petioles, glossy dark green above with some brown indumentum below. Discovered by Robert Eiland in the 1960s near Wetumpka, Alabama.

Propagated by Monrovia Nursery, Azusa, California, and named by them ca. 1980. The name 'Monland' comes from *Mon*rovia and Ei*land*; Timeless Beauty is the trademark name. This form appears to be sterile, suggesting that it may be a hybrid, perhaps with *Magnolia virginiana* as the other parent. Registered in 1988. Plant Patent #6178. Hardy to Zones (6)7.

'Nannetensis'   Flowers double. Cultivated since 1865.

'Ocean Wave'   Leaf margins undulate. Selected from a tree on the campus of the University of Florida, Gainesville. Registered by Harold Hume of Gainesville, Florida, in 1961.

'Opal Haws'   Original tree in Boise, Idaho; hardy to −24°F (−31°C). Small flowers and leaves. Originally published as 'Suzette'. Registered by Steven Gossett, Boise, Idaho, ca. 1980.

'Orbit'   A dwarf, compact, rounded form selected by J. C. McDaniel of Urbana, Illinois. Date of selection unknown.

'Pioneer'   Hardy form from Oregon, similar to 'Victoria', but shrubbier, with leaves less dark green. Hardy to Zones (5)6. Registered by William Curtis, Wil-Chris Nurseries, Sherwood, Oregon, ca. 1965.

'Praecox'   Early-flowering form selected in France in the early 1800s and extensively cultivated in England. This cultivar is not as commonly grown now but is still available.

'Praecox Fastigiata'   Seedling of 'Praecox' with fastigiate growth habit; long-flowering. Introduced ca. 1961 by Henry Hohman of Kingsville Nursery, Kingsville, Maryland.

'Pyramidata'   Pyramidal form with large flowers. Introduced by Henry Nehrling, Gotha, Florida, in 1944.

'Reflexa'   Older leaves reflexed. Flowers often double. Published by Henry Nehrling, Gotha, Florida, in 1933.

'Robert Reich'   Leaves and flowers larger than those of species. Registered by Ken Durio, Louisiana Nursery, Opelousas, Louisiana, ca. 1980.

'Ruff'   Leaves deep green above, brick red tomentum below; foliage similar to that of 'Satin Leaf'. Selected from a plant in the garden of Wallace Ruff, Eugene, Oregon. Introduced in 1973 by Gossler Farms Nursery, Springfield, Oregon, 1973. **Plate 27.**

'Russet'   Branchlets, buds, petioles, and leaf undersurfaces are covered with orange-brown tomentum; compact form. Similar to 'Little Gem' but hardier. Introduced by the Saratoga Horticultural Foundation, Saratoga, California, in 1966. Plant Patent #2617. **Plate 35.**

'Saint George'   Leaves with dark indumentum beneath; hardy to −15°F (−26°C) with some foliage burn. Flowers with extra tepals. Registered in 1968 by A. A. Pickard, Magnolia Gardens, Canterbury, Kent, England.

'Saint Mary'   Leaves dark brown beneath, margins wavy; flowers when young. Originated before 1930 at Glen St. Mary Nursery, Glen St. Mary, Florida, from a seedling received from Joseph Vestal and Son, Little Rock, Arkansas. Introduced 1941 and named by W. B. Clarke and Company, San José, California. **Plate 27.**

'Samuel Sommer'   Leaves larger and glossier than those of the species, veins prominent, rusty tomentum beneath; flowers 4–6 in. (10–15 cm) across. Introduced in 1961 by the Saratoga Horticultural Foundation, Saratoga, Florida. Plant Patent #2015. Hardy to Zones (5)6.

'San Marino'   Compact habit; leaves with light indumentum, wavy margins. Introduced in 1970 by the Saratoga Horticultural Foundation, Saratoga, California, and

Gossler Farms Nursery, Springfield, Oregon. Plant Patent #2830.

'Satin Leaf'   Leaves and flowers larger than species; leaf underside covered with deep red-brown tomentum. Original tree native near Tallahassee, Florida. Introduced by Southern States Nursery Company, Macclenny, Florida, ca. 1950.

'Silver Tip'   Upright form. Leaves with silvery gray indumentum. Registered by Ken Durio of Louisiana Nursery, Opelousas, Louisiana, ca. 1980.

'Smitty'   Flowers 8–10 in. (20.3–25.4 cm) across. Introduced by Shady Grove Nursery, Orangeburg, South Carolina, in 1988.

'Spring Hill'   Selected for excellent foliage characteristics by Magnolia Nursery, Chunchula, Alabama, ca. 1985.

'Stalwart'   Narrow pyramidal tree flowering at an early age; pink sheaths surrounding leaf buds. Introduced by the Saratoga Horticultural Foundation, Saratoga, California, in 1958.

'Sunset'   Leaves with light yellow variegation on upper surface; lower surface yellowish green. Originated from a seedling native near Glen St. Mary, Florida. Registered by Harold Hume of Gainesville, Florida, 1961.

'Suzette'   See 'Opal Haws'

'Symmes Select'   Compact form with dark green leaves and slightly undulating margins. Dark brown tomentum on undersurfaces of leaves. Flowers typical. Selected from seedling in 1966 by John Symmes of Cedar Lane Farm, Madison, Georgia. Registered 1988.

Timeless Beauty™   See 'Monland'

'Tulsa'   Vigorous, hardy form growing with only minor injury in Winchester, Massachusetts. Flowers 6–8 in. (15–20.3 cm) in diameter. Seedling from a tree in the Tulsa Rose Garden, Tulsa, Oklahoma. Registered by L. C. Case of Winchester, Massachusetts, in 1988.

'24 Below'   Selected by Frank Galyon, Knoxville, Tennessee, from a tree that survived −24°F (−31°C) in 1985. Registered by Galyon in 1991.

'Undulata'   Leaves with undulated margins. Cultivated since 1844.

'Variegata'   Leaves with yellow variegation. Introduced 1909; doubtfully the same form as sold under this name now.

'Victoria'   Hardy cultivar from Victoria, British Columbia, with glossy foliage and dark indumentum. First published by L. H. Bailey in *Hortus*, 1930, and grown extensively in the northwestern United States. Hardy to Zones (5)6.

'Workman'   Smaller than the species in all ways. Flowers 4–5 in. (10–12.7 cm) across; growth compact; flowers April to mid November; leaves with wavy margins. Registered by Ken Durio of Louisiana Nursery, Opelousas, Louisiana, ca. 1980.

HYBRIDS

Hybrids between *Magnolia grandiflora* and *M. guatemalensis* were made by J. C. McDaniel ca. 1968. The hybrids most resembled the hexaploid ($2n = 114$) *M. grandiflora*, this species showing dominance over the diploid ($2n = 38$) *M. guatamalensis*. Dr. Frank Santamour (1981) crossed *Magnolia liliiflora* 'Darkest Purple' onto *M. grandiflora* in hopes of obtaining a pink or purple-flowered, evergreen magnolia. The hybrids were more vigorous than *M. grandiflora* but had white flowers and were almost indistinguishable from that species. There is a possibility, however, that *M. liliiflora* provided the hybrids with more cold-hardiness. More information on this cross is included with the description of *M. liliiflora*.

Santamour had previously (1979) made a similar cross with *Magnolia acuminata* onto *M. grandiflora* in hopes of obtaining a yellow-flowered, evergreen magnolia. The hybrids had white flowers, and the degree of "evergreenness" varied among the seedlings. For a more complete discussion of this cross, see description of *M. acuminata*. Cultivated hybrids involving *Magnolia grandiflora* include the following:

M. coco × *M. grandiflora* = 'Shirley Curry'

M. grandiflora × *M. virginiana* = 'Freeman', 'Griffin', 'Maryland'

M. grandiflora 'Samuel Sommer' × *M. virginiana* var. *australis* = 'Sweet Summer'

Magnolia schiedeana Schlecht.

Botanische Zeitung 22:144. 1864. Schiede's Magnolia, Corpus, Laurel. Type specimen: destroyed in World War II. Lectotype: *Schiede 296*. British Museum, London. **Plate 70.**

Magnolia schiedeana is an evergreen tree reaching 100 ft. (30 m) tall. The branches are pubescent or glabrous and are usually ringed with stipular scars. The coriaceous oval to elliptic leaves are 4–8 in. (10–20 cm) long and 2–4 in. (5–10 cm) wide, with an acute tip and rounded or cuneate base.

Petioles are 1–1.5 in. (2.5–4 cm) long and glabrous to pubescent. The creamy white flowers are approximately 9–10 in. (23–25.4 cm) across with 9–12 obovate tepals 2–3 in. (5–8 cm) long. A decoction of the flowers is sometimes used as a remedy for scorpion stings. The outer 3 tepals are sepaloid. The stamens are 0.3–0.5 in. (0.8–1.2 cm) long. The pedicels are 0.4–1 in. (1–2.5 cm) long and glabrous to pubescent. The follicetum is ellipsoid and 1.5–2 in. (4–5 cm) long. The outer seed coats are bright red. Chromosome number: $2n = 114$.

This species is native along cool, moist streamsides at 4600–7200 ft. (1380–2160 m) altitude in the mountains of Mexico from Sinaloa and Nayarit to Veracruz. It was collected by Schiede near Jalapa, Veracruz, in 1829 and was named in his honor by Schlechtendahl in 1864. Schiede's type specimen in Berlin was destroyed in World War II, and the lectotype was selected by Dandy.

Magnolia schiedeana is the most northern of the Mexican magnolias; its range, therefore, is the closest to that of *M. grandiflora*, and it is the most likely tropical species for increased cultivation in the United States. Botanically the two species are close as well. They are the only two hexaploid ($2n = 114$) species in this section, and they look so much alike that they are easily confused. The main difference is that *M. schiedeana* is less pubescent and has smaller flowers than *M. grandiflora*. It is possible that *M. schiedeana* should be combined with *M. grandiflora*.

Dandy recognized two varieties of *Magnolia schiedeana*. The typical variety occurs at 4600–5900 ft. (1380–1770 m) altitude and shows little or no pubescence, while the second variety grows at 6900–7200 ft. (2070–2160 m) and is pubescent to tomentose on branchlets and leaves. These varieties were never actually published and are therefore not valid. The recognition of two forms of this species does not appear to be warranted. Dandy himself reported in 1970 that *M. schiedeana* is the "most widely distributed of Mexican magnolias and also the most variable. The indumentum varies greatly."

Magnolia schiedeana is in limited cultivation in the United States. It will grow in southern California and in the extreme southeastern United States. A plant collected by Dr. Frederick G. Meyer in the Sierra Madre Oriental as *M. schiedeana* has proven hardy in the University of Washington Arboretum at Seattle. The identity of this plant, however, has yet to be confirmed. *Magnolia schiedeana* is not considered hardy enough to be grown in England. Flowers March–May (July); fruits the following March–April. Zones 9–10. No cultivars have been selected from nor hybrids made with this species.

Magnolia guatemalensis Donn. Sm.

> *Botanical Gazette* 47:253. 1909. Guatemala Magnolia. Type specimen: *Von Tuerckheim II.2165.*
> British Museum, London. **Plate 37.**

Magnolia guatemalensis is a large evergreen tree 25–75 ft. (7.6–22.6 m) tall with a pyramidal shape and glabrous to pubescent twigs. The elliptic leaves are 5–8 in. (12.7–20.3) cm long by 2–4 in. (5–10 cm) wide. Leaves are coriaceous, glabrous to pubescent above, and pubescent beneath especially on the midrib, becoming glabrous as they mature. The leaf apex is acuminate, and the base is cuneate to obtuse. The petiole is 1 in. (2.5 cm) long and varies from glabrous to pubescent. The fragrant flowers are white with 9 oblong tepals, the outer 3 tepals sepaloid and 1–3 in. (2.5–7.6 cm) long, and the inner 6 tepals 3 in. (7.6 cm) long and 1–2 in. (2.5–5 cm) wide. The stamens are 0.5–0.6 in. (1.3–1.5 cm) long, and the pedicel is about 1 in. (2.5 cm) long and glabrous to pubescent. The follicetum is ellipsoid, 1–2 in. (2.5–5 cm) long. Chromosome number: $2n = 38$.

This species is native to swampy areas in forests at 4600–6500 ft. (1380–1950 m) altitude in Guatemala from Alta Verapaz to Santa Rosa departments. It has also been reported as fairly frequent in the forests of Santa Ana Department, El Salvador, and Intibuca Department, Honduras. The specimen from Honduras probably belongs to *Magnolia hondurensis*, as Molina (1974) reports its abundance in that locality.

Magnolia guatemalensis was first collected by Von Tuerckheim in 1908 in a swamp near Tactic in Alta Verapaz Department of Guatemala at 5000 ft. (1500 m). It was named by Donnell Smith in 1909.

This species has not been successfully cultivated in the United States except in southern California, Louisiana, and Florida. McDaniel (1968) collected seeds from plants near the type locality in Guatemala, and seedlings were distributed by the United States National Arboretum. The flowers are smaller and less fragrant than *Magnolia grandiflora*, and the latter species is preferred for cultivation, even in Central America. Flowers January–May (July); fruits July–September. No cultivars of this species have been selected.

HYBRIDS

Hybrids between *Magnolia guatemalensis* and *M. grandiflora* were made by J. C. McDaniel of Urbana, Illinois, ca. 1968. The hybrids most resembled the hexaploid *M. grandiflora*, which showed dominance over the diploid *M. guatemalensis*. William Kosar of the U.S. National Arboretum crossed *M. guatemalensis* with pollen of *M. virginiana* var. *virginiana* in 1965 and 1966, resulting in several vigorous hybrids to be used in further breeding work.

Magnolia hondurensis A. Molina.

> *Ceiba* 18(1–2):95–106. 1974. Honduran Magnolia. Type specimen: *Molina & Molina 24379.*
> Escuela Agricola Panamericana, Tegucigalpa, Honduras.

Magnolia hondurensis is an upright tree reaching 100 ft. (30 m) in height and 1.5 ft. (0.5 m) in trunk diameter. Bark is smooth, greenish to brownish gray, and the black twigs are fragile. Leaves are coriaceous, elliptic, oblanceolate or lanceolate, 3–7.5 in. (7.6–19.1 cm) long, and 1.5–3 in. (3.8–7.6 cm) wide. Leaf apex is acute or acuminate, and leaf base is acute or sometimes obtuse. Leaves are glossy green above and pale green and rusty tomentose below, becoming glabrous with age. The fragrant flowers are borne on densely pubescent pedicels. Flowers consist of 3 (rarely 4) white and fleshy sepaloid tepals. The 6–9 petaloid tepals are white, concave, obovate or spatulate, and about 2 in. (5.1 cm) long, sometimes appearing bilobed. Stamens are chestnut-colored or brown. The follicetum is 1–2 in. (2.5–5.1 cm) long and chestnut-colored. Seed coats are reddish brown.

This magnolia is native to the cloud forests in the mountains of Honduras. In his original paper, Molina (1974) describes the destruction of populations of *Magnolia hondurensis* when land is cleared for agriculture, reporting that trees are "being exterminated without much hope of saving them." He also advocated the selection of this species for the national flower of Honduras. The flowers are reported by Molina to be boiled for use as a condiment or medicinal. The wood is used for tool handles.

Magnolia yoroconte Dandy.

Journal of Botany 68:147. 1930. Yoroconte Magnolia. Type specimen: *Whitford and Stadtmiller 501.* U.S. National Herbarium, Washington, D.C.

Magnolia yoroconte is a large evergreen tree reaching 125 ft. (38 m) in height, with glabrous or slightly pubescent branchlets. The thick, coriaceous leaves are elliptic to oblong, 4–5.5 in. (10.2–14 cm) long and 1.5–2 in. (4–5 cm) wide, and glabrous or slightly pubescent with an acute apex and obtuse or rounded base. The glabrous petiole is slender and about 1 in. (2.5 cm) long. The white flowers have 9 obovate tepals about 1 in. (2.5 cm) long, of which the outer 3 are sepaloid. The stamens are about 0.5 in. (1.3 cm) long. The fruit is ellipsoid. The outer seed coats are red.

This species is native only to the Tarros Mountains in the Copan Department of Honduras at about 4000 ft. (1200 m). The species was named by Dandy, apparently based only on this one specimen, and it is doubtful that separate species status is warranted. It is very similar to *Magnolia poasana* from Costa Rica and perhaps should be merged with that species. The vernacular name for this plant is Yoroconte, from which the epithet is derived. Its wood is used for posts and in construction of homes. No cultivars have been selected from nor hybrids made with this species.

Magnolia poasana (Pittier) Dandy.

[*Talauma poasana* Pittier.] *Bulletin of Miscellaneous Information, Kew* 1927: 263. 1927. Poas Magnolia. Type specimen: *Pittier 2043.* U.S. National Herbarium, Washington, D.C. **Plate 64.**

Magnolia poasana is an evergreen tree to 70 ft. (21 m) tall with glabrous branchlets. The leaves are elliptic or ovate to obovate, 4–8 in. (10–20 cm) long and 2–3.5 in. (5–9 cm) wide with an acuminate apex and cuneate or rounded base. The leaves are thick and coriaceous, glabrous above, and usually somewhat glaucescent beneath. The petiole is glabrous, 1–2 in. (2.5–5 cm) long. The flowers are fragrant and creamy white with 9 obovate tepals. The outer 3 tepals are sepaloid, greenish outside and white inside, and about 2 in. (5 cm) long. The inner 6 tepals are white, up to 3 in. (7.6 cm) long. The stamens are 0.4–0.5 in. (1.2–1.3 cm) long. The fruit is oblong, about 2 in. (4–5 cm) in length, bearing red seeds.

This species is native to forests at 6000–8500 ft. (1800–2550 m) in Costa Rica from Alajuela Province to Cartago and San Jose provinces and is particularly prevalent on the volcanoes Poas and Barba. This species was considered by Donnell Smith to be identical to *Talauma cespedesii* from Colombia. In 1910 Pittier collected plants on Poas Volcano in Alajuela Province, Costa Rica, at an altitude of 7500 ft. (2250 m). He gave the plant species status, naming it *Talauma poasana*. In 1927 Dandy transferred the species from *Talauma* to *Magnolia*. *Magnolia poasana* is known locally as Candelillo and is reportedly used for medicinal purposes. It has not come into widespread cultivation as its hardiness is questionable. It differs from many of the species in this section in having leaves that are glabrous on the underside. Leaves and flowers are much smaller than those of *M. grandiflora* and even *M. guatemalensis*. Flowers May–June; fruits September–October. No cultivars have been selected nor hybrids made with this species.

Magnolia sororum Seibert.

> *Annals of the Missouri Botanical Garden* 25:828. 1938. Sister Magnolia. Type specimen: *Gene and Peggy White 21.* Missouri Botanical Garden Herbarium, St. Louis. **Fig. 7.1.**

Magnolia sororum is an evergreen tree reaching 100 ft. (30 m) in height. The twigs are covered in tawny pubescence when young, becoming nearly glabrous with age. Leaves are elliptic or obovate to oblong, 4–8 in. (10–20 cm) long and 3–4 in. (7.6–10 cm) wide. The leaves are coriaceous and pubescent above and densely pubescent beneath, with an obtuse to acute apex and obtuse to rounded base. The petiole is up to 1 in. (2.5 cm) long, usually densely pubescent but sometimes glabrescent. The flowers are white and very fragrant with 9 obovate tepals, of which the outer 3 are sepaloid and 2.5–3.0 in. (6.4–7.6 cm) long. The inner 6 tepals are about 3 in. (7.6 cm) long, surrounding stamens which are 0.5–0.6 in. (1.3–1.6 cm) long. The fruit is ellipsoid, 1.0–2.5 in. (2.5–6.4 cm) in length, and the carpels have short beaks. The outer seed coats are red.

This species is native to damp woods at 5200–7100 ft. (1560–2130 m) altitude and is quite abundant on the east and northwest side of Volcan de Chiriqui in Panama. It is also known from the province of San José, Costa Rica. The plant was named by Seibert in 1938 based on specimens collected by the sisters Gene and Peggy White in the valley of upper Rio Chiriqui Veijo, Chiriqui Province, Panama, at 5900 ft. (1770 m). Seibert assigned the epithet from the Latin word *soror* (sister) in honor of the White sisters, who, according to Seibert, "made a special effort to recollect the plant" after Seibert's own specimens were destroyed by fire. The local vernacular name is Vaco.

Woodson and Seibert (1938) and Dandy (1962) report that this species is differentiated from the closely related *Magnolia poasana* and *M. guatemalensis* in having dense pubescence and fewer carpels in the follicetum. Vazquez (1990) proposed a subspecies *lutea* for those plants having yellowish pubescence. Degree of pubescence, as often noted in this book, is a variable character and should be used only with extreme care in delineating species. *Magnolia sororum* may perhaps be best considered as a variety of *M. poasana*. No cultivars have been selected nor hybrids made with this species.

Magnolia sharpii Miranda.

> *Anales del Instituto de Biología, Universidad de México* 26:79, figs. 1–3. 1955. Sharp's Magnolia. Type specimen: *Thomas MacDougall (1955).* University of Mexico Institute of Biology, Mexico, D.F. **Plates 68–69.**

Plants of *Magnolia sharpii* are evergreen trees reaching 100 ft. (30 m) in height, with grayish brown, pubescent twigs. The coriaceous leaves are ovate to obovate, 6–9 in. (15–23 cm) long and 4–7 in. (10–18 cm) wide, with a rounded or acuminate apex and rounded or cordate base. The upper surface is glossy dark green, and the lower surface is covered with grayish pubescence, which may have a yellowish or brownish tinge near the veins. The leaf apex is rounded or acuminate, the base rounded or cordate. The petioles are 1–2 in. (2.5–5 cm) long and pubescent. The fragrant flowers are 10–11 in. (25.4–28 cm) across and white with 9 obovate tepals. The outer 3 are sepaloid, greenish white, 4 in. (10 cm) long and 2 in. (5 cm) wide, while the inner 6 tepals are white, 4–5 in. (10–12.7 cm) long, and 2–3 in. (5–7.6 cm) wide. The stamens are purple at the base and

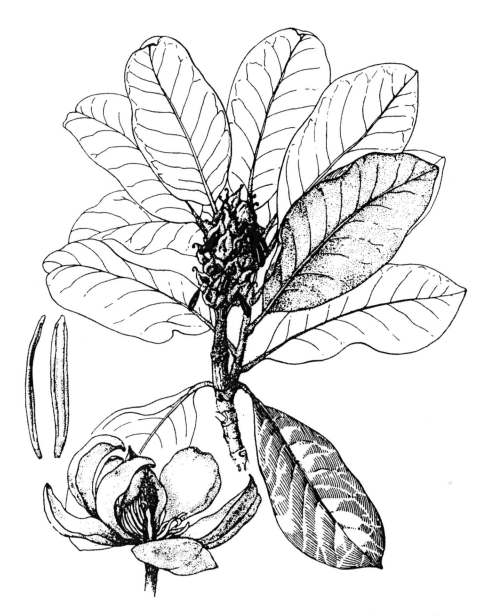

Fig. 7.1. *Magnolia sororum* from Dandy 1962. Reprinted with permission from the Missouri Botanical Garden.

0.5–0.6 in. (1.3–1.6 cm) long, and the pedicel is about 1 in. (2.5 cm) long. The follicetum is ovoid, 3.5–5 in. (9–12.7 cm) in length, and covered with yellow pubescence.

This species is native to evergreen forests of the Central Plateau in Chiapas, Mexico, at elevations of 6500–9800 ft. (1950–2940 m). It was named by Dr. Faustino Miranda in honor of Dr. A. J. Sharp, former head of the University of Tennessee Botany Department. The vernacular name is simply Magnolia or, in Tzetal, Tajchac. It differs from *Magnolia grandiflora* and others in having gray or yellowish pubescence rather than tawny-colored pubescence.

Several attempts have been made to bring *Magnolia sharpii* into cultivation. Frank Galyon (1966) collected material in Chiapas which he left with a nurseryman in Mexico City for grafting. The grafts were not successful. Scion wood also was collected by a team from Stanford University and distributed to the Strybing Arboretum and the U.S. National Arboretum for propagation; again, grafting attempts failed. Seeds collected in Chiapas by Thomas MacDougall in 1962 produced five seedlings, one of which succumbed to 12°F (−11°C) temperatures at Fort Bragg, California. The other plants survived, and J. C. McDaniel obtained scions from two of the plants in 1969. These were grafted onto *M. grandiflora,* and one of these plants was given to the Strybing Arboretum. Strybing staff member Sullivan took three cuttings in 1970, and all rooted easily. McDaniel also attempted to bud *M. sharpii* onto *M. guatemalensis* and *M. grandiflora,* but these efforts failed. It is doubtful that this species can be widely cultivated since it lacks hardiness. The plants cultivated at Fort Bragg, California, bloom in September. No cultivars have been selected nor hybrids made with this species.

Magnolia ptaritepuiana Steyerm.

[*Magnolia roraimae* Steyerm.] *Fieldiana: Botany* 28:233. 1951. Ptari-tepui Magnolia. Type specimen: *Steyermark 59547.* Field Museum of Natural History, Chicago.

Magnolia ptaritepuiana is an evergreen tree 30–40 ft. (9–12 m) in height with pubescent or tomentose twigs. Leaves are obovate, 7–8 in. (18–20.3 cm) long and 5–6 in. (12.7–15 cm) wide, and coriaceous. They are glabrous above and tomentose beneath when young, becoming almost glabrous at maturity. The leaf apex is rounded, the base cuneate or obtuse. The petioles are about 1 in. (2.5 cm) long and may be pubescent or glabrous. The fragrant flowers have 9 creamy white tepals of which the outer 3 are 1.5–2 in. (3.8–5 cm) long, and the inner 6 are 2–2.5 in. (5–6.4 cm) long. The follicetum is ellipsoid and glabrous.

This species is native to forest slopes at 6500–7500 ft. (1950–2250 m) in the mountains of southeastern Venezuela. It was collected by Steyermark in 1944 in Ptari-tepui in Bolivar Province of Venezuela and was named by him in 1951. This species differs from most others in this section in having glabrous carpels. McDaniel (1976a) reported that this species may belong to the genus *Dugandiodendron,* showing some characters of *Magnolia* and some of *Talauma.* This species is not known to be in cultivation.

Magnolia cubensis Urb.

[*M. cubensis* subsp. *acunae* Imch. *M. cubensis* var. *baracoensis* Imch.] *Symbolae Antillanae* 1:307. 1899. Cuban Magnolia. Type specimen: *Linden 2040.* Berlin Herbarium. **Fig. 7.2.**

Magnolia cubensis is an evergreen tree 40–65 ft. (12–20 m) in height with glabrous twigs. The oblong to elliptic leaves are 3–6 in. (7.6–15.2 cm) long and 1–3 in. (2.5–7.6 cm) wide. They are coriaceous and glabrous on both upper and lower surfaces and have an acute or acuminate leaf apex and rounded or acute base. The petioles are 0.5–1 in. (1–2.5 cm) long and glabrous. Flowers are creamy white with 6–9 tepals, of which the outer 3 are obovate, approximately 1 in. (2.5 cm) long and 0.5 in. (1.3 cm) wide. The inner 6 tepals are slightly smaller. The follicetum is ovoid and 1–2 in. (2.5–5.1 cm) long. The outer seed coat is reddish orange.

This species is native to the forests of the Sierra Maestra in Oreinte Province of Cuba at 2300–5900 ft. (690–1770 m) altitude. It was named by Urban in 1899. The vernacular name is Maranon de la Maestra (Cashew of the Maestra). It differs from the other members of this section in having only 5–8 carpels. Flowers May–June; fruits July–January. Zone 10. No cultivars have been selected from nor hybrids made with this species.

Fig. 7.2. *Magnolia cubensis* from Howard 1948. Reprinted with permission from the Bulletin of the Torrey Botanical Club.

Fig. 7.3. *Magnolia domingensis* from Howard 1948. Reprinted with permission from the Bulletin of the Torrey Botanical Club.

Magnolia domingensis Urb.

In Fedde, *Repertonium Specierum Novarum Regni Vegetabilis* 13:447. 1914. Dominican Magnolia. Type specimen: *Nash & Taylor 1081*. New York Botanical Garden. **Fig. 7.3.**

Magnolia domingensis is a small, evergreen tree 10–15 ft. (3–4.6 m) tall, with spreading branches which are covered in white or yellowish hairs when young, becoming brown and glabrous with age. Leaves are obovate, 3–5.5 in. (7.6–14 cm) long and 1.5–3 in. (4–8 cm) wide, and coriaceous with a rounded or sometimes slightly emarginate apex and an acute base. The upper surface is glabrous; the lower surface is pubescent when young, becoming almost glabrous. The petioles are about 0.5 in. (1.3–1.5 cm) long and pubescent when young, becoming glabrous. The creamy white flowers are little known, but have 9–12 tepals. The follicetum is oval and about 1 in. (2.5 cm) long. The outer seed coats are red.

This species is native to forests at 2800–4900 ft. (840–1470 m) in northern Haiti. It was named by Urban in 1914. In a report in 1928, Urban referred to Ekman's collection number H4339 in the Riksmuseum Herbarium in Stockholm, Sweden, in flower, as being representative of this species. But in 1931 this specimen was used as the basis of the new species *Magnolia emarginata*, leaving no flowering specimens for *M. domingensis*. It is partly because of this that the flowers are so little known. This species differs from others in the West Indies group in having pubescent carpels. Fruits in July. This species is not known to be in cultivation.

Magnolia ekmanii Urb.

Arkiv för botanik 23, A, 11:12, fig. 1(4). 1931. Ekman Magnolia. Type specimen: *Ekman H10395.* Riksmuseum Herbarium, Stockholm. **Fig. 7.4.**

Magnolia ekmanii is an evergreen tree with glabrous branches and oval, glabrous, coriaceous leaves. The leaves are 3.5–5 in. (9–12.7 cm) long by 2–3 in. (5–7.6 cm) wide with a rounded apex and base. The petioles are about 1 in. (2.5 cm) long. The flowers are fragrant with 6–9 obovate tepals, the outer 3 being 1–1.5 in. (2.5–3.8 cm) long and reflexed at maturity. The inner tepals are 1–1.5 in. (2.5–3.8 cm) long. The follicetum is oblong and about 1 in. (2.5 cm) long.

This species is native to the edges of pine forests at about 3900 ft. (1170 m) altitude in southwestern Haiti. It was named by Urban in 1931. This specimen was previously thought by Urban (in 1928) to be *Magnolia domingensis,* and the two are very similar, differing mainly in pubescence of the carpels (*M. ekmanii* lacking the pubescence). Flowers July; fruits November. This species is not known to be in cultivation.

Magnolia emarginata Urb. & Eckm.

In Urb. *Arkiv för botanik* 23, A, 11:11, fig. 1(3). 1931. Emarginate Magnolia. Type specimen: *Ekman H4339.* Riksmuseum Herbarium, Stockholm.

Magnolia emarginata is an evergreen tree 30–50 ft. (9–15 m) tall with glabrous branches and obovate leaves, 3–4 in. (7.6–10.2 cm) long and 2–3 in. (5–7.6 cm) wide. They are coriaceous and glabrous on both surfaces, with a rounded to truncate, more or less emarginate, apex; the base is cuneate or rounded. The petioles are about 0.5 in. (1–1.5 cm) long and glabrous. The creamy white flowers have 9 obovate tepals, the outer 3 of which are about 1 in. (2.5 cm) long; the inner 6 are slightly longer. The follicetum is oblong, 1–1.5 in. (2.5–3.8 cm) long, and glabrous.

This species is native to forests at 2900–4300 ft. (900–1300 m) in northern Haiti. It was named by Urban and Ekman in 1931 and is very similar to the other two species from Haiti, *Magnolia domingensis* and *M. ekmanii.* It differs from the former in having glabrous carpels. It differs from the latter in leaf shape and in having more carpels. It differs from both in having emarginate leaves. Flowers June; fruits March. This species is not known to be in cultivation.

Magnolia pallescens Urb. & Eckm.

In Urb. *Arkiv för botanik* 23, A, 11:10, fig. 1(2). 1931. Pale Magnolia. Type specimen: *Ekman H13884.* Riksmuseum Herbarium, Stockholm. **Fig. 7.5.**

Magnolia pallescens is a medium-sized tree to 65 ft. (19.6 m) tall. The branchlets are covered with short, golden-colored hairs, especially when young. The leaves are obovate to orbicular, 2–3.5 in. (5–9 cm) long, and 2–3 in. (5–7.6 cm) wide. They are thick and coriaceous, glabrous above, and covered with yellowish pubescence beneath. The leaf tip is rounded or truncate, and the base is obtuse or rounded. The petioles are about 0.5 in. (1.3 cm) long and densely tomentose. The flowers are creamy white with 9 ovate tepals. Of these, the outer 3 are about 1 in. (2.5 cm) long, while the inner 6 are slightly longer. The follicetum is ovoid to globose, about 1 in. (2.5 cm) in length. The outer seed coats are dark red.

This species is native to forests at about 5200–6900 ft. (1560–2070 m) in western Dominican Republic. It was named by Urban and Ekman in 1931. The epithet *pallescens* means "pale" and probably refers to the yellowish pubescence. The vernacular name is Evano Verde, and the wood is reportedly used for cabinet work. Flowers October; fruits August. This species is not known to be in cultivation.

Plate 1. Typical fruit of *Magnolia acuminata,* often appearing misshapen. Photo by Richard Figlar, Pomona, New York.

Plate 2. *Magnolia acuminata* var. *subcordata.* Photo by the author.

Plate 3. *Magnolia acuminata* var. *subcordata* 'Skyland's Best'. Photo by Richard Figlar, Pomona, New York.

Plate 4. *Magnolia biondii* summer habit. Photo by the author.

Plate 6. *Magnolia biondii*. Photo by Peter Del Tredici, Arnold Arboretum, Jamaica Plain, Massachusetts.

Plate 5. *Magnolia biondii* in flower at the Arnold Arboretum, Jamaica Plain, Massachusetts. Photo by Peter Del Tredici, Arnold Arboretum, Jamaica Plain, Massachusetts.

Plate 7. *Magnolia campbellii* var. *alba* flowering habit. Photo by the author.

Plate 8. *Magnolia campbellii* var. *alba*. Photo by the author.

Plate 9. *Magnolia campbellii* var. *alba* 'Stark's White'. Photo by Richard Figlar, Pomona, New York.

Plate 10. *Magnolia campbellii* var. *alba* 'Strybing White'. Photo from the collections of the Strybing Arboretum Society's Helen Crocker Russell Library of Horticulture, Strybing Arboretum, San Francisco, California.

Plate 11. *Magnolia campbellii* var. *mollicomata* 'Lanarth'. Photo by Richard Figlar, Pomona, New York.

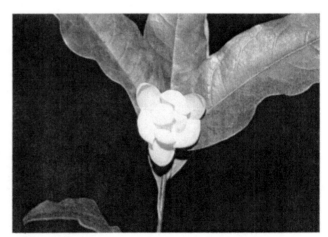

Plate 12. *Magnolia coco*. Photo by Kenneth Durio, Opelousas, Louisiana, from the collections of the Magnolia Society.

Plate 13. *Magnolia cylindrica*. Photo from the collections of the Magnolia Society.

Plate 14. *Magnolia dawsoniana*. Photo from the collections of the Strybing Arboretum Society's Helen Crocker Russell Library of Horticulture, Strybing Arboretum, San Francisco, California.

Plate 15. *Magnolia delavayi* habit. Photo by Richard Figlar, Pomona, New York.

Plate 16. *Magnolia delavayi* flower. Photo by Richard Figlar, Pomona, New York.

Plate 18. *Magnolia denudata.* Photo by the author.

Plate 17. *Magnolia delavayi* flower. Photo from the collections of the Magnolia Society.

Plate 19. *Magnolia denudata.* Photo by the author.

Plate 20. *Magnolia denudata.* Photo by the author.

Plate 21. *Magnolia fraseri.* Photo by John Tobe, Clemson, South Carolina.

Plate 22. Fruit of *Magnolia fraseri.* Photo by John Tobe, Clemson, South Carolina.

Plate 23. *Magnolia fraseri* var. *pyramidata.* Photo by Richard Figlar, Pomona, New York.

Plate 24. *Magnolia grandiflora.* Photo by the author.

Plate 25. *Magnolia grandiflora*. Photo by the author.

Plate 26. *Magnolia grandiflora* leaves often have a rust-colored undersurface which adds to their ornamental value. Photo by the author.

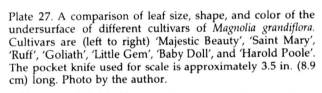

Plate 27. A comparison of leaf size, shape, and color of the undersurface of different cultivars of *Magnolia grandiflora*. Cultivars are (left to right) 'Majestic Beauty', 'Saint Mary', 'Ruff', 'Goliath', 'Little Gem', 'Baby Doll', and 'Harold Poole'. The pocket knife used for scale is approximately 3.5 in. (8.9 cm) long. Photo by the author.

Plate 28. *Magnolia grandiflora* espaliered. Photo by Joseph Hickman, Benton, Illinois.

Plate 29. Fruit of *Magnolia grandiflora* 'Charles Dickens'. Photo by J. C. McDaniel from the collections of the Magnolia Society.

Plate 30. Dark brown tomentum on the undersurface of the leaves is a characteristic of *Magnolia grandiflora* 'Bracken's Brown Beauty'. Photo by the author.

Plate 31. *Magnolia grandiflora* 'Hasse' has small, compact leaves. Photo by the author.

Plate 32. *Magnolia grandiflora* 'Little Gem' has flowers typical of the species, but 'Little Gem' is smaller than typical in all aspects. Photo by the author.

Plate 33. *Magnolia grandiflora* 'Majestic Beauty' has light tomentum on the leaf undersurfaces. Photo by the author.

Plate 34. *Magnolia grandiflora* 'Majestic Beauty'. Photo by the author.

Plate 35. *Magnolia grandiflora* 'Russet' has reddish brown tomentum. Photo by the author.

Plate 36. An unnamed variegated form of *Magnolia grandiflora* in Lafayette, Louisiana. Photo by Richard Figlar, Pomona, New York.

Plate 37. *Magnolia guatemalensis*. Photo by J. C. McDaniel from the collections of the Magnolia Society.

Plate 38. *Magnolia hypoleuca*. Photo by Gerald Straley, University of British Columbia, from the collections of the Magnolia Society.

Plate 39. *Magnolia hypoleuca*. Photo by Richard Figlar, Pomona, New York.

Plate 40. *Magnolia hypoleuca* fruit. Photo from the author's collection.

Plate 41. *Magnolia kobus*. Photo by the author.

Plate 42. *Magnolia kobus*. Photo by Richard Figlar, Pomona, New York.

Plate 43. *Magnolia kobus* 'Borealis'. Photo by Larry Langford, Gibson, Tennessee.

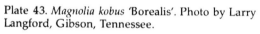

Plate 44. *Magnolia kobus* var. *loebneri*. Photo from the collections of the Magnolia Society.

Plate 45. *Magnolia kobus* var. *loebneri* 'Spring Snow'. Photo from the collections of the Magnolia Society.

Plate 46. *Magnolia kobus* var. *stellata*. Photo by the author.

Plate 47. *Magnolia kobus* var. *stellata*. Photo by the author.

Plate 48. *Magnolia kobus* var. *stellata* 'Centennial'. Photo by the author.

Plate 49. *Magnolia kobus* var. *stellata* 'Jane Platt'. Photo by the author.

Plate 50. *Magnolia kobus* var. *stellata* 'Royal Star'. Photo by the author.

Plate 51. *Magnolia kobus* var. *stellata* 'Rubra' at the Morris Arboretum, Philadelphia, Pennsylvania. Photo by the author.

Plate 52. *Magnolia liliiflora*. Photo by the author.

Plate 53. *Magnolia liliiflora* 'Holland Red'. Photo by Joseph Hickman, Benton, Illinois.

Plate 54. *Magnolia liliiflora* 'O'Neill'. Photo by the author.

Plate 55. *Magnolia macrophylla* habit at Longwood Gardens, Philadelphia, Pennsylvania. Photo by the author.

Plate 56. *Magnolia macrophylla*. Photo by the author.

Plate 57. *Magnolia macrophylla.*
Photo by the author.

Plate 58. *Magnolia macrophylla.*
Photo by the author.

Plate 59. *Magnolia macrophylla* var. *ashei*
blooms at an early age. This plant is
only four years old and 14 in. (35.6 cm)
tall, with a flower 10 in. (25.4 cm) in
diameter. Photo by Richard Figlar,
Pomona, New York.

Plate 60. *Magnolia macrophylla* var. *dealbata*. Photo by Martin Grantham, Horticulturist, University of California Botanical Garden, Berkeley, California.

Plate 61. *Magnolia macrophylla* var. *dealbata*. Photo by Martin Grantham, Horticulturist, University of California Botanical Garden, Berkeley, California.

Plate 62. *Magnolia officinalis*. Photo by the author.

Plate 63. *Magnolia officinalis* 'Biloba' has leaves with an emarginate (bilobed) apex. Photo by Richard Figlar, Pomona, New York.

Plate 64. *Magnolia poasana*. Photo by Martin Grantham, Horticulturist, University of California Botanical Garden, Berkeley, California.

Plate 65. *Magnolia sargentiana* var. *robusta.* Photo by Richard Figlar, Pomona, New York.

Plate 66. *Magnolia sargentiana* var. *robusta.* Photo by the author.

Plate 67. *Magnolia sargentiana* var. *robusta* 'Blood Moon'. Photo from the collections of the Magnolia Society.

Plate 68. *Magnolia sharpii*. Photo by Martin Grantham, Horticulturist, University of California Botanical Garden, Berkeley, California.

Plate 69. *Magnolia sharpii*. Photo by Holly Forbes, Assistant Curator, University of California Botanical Garden, Berkeley, California.

Plate 70. *Magnolia schiedeana*. Photo by Martin Grantham, Horticulturist, University of California Botanical Garden, Berkeley, California.

Plate 71. *Magnolia sieboldii* flowering habit. Photo by the author.

Plate 72. *Magnolia sieboldii*. Photo by Richard Figlar, Pomona, New York.

Plate 73. *Magnolia splendens*. Photo by Richard Figlar, Pomona, New York.

Plate 74. *Magnolia sprengeri* 'Diva'. Photo from the collections of the Strybing Arboretum Society's Helen Crocker Russell Library of Horticulture, Strybing Arboretum, San Francisco, California.

Plate 75. *Magnolia tripetala*. Photo by the author.

Plate 76. *Magnolia tripetala*. Photo by the author.

Plate 77. *Magnolia virginiana* has a rounded, shrubby habit. Photo by the author.

Plate 78. Leaf undersurfaces of *Magnolia virginiana* are silvery white. Photo by the author.

Plate 79. *Magnolia virginiana*. Photo by the author.

Plate 80. *Magnolia virginiana*. Photo by the author.

Plate 81. *Magnolia virginiana* var. *australis* has an upright, open habit, contrasting with the rounded, shrubby habit of typical *M. virginiana*. Photo by the author.

Plate 82. *Magnolia zenii* at the Arnold Arboretum, Jamaica Plain, Massachusetts. Photo by Peter Del Tredici, Arnold Arboretum.

Plate 83. *Magnolia zenii*. Photo by Peter Del Tredici, Arnold Arboretum, Jamaica Plain, Massachusetts.

Plate 84 *Magnolia* × *brooklynensis* 'Evamaria'. Photo by the author.

Plate 85. *Magnolia × brooklynensis* 'Evamaria'. Photo by the author.

Plate 86. *Magnolia × brooklynensis* 'Woodsman'. Photo by J. C.
McDaniel from the collections of the Magnolia Society.

Plate 87. *Magnolia × brooklynensis* 'Woodsman'. Photo by J. C.
McDaniel from the collections of the Magnolia Society.

Plate 88. *Magnolia* × *kewensis* 'Wada's Memory'. Photo by the author.

Plate 89. *Magnolia* × *proctoriana*. Photo by the author.

Plate 90. *Magnolia* × *soulangiana*. Photo by the author.

Plate 91. *Magnolia* × *soulangiana* 'Alexandrina'. Photo by the author.

Plate 93. *Magnolia* × *soulangiana* 'Lennei'. Photo by the author.

Plate 92. *Magnolia* × *soulangiana* 'Deep Purple Dream'. Photo by David Ellis, Chunchula, Alabama.

Plate 94. *Magnolia* × *soulangiana* 'Lennei'. Photo by the author.

late 95. *Magnolia* × *soulangiana* 'Lombardy Rose'. Photo by
ie author.

Plate 96. *Magnolia* × *soulangiana* 'Niemetzii'. Photo
by the author.

Plate 97. *Magnolia* × *soulangiana* 'Rustica Rubra'. Photo by the author.

Plate 98. *Magnolia* × *soulangiana* 'Rustica Rubra'. Photo by the author.

Plate 99. *Magnolia* × *soulangiana* 'Veitchii Rubra'. Photo by the author.

Plate 100. *Magnolia* × *soulangiana* 'Verbanica'. Photo by the author.

Plate 101. *Magnolia* × *thompsoniana* 'Urbana'. Photo by J. C. McDaniel from the collections of the Magnolia Society.

Plate 102. *Magnolia* × *veitchii.* Photo by the author.

Plate 103. *Magnolia* × *veitchii.* Photo by the author.

Plate 105. *Magnolia × wieseneri*. Photo from the collections of the Magnolia Society.

Plate 104. *Magnolia × veitchii* 'Isca'. Photo by the author.

Plate 106. *Magnolia* 'Albatross'. Photo by Joseph Hickman, Benton, Illinois.

Plate 107. *Magnolia* 'Ann'. Photo by the author.

Plate 108. *Magnolia* 'Ann'. Photo by the author.

Plate 109. *Magnolia* 'Athene'. Photo from Mark Jury Nursery, North Taranaki, New Zealand.

Plate 110. *Magnolia* 'Atlas'. Photo from Mark Jury Nursery, North Taranaki, New Zealand.

Plate 111. *Magnolia* 'Big Dude'. Photo by Richard Figlar, Pomona, New York.

Plate 112. *Magnolia* 'Butterflies'. Photo by Philip Savage, Bloomfield Hills, Michigan, from the collections of the Magnolia Society.

Plate 113. *Magnolia* 'Butterflies'. Photo by the author.

Plate 114. *Magnolia* 'Darrell Dean'. Photo by Kenneth Durio, Opelousas, Louisiana, from the collections of the Magnolia Society.

Plate 115. *Magnolia* 'Elizabeth'. Photo by Lola Koerting from the Brooklyn Botanic Garden, Brooklyn, New York.

Plate 116. *Magnolia* 'Elizabeth'. Photo by Lola Koerting from the Brooklyn Botanic Garden, Brooklyn, New York.

Plate 117. *Magnolia* 'Galaxy'. Photo by the author.

Plate 118. *Magnolia* 'Iolanthe'. Photo from the collections of the Magnolia Society.

Plate 119. *Magnolia* 'Jersey Belle'. Photo from the collections of the Magnolia Society.

Plate 120. *Magnolia* 'Judy'. Photo by the author.

Plate 121. *Magnolia* 'Legacy'. Photo by David Leach, North Madison, Ohio, from the collections of the Magnolia Society.

Plate 122. *Magnolia* 'Lotus'. Photo from Mark Jury Nursery, North Taranaki, New Zealand.

Plate 123. *Magnolia* 'Marillyn'. Photo by Lola Koerting from the Brooklyn Botanic Garden, Brooklyn, New York.

Plate 124. *Magnolia* 'Maryland'. Photo by the author.

Plate 125. *Magnolia* 'Maryland'. Photo by the author.

Plate 126. *Magnolia* 'Maryland'. Photo by the author.

Plate 127. *Magnolia* 'Nimbus'. Photo by Richard Figlar, Pomona, New York.

Plate 128. *Magnolia* 'Nimbus'. Photo by Richard Figlar, Pomona, New York.

Plate 129. *Magnolia* 'Orchid'. Photo by the author.

Plate 130. *Magnolia* 'Paul Cook'. Photo by Kenneth Durio, Opelousas, Louisiana, from the collections of the Magnolia Society.

Plate 131. *Magnolia* 'Pickard's Maime'. Photo from the collections of the Magnolia Society.

Plate 132. *Magnolia* 'Pickard's Sundew'. Photo by the author.

Plate 133. *Magnolia* 'Randy'. Photo by the author.

Plate 134. *Magnolia* 'Raspberry Ice'. Photo by the author.

Plate 135. *Magnolia* 'Ricki'. Photo by the author.

Plate 136. *Magnolia* 'Royal Crown'. Photo by the author.

Plate 137. *Magnolia* 'Sangreal'. Photo by David Ellis, Chunchula, Alabama.

Plate 138. *Magnolia* 'Spectrum'. Photo by Richard Figlar, Pomona, New York.

Plate 139. *Magnolia* 'Tina Durio'. Photo by Richard Figlar, Pomona, New York.

Plate 140. *Magnolia* 'Tina Durio'. Photo by Kenneth Durio, Opelousas, Louisiana, from the collections of the Magnolia Society.

Plate 141. *Magnolia* 'Todd Gresham'. Photo by David Ellis, Chunchula, Alabama.

Plate 142. *Magnolia* 'Vulcan'. Photo from Mark Jury Nursery, North Taranaki, New Zealand.

Fig. 7.4. *Magnolia ekmanii* from Howard 1948. Reprinted with permission from the Bulletin of the Torrey Botanical Club.

Fig. 7.5. *Magnolia pallescens* from Howard 1948. Reprinted with permission from the Bulletin of the Torrey Botanical Club.

Magnolia hamorii Howard.

> *Bulletin of the Torrey Botanical Club* 75(4):335–357. 1948. Hamor's Magnolia. Type specimen: *Howard 8484.* Gray Herbarium, Harvard University, Cambridge, Massachusetts. **Fig. 7.6.**

Magnolia hamorii is a tree 40–50 ft. (12–15 m) tall with smooth, dark gray bark. The branches are covered in yellowish pubescence. The leaves are oval, 3–4 in. (7.6–10 cm) long, 2–3 in (5–8 cm) wide, and glabrous on both surfaces. The leaf apex is emarginate (bilobed), with one lobe smaller than the other; the base of the leaf is rounded, and the petiole is about 1 in. (2.5 cm) long and pubescent. The fragrant flowers are creamy white with 9–12 tepals, the outer 3 of which are oblong, about 2 in. (5 cm) long, and emarginate at the apex with unequal lobes. The inner tepals are about 2 in. (5 cm) long with a rounded apex. The follicetum is oblong, 1–2 in. (2.5–5 cm) long. The outer seed coats are pink to orange-red. Chromosome number: $2n = 38$.

This species is native to Barahona Province of the Dominican Republic. It has been found in small stands on limestone hilltops in wet cloud forests at about 4400 ft. (1320 m) altitude. It was named by R. A. Howard in 1948 in honor of Mr. George Hamor of the Barahona Sugar Company. The vernacular name is Caimoni. Flowers and fruits were collected in August. *Magnolia hamorii* is probably most closely related to *M. emarginata* but differs from that species in having unequal lobes at the leaf apex. In addition, shoots and petioles of *M. emarginata* are glabrous, whereas they are pubescent in *M. hamorii.* This species is not known to be in cultivation.

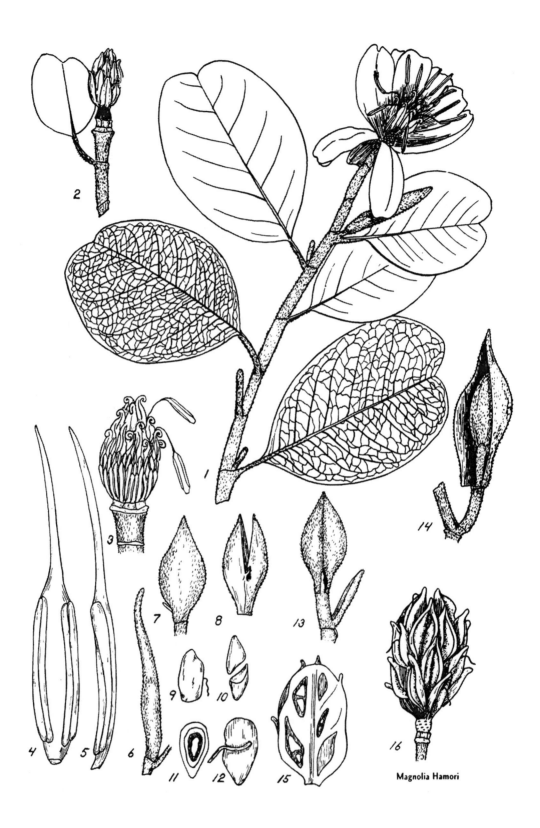

Fig. 7.6. *Magnolia hamorii* from Howard 1948. Reprinted with permission from the Bulletin of the Torrey Botanical Club.

Magnolia portoricensis Bello.

Anales de Sociedad Española de Historia Natural 10:233. 1881. Puerto Rican Magnolia. Type specimen: not designated.

Magnolia portoricensis is an evergreen tree 25–80 ft. (7.6–24 m) tall with glabrous branchlets. The leaves are oval to slightly obovate, 3–9.5 in. (7.6–24 cm) long and 2–5 in. (5–12.7 cm) wide, glabrous, and coriaceous, with an acute apex and a rounded base. The petioles are about 1 in. (2.5 cm) long and glabrous. The fragrant flowers are creamy white with 9–12 tepals, of which the outer are greenish, concave, and about 1.5 in. (3.8 cm) long. The inner tepals are ovate and 1–2 in. (2.5–5 cm) long, and the pedicels are 1–2 in. (2.5–5 cm) in length. The follicetum is oblong and 1–1.5 in. (2.5–3.8 cm) long. The outer seed coats are reddish orange.

This species is native to montane forests at 2600–3200 ft. (780–960 m) in western Puerto Rico. It was named by Bello in 1881 based on collections made by him in the Mayaguez District. *Magnolia portoricensis* differs from *M. splendens* of eastern Puerto Rico in being completely glabrous. Vernacular names include Burro, Mauricio, and Jaqilla. Flowers May–June; fruits September. This species is not known to be in cultivation.

Magnolia splendens Urb.

[*Talauma mutabilis* var. *splendens* Urb. ex McLaughlin, *T. splendens* (Urb.) McLaughlin.] *Symbolae Antillanae* 1:306. 1899. Shining Magnolia. Type specimen: not designated. **Plate 73.**

Magnolia splendens is an evergreen tree 16–80 ft. (5–24 m) tall with grayish to yellowish pubescence on the branchlets. The ovate leaves are 4–7 in. (10–17.8 cm) long by 1.5–3.5 in. (3.8–9 cm) wide. They are coriaceous, glabrous above, and pubescent below, with an acute or acuminate apex and a rounded base. The petioles are 1–1.5 in. (2.5–3.8 cm) long and tomentose, becoming almost glabrous with age. The flowers are creamy white and small, about 1 in. (2.5 cm) across. They are made up of 9–12 tepals, of which the inner 6–9 are obovate and 1–2 in. (2.5–5 cm) long. The outer 3 tepals are oblong and 1–1.5 in. (2.5–3.8 cm) long. The pedicels are 1–2 in. (2.5–5 cm) long and tomentose. The follicetum is oblong, 1–1.5 in. (2.5–3.8 cm) in length, bearing red seeds.

This species is native to montane forests of eastern Puerto Rico. It was named by Urban based on three collections: *Eggers 1070*, *Sintenis 1423*, and *Sintenis 5324*. The epithet *splendens* means "shining" or "brilliant." Vernacular names include Sabino and Laurel Sabino. The wood is used for furniture and in construction. McLaughlin (1933) transferred this species to *Talauma* on the basis of wood characteristics, but flower and fruit characteristics do not support this change. Flowers June–August; fruits October. This species is not known to be in cultivation.

Magnolia chimantensis Steyerm. & Maguire.

Memoirs of the New York Botanical Garden 17:443. 1967. Chimanta Magnolia. Type specimen: *Steyermark 75840*. Instituto Botanico, Caracas, Venezuela.

Magnolia chimantensis is an evergreen tree reaching 50 ft. (15 m) tall, with branchlets that are pubescent above and glabrous below. The leaves are oblong to elliptic, 3–6 in. (7.6–15 cm) long and 1.5–3 in. (3.8–7.6 cm) wide. The leaf apex is acute, the base obtuse. The upper surface is glabrous, the lower surface covered with greenish yellow hairs. The petioles are about 1 in. (2.5 cm) long and pubescent to glabrous. The fragrant flowers are creamy white with 9–12 tepals, each of which is 1–1.5 in. (2.5–3.8 cm) long. The follicetum is green.

This species is native to the Chimanta Massif region of Venezuela at altitudes of 4900–7500 ft. (1470–2250 m). It was named by Steyermark and Maguire in 1967. The type locality is not far from that of *Magnolia ptaritepuiana*, and, in fact, these two species may be one and the same. Steyermark and Maguire, however, note that *M. chimantensis* differs in having smaller tepals, narrower leaves, and more pubescence on the lower leaf surface. This species is not known to be in cultivation.

Magnolia subgenus *Magnolia* section *Gwillimia* DC.

[*Gwillimia* Rottl. ex Sims.] *Regni Vegetabilis Systema Naturale* 1:455, 548. 1817. TYPE: *Magnolia coco.*

Magnolias in section *Gwillimia* are evergreen shrubs or small trees. The stipules are adnate to the petioles, leaving scars on the petioles when falling. The flowers are creamy white with 9(10–11) tepals. The gynoecium is sessile. The carpels have short beaks. The species for which chromosome numbers have been determined are diploid with $2n = 38$ chromosomes.

Gwillimia was first described by de Candolle in 1817. The section name is taken from Rottler's genus name published in 1806 and included all nine Asiatic species of *Magnolia* known at that time. Since then, most of the species have been moved to other sections. *Magnolia coco*, the type species of section *Gwillimia*, remains from de Candolle's original group, while new species have been added to the section. It now contains eighteen species from tropical and subtropical Asia from southern China to Siam, Borneo, and the Philippines.

Each of the eighteen species described in the literature is discussed here, although it is doubtful that they all deserve specific status. Six of them were named by Dandy based on a single, poor specimen of each, and many are so closely related that it is difficult to differentiate them. Furthermore, this section strongly resembles the genus *Talauma*, the main difference in the two being fruit characteristics. As some species have been assigned to section *Gwillimia* without knowledge of the fruit, it is possible that they will be transferred to *Talauma* once fruit is found. Thus, while *Gwillimia* is currently one of the largest magnolia sections, it is doubtful that it will remain so with continued study. Until such study is undertaken, the species currently described in the literature are presented together with a discussion of the presently available information. Two species in this section, *Magnolia coco* and *M. delavayi*, are cultivated. Due to their lack of hardiness, however, cultivation is limited.

Magnolia albosericea Chun & Tsoong.

Acta Phytotaxonomica Sinica 9:117. 1964. Type specimen: *How 72740*. Herbarium of the Canton Botanic Garden.

Magnolia albosericea is an evergreen shrub or small tree 15–25 ft. (4.6–7.6 m) tall. The bark is gray, and the branches are first covered in white tomentum, becoming glabrous later. The leaves are elliptic, 10–12 in. (25.4–30.5 cm) long, and 2–4 in. (5–10 cm) wide. They are coriaceous, pubescent at first, becoming glabrous, glaucescent beneath. The leaf apex is obtuse or acute, the base cuneate. The petiole is pubescent at first, becoming glabrous, about 1 in. (2.5 cm) long. The flowers are white with 9 tepals. The stamens are about 0.5 in. (1.3 cm) long. The gynoecium is ellipsoid and pubescent. The follicetum is ellipsoid, about 2 in. (5 cm) long.

This species is native to forests at 1000–1300 ft. (300–390 m) in Hainan Province of China. It was named by Chun and Tsoong in 1964. The epithet *albosericea* means "white silky hairs," referring to the tomentum on the branchlets and petioles. This species is not known to be in cultivation.

Magnolia champacifolia Gagnep.

In Humbert, *Supplément de la Flore Générale de l'Indo-Chine* 1:39. 1938. Type specimen: *Poilane (Institut des Sciences) 6472.* Herbarium of the Musée National d'Histoire Naturelle, Paris.

Magnolia champacifolia is an evergreen shrub or small tree 10–30 ft. (3–9 m) tall. The branchlets are tomentose, becoming almost glabrous. The leaves are lanceolate to oblong and coriaceous. They are 8–12 in. (20.3–30.5 cm) long and 3–5 in. (7.6–12.7 cm) wide. Young leaves are pubescent to tomentose on both surfaces. The leaf apex is acuminate, the base is cuneate. The petiole is densely pubescent at first, becoming glabrous, about 2 in. (5 cm) long. The flowers are white with 9 tepals, the outer 3 tepals densely pubescent outside at the base and along the middle. The fruit is unknown.

This species is native to forests at 4200–5000 ft. (1260–1500 m) in Nha-trang Province of Annam. It was named by Gagnepain in 1938 based on specimens collected by Poilane in 1923. The epithet *champacifolia* refers to the resemblance of its leaves to those of *Michelia champaca.* This species is not known to be in cultivation.

Magnolia championii Benth.

[*M. liliifera* var. *championi* (Benth.) Pampan., *M. pumila* var. *championi* (Benth.) Finet & Gagnep.] *Flora of Hong Kong* 8. 1861. Champion's Magnolia. Type specimen: *Champion 37.* Herbarium of the Royal Botanic Gardens, Kew.

Magnolia championii is an evergreen shrub or small tree 3–9 ft. (0.9–2.7 m) high. The branchlets are densely pubescent, becoming glabrous with age. The leaves are oblong or elliptic, 3–6 in. (7.6–15.2 cm) long and 1–2 in. (2.5–5 cm) wide. The leaves are coriaceous, glabrous above, glaucescent and pubescent at first, becoming glabrous beneath. The leaf apex is acuminate, the base cuneate. The petioles are about 0.5 in. (1.3 cm) long, pubescent at first, becoming almost glabrous. The white flowers are very fragrant, especially at night, and about 1.5 in. (4 cm) in diameter. The flowers have 9 tepals, the outer 3 sepaloid. The stamens are 0.3–0.5 in. (0.9–1.3 cm) long. The gynoecium is pubescent. The pedicels are densely pubescent and 1–2 in. (2.5–5 cm) long. The follicetum is ellipsoid, about 2 in. (5 cm) long. The outer seed coats are red.

This species is native to woods and grassy slopes at 500–800 ft. (150–240 m) altitude in Hong Kong and Ting Wu Shan, Guangdong Province, China. It was first collected near Hong Kong by Champion, who identified it as *Magnolia coco* (then referred to as *M. pumila* or *M. liliifera*). Ten years later, Bentham described the plant as a new species, naming it in honor of Champion. In 1886 it was again reduced to *M. coco* by Forbes and Hemsley. Subsequently, Finet and Gagnepain, and later Pampanini, recognized the plant as a variety of *M. coco,* from which it differs in having pubescent young branchlets, pedicels, and carpels. Although Dandy (unpublished) considered it to be truly a distinct species, it will probably be considered a variety of *M. coco* when the plants are more completely understood. *Magnolia championii* is also closely related to *M. fistulosa* and *M. paenetalauma.* This species is not known to be in cultivation.

Magnolia clemensiorum Dandy.

Journal of Botany 68:207. 1930. Clemens Magnolia. Type specimen: *Clemens 339967.* University of California (Berkeley) Herbarium.

This species is an evergreen tree about 10 ft. (3 m) tall and entirely glabrous. The leaves are elliptic to oblong, coriaceous, and up to 8 in. (20 cm) long and 2 in. (5 cm) wide. The leaf apex is acuminate, the base cuneate. The petiole is about 0.5 in. (1.3 cm) long. The flowers are white with 9 tepals, the outer 3 somewhat sepaloid. The stamens are about 0.5 in. (1.3 cm) long. The fruit is unknown.

Magnolia clemensiorum is native to forests of Tourane Province of Annam, the only locality from which it is known. The species was named by Dandy in 1930 based on a single specimen collected by J. and M. S. Clemens in 1927. Dandy notes that its closest relative is *M. coco*, from which it differs in having more reticulate leaves, a more elongated pedicel, and slightly longer stamens. The single known collection has the flowers terminating in short axillary branchlets, making the flowers appear axillary. A similar habit occasionally occurs in *M. coco*, *M. fistulosa*, and some species of *Talauma*. This species is not known to be in cultivation.

Magnolia coco (Lour.) DC.

[*Gwillimia indica* Rottl. ex DC., *Liriodendron coco* Lour., *Liriopsis pumila* Spach ex Baill., *Magnolia liliifera* Baill., *M. pumila* Andr., *Talauma coco* (Lour.) Merr., *T. longifolia* (not Ridl.) Craib., *T. pumila* (Andr.) Bl.] *Regni Vegetabilis Systema Naturale* 1:459. 1817. Coconut Magnolia. Type specimen: unknown. **Plate 12.**

Plants of *Magnolia coco* are evergreen shrubs or small trees 7–13 ft. (2–4 m) tall with glabrous branchlets. The leaves are elliptic to oblong, 4–8 in. (10.2–20.3 cm) long and 1–3 in. (2.5–7.6 cm) wide. The leaves are coriaceous, dark green and glabrous above, pale green and glabrous beneath. The leaf apex is acuminate, the base cuneate. The petioles are about 0.5 in. (1.3 cm) long, stout and glabrous. The flowers are very fragrant (especially at night), about 1 in. (2.5 cm) in diameter, with 9(10–11) tepals. The outer 3 tepals are sepaloid, greenish, glaucescent, and about 1 in. (2.5 cm) long. The inner 6 tepals are white, about 1.5 in. (3.8 cm) long, and obovate. The stamens are white, about 0.2–0.3 in. (0.5–0.9 cm) long. The gynoecium is glabrous. The pedicels are glabrous, about 0.5 in. (1.3 cm) long. The follicetum is oblong, and 1–2 in. (2.5–5 cm) long; the carpels are shortly beaked. The outer seed coats are red. Chromosome number: $2n = 38$.

Magnolia coco is native only to open slopes of Guangdong Province, China. It has long been considered to be native to southeastern China, an assumption apparently based on cultivated specimens, for the species has been widely grown throughout southeast Asia. It was first described in 1790 by Loureiro as *Liriodendron coco*, yet the plant was known before that time. In 1817 de Candolle transferred the species from *Liriodendron* to *Magnolia*. In the meantime, Andrews (1802) described the same species under the name *Magnolia pumila*, and it has commonly been known by this name since. Hance (1866) was the first to point out that *M. coco* and *M. pumila* were the same plant. Although Andrews' *M. pumila* was published before de Candolle's change in genus in 1817, de Candolle's *M. coco* was based on Louriero's original naming of the species. Thus, the correct name is *Magnolia coco*. The epithet *coco* was derived by Loureiro from the vernacular name Fula Coco (coconut flower) and refers to the resemblance of the unopened flowers to the fruit of the coconut, *Cocos nucifera*.

This species has caused considerable confusion for botanists over the years. Early specimens were collected without fruit, for plants seldom fruit in cultivation. In the absence of fruit, the species much resembles *Talauma* and was placed in that genus by some authors. Others described new genera for the species. Within *Magnolia*, it is thought to be most closely related to *M. championii*. It differs from the latter species in being entirely glabrous, while the young branchlets, flower buds, pedicels, and carpels of *M. championii* are covered in yellowish indumentum. *Magnolia coco* and *M. championii* have been combined, perhaps rightly so, by some botanists.

Magnolia coco was introduced into England in 1786 by Lady Amelia Hume. It has not proven hardy enough for widespread cultivation. The flowers are small and very fragrant. They usually last only a day and open in the evening, the tepals falling by morning. *Magnolia coco* is sometimes grown as a houseplant or greenhouse plant. Figlar (1986) reported it to be a good houseplant, a use for which it is especially suited thanks to its small size and slow growth rate. This species is easily propagated by layers or cuttings. Hardy to Zones 9–10. No cultivars of this species have been selected.

HYBRID

M. coco × *M. grandiflora* = 'Shirley Curry'

Magnolia craibana Dandy.

Bulletin of Miscellaneous Information, Kew 1929: 105. 1929. Craib's Magnolia. Type specimen: *Kerr 15537.* Herbarium of the Royal Botanic Gardens, Kew.

Magnolia craibana is a small, evergreen tree 15–30 ft. (4.6–9.1 m) tall and entirely glabrous. The leaves are elliptic to oblong, about 6 in. (15 cm) long, 2 in. (5 cm) wide, and thinly coriaceous. The leaf apex is obtuse to rounded, the base is cuneate. The petiole is about 1 in. (2.5 cm) long. The flowers are white with 9 tepals. The fruit is unknown.

This species is native to evergreen forests at about 4600 ft. (1380 m) in peninsular Siam. The type specimen lacks fruit, as do the other twenty or so specimens examined by Dandy. He reported that without the fruit, it is impossible to tell whether this species belongs to *Magnolia* or *Talauma*. In *Magnolia*, it is distinguished from *M. coco* and *M. pulgarensis* in having a rounded or obtuse rather than acuminate leaf apex. The species is named in honor of William Grant Craib, a botanist known for his work on the flora of Siam. It is not known to be in cultivation.

Magnolia delavayi Franch.

Plantae Delavayanae 33, t 9–10. 1889. Delavay's Magnolia. Type specimen: *Delavay 2231.* Herbarium of the Musée National d'Histoire Naturelle, Paris. **Plates 15–17.**

Magnolia delavayi is an evergreen tree or large shrub to 40 ft. (12.2 m) tall. The bark is rough and dark brown. The twigs are gray or brown, slightly pubescent at first, becoming almost glabrous. The leaves are usually ovate, rarely elliptic, 4–12 in. (10–30.5 cm) long, and 2–8 in. (5–20.3 cm) wide. Leaves are coriaceous, dark green above, pale green beneath, and slightly pubescent, becoming glabrous. The leaf apex is obtuse to acute, the base rounded to obtuse. The petioles are 2–4 in. (5–10 cm) long, and pubescent, becoming glabrous. The flowers are about 7 in. (18 cm) across, fragrant, and creamy white, with 9 glabrous tepals. The outer 3 tepals are greenish, becoming reflexed; the inner tepals are spatulate, 2–4 in. (5–10 cm) long, and 1–2 in. (2.5–5 cm) wide. The stamens are creamy yellow and 0.5 in. (1.3 cm) long. The pedicels are about 0.5–1 in. (1.3–2.5 cm) long and glabrous or slightly pubescent. The follicetum is oblong, 4–5 in. (10–12.7 cm) long, and brown. The outer seed coats are reddish orange. Chromosome number: $2n = 38$.

This species is native to the forests at about 4500–7500 ft. (1350–2250 m) in Yunnan and southern Sichuan provinces of China. It was named by Franchet in 1889 based on a specimen collected by French missionary Père Jean M. Delavay in 1886. The specific epithet honors the collector. This species has no close affinities within the section. It resembles American species of *Talauma*, whereas other members of the section resemble Asiatic *Talauma* species.

Magnolia delavayi was introduced into cultivation in England by E. H. Wilson, who collected seeds in 1889 and sent them to Veitch. Wilson later considered this species "one of the very first good finds I made in my plant hunting career." Like *Magnolia grandiflora*, this species is cultivated for its evergreen foliage. The flowers are not as abundant or as large as those of *M. grandiflora*, and they are sparse, usually open at night, and often last only a few hours. *Magnolia delavayi* is probably more tolerant of winds than most magnolias and nearly as hardy as *M. grandiflora*. It reaches flowering age in 9–10 years from seed, although it seldom fruits in cultivation. It is easily propagated by layering, the layers taking about two years to root. It differs from *M. grandiflora* in having stipules adnate to the petiole and in lacking the rufous tomentum on the undersides of the leaves. Flowers July–August; fruits October–November. Zones (8)9.

CULTIVARS

'Multitepal'   Large flowers 8 in. (20.3 cm) across with 33 tepals. Hardy to 10°F (−12°C). Offered by Louisiana Nursery, Opelousas, Louisiana, in 1990.

'Treseder'   More tepals than typical, leaves large. Hardy to 10°F (−12°C). Offered by Louisiana Nursery, Opelousas, Louisiana, in 1990.

HYBRIDS

No hybrids have been made with this species.

Magnolia eriostepta Gagnep.

In Humbert, *Supplément de la Flore Générale de l'Indo-Chine* 1:39, fig. 5(4). 1938. Woolly Magnolia. Type specimen: *Poilane (Institut des Sciences) 7143*. Herbarium of the Musée National d'Histoire Naturelle, Paris.

Magnolia eriostepta is a small evergreen tree 15–30 ft. (4.6–9.1 m) tall. The branchlets are densely pubescent at first, becoming almost glabrous. The leaves are elliptic, somewhat coriaceous, 6–8 in. (15–20 cm) long, and 3–4 in. (7.6–10 cm) wide. They are tomentose above when young, becoming almost glabrous; the undersurface is glaucous. The leaf apex is rounded to obtuse; the base is cuneate. The petiole is pubescent at first, becoming glabrous, about 2 in. (5 cm) long. The flowers are white with 9 tepals, the outer 3 densely pubescent on the outside. The gynoecium is also densely pubescent.

This species is native to forests at 4000–5000 ft. (1200–1500 m) in Tourane Province of Annam. It was named by Gagnepain in 1938. The epithet *eriostepta* refers to the woolliness of the species (*erio* is Greek for "woolly"). The vernacular name is Tanon. This species is not known to be in cultivation.

Magnolia fistulosa (Finet & Gagnep.) Dandy.

[*Talauma fistulosa* Finet & Gagnep.] *Notes from the Royal Botanic Garden, Edinburgh* 16:124. 1928. Type specimen: *Balansa 384*. Herbarium of the Musée National d'Histoire Naturelle, Paris.

Magnolia fistulosa is a small, evergreen tree to 16 ft. (5 m) tall. The branchlets may be pubescent at first, becoming glabrous. The leaves are obovate to oblong, thinly coriaceous, 8–12 in. (20–30.5 cm) long, and about 4 in. (10 cm) wide. The upper surface is glabrous, and the lower surface is pubescent, becoming glabrous. The leaf apex is acuminate to cuspidate, the base cuneate. The petiole is pubescent when young, becoming glabrous, and about 1 in. (2.5 cm) long. The flowers are white with 9 tepals, the outer 3 about 2 in. (5 cm) long. The stamens are about 0.5–1 in. (1.3–2.5 cm) long. The gynoecium is pubescent. The pedicel is about 0.5–1 in. (1.3–2.5 cm) long and glabrous to pubescent. The fruit is unknown.

This species is native to forests at 1600–4000 ft. (450–1200 m) altitude in Tongking. It was named by Finet and Gagnepain in 1907 as *Talauma fistulosa* and transferred to *Magnolia* by Dandy in 1928. Like *Magnolia coco* and *M. clemensiorum*, flowers are occasionally produced on short lateral branchlets, thus appearing axillary. The epithet *fistulosa* means "hollow" or "tubular." It is not known to what character in this species this refers.

Magnolia henryi Dunn.

[*Manglietia wangii* Hu, *Talauma kerrii* Craib.] *Journal of the Linnean Society of London, Botany* 35:484. 1903. Henry's Magnolia. Type specimen: *Henry 12782A.* Herbarium of the Royal Botanic Gardens, Kew.

Magnolia henryi is a small, evergreen tree of 20–50 ft. (6–15 m) in height. The branchlets are initially slightly pubescent, becoming glabrous. The leaves are oblong, about 8–25 in. (20–63.5 cm) long, 3–9 in. (7.6–22.9 cm) wide, and coriaceous. They may be slightly pubescent when young and become glabrous with age. The leaf apex is acuminate to obtuse or rounded; the base is cuneate or obtuse. The petiole is slightly pubescent becoming glabrous, and about 3 in. (7.6 cm) long. The flowers are white and very fragrant with 9 glabrous tepals. The stamens are about 0.5 in. (1.3 cm) long. The gynoecium is oblong and glabrous. The pedicels are stout, curved, and 4–5 in. (10–12.7 cm) long.

This species is native to forests at about 2000–5000 ft. (600–1500 m) altitude from southern Yunnan Province of China to northern Siam. It was first discovered by A. Henry in Yunnan Province and was named in his honor by Dunn in 1903. The plant was later independently described in the genera *Talauma* and *Manglietia*, yet characteristics of the fruit justify its placement in *Magnolia*. This species is easily recognizable by its large leaves with very prominent venation on the lower surface and by its cylindric fruit. The species is not known to be in cultivation.

Magnolia nana Dandy.

Journal of Botany 68:207. 1930. Dwarf Magnolia. Type specimen: *Poilane (Institut des Sciences) 5066.* Herbarium of the Musée National d'Histoire Naturelle, Paris.

Magnolia nana is a small, evergreen shrub about 3 ft. (0.9 m) tall and entirely glabrous. The leaves are lanceolate to oblong, 4–6 in. (10–15 cm) long, and 1–2 in. (2.5–5 cm) wide. The leaves are coriaceous; the upper surface is shiny, the lower surface rather dull. The leaf apex is obtuse; the base is obtuse to cuneate. The petiole is about 1 in. (2.5 cm) long. The flowers are white with 9 tepals. The follicetum is oblong, about 2 in. (5 cm) long.

This species is native to forests of 4300 ft. (1290 m) in Nha-trang Province of Annam. It was named by Dandy in 1930 based on a single specimen. The epithet *nana* means "dwarf" and refers to the plant's small stature. Dandy reports this as a "very distinct species," yet it is known from only one specimen and differs from *Magnolia clemensiorum* only in having a smaller habit and slightly different leaf shape. This species is not known to be in cultivation.

Magnolia pachyphylla Dandy.

Bulletin of Miscellaneous Information, Kew 1928: 186. 1928. Thick-leaved Magnolia. Type specimen: *Curran (Forestry Bureau 3864.)* Herbarium of the Royal Botanic Gardens, Kew.

Magnolia pachyphylla is a small evergreen tree. The branchlets are tomentose becoming slightly pubescent. Leaves are elliptic to oblong, 4–6 in. (10–15 cm) long, and 2–3 in. (5–7.6 cm) wide. They are coriaceous, glabrous above, and tomentose below, becoming glabrous. The leaf apex is rounded to obtuse; the base is cuneate to obtuse. The petiole is tomentose at first, becoming glabrous, and about 1 in. (2.5 cm) long. The flowers are white with 9 tepals, the outer 3 pubescent at the base. The fruits are unknown.

This species is native to the Philippine island of Palawan and was named by Dandy based on a single specimen. Dandy reports that the species differs from others in this section in having rigidly coriaceous leaves which are broadest below or at the middle, rounded to obtuse at the apex, and tomentose beneath at first. The bracts and young branchlets are covered with dense tomentum. The epithet *pachyphylla* means "thick-leaved." This species is not known to be in cultivation.

Magnolia paenetalauma Dandy.

Journal of Botany 68:206. 1930. Type specimen: *Tsang & Fung 538 in Herbarium Lingnan University (18072).* Herbarium of the British Museum, London.

Magnolia paenetalauma is an evergreen shrub or small tree reaching 35 ft. (11 m) in height. The branchlets are pubescent when young, becoming glabrous. The leaves are oblong to oblanceolate, 6–9 in. (15–23 cm) long, and 2–3 in. (5–7.6 cm) wide. They are thinly coriaceous, glabrous above, and pubescent beneath, becoming glabrous. The leaf apex is gradually acuminate, the base is cuneate to attenuate. The petiole is pubescent at first, becoming glabrous, and about 0.5 in. (1.3 cm) long. The flowers are fragrant, white, and with 9 tepals. The stamens are 0.5 in. (1.3 cm) long. The gynoecium is pubescent and oblong. The pedicel is slender, pubescent, slightly curved, and about 1 in. (2.5 cm) long. The follicetum is ellipsoid to oblong and 1–2 in. (2.5–5 cm) long.

This species is native to forests and thickets, usually near streams, at altitudes up to 1400 ft. (420 m) in southwestern Guangxi and western Guangdong Province of China. It was named by Dandy in 1930. In the absence of fruit the species resembles *Talauma gitingensis*. The epithet *paenetalauma* means "nearly talauma." It most closely resembles *Magnolia championii* but differs in having smaller flowers, more carpels, a curved peduncle at flowering time, and leaves lacking the slightly glaucous undersurfaces. This species is not known to be in cultivation.

Magnolia persuaveolens Dandy.

Bulletin of Miscellaneous Information, Kew 1928: 186. 1928. Fragrant Magnolia. Type specimen: collections by *Low,* Herbarium of the Royal Botanic Gardens, Kew.

Magnolia persuaveolens is a small evergreen tree. The branchlets are pubescent when young, becoming more glabrous. The leaves are obovate, 2–4 in. (5–10 cm) long by 1 in. (2.5 cm) wide. They are coriaceous, glabrous above, pubescent beneath. The leaf apex is rounded and slightly emarginate, the base cuneate to obtuse. The petiole is pubescent at first, becoming almost glabrous, and 0.5 in. (1.3 cm) long. The flowers are very fragrant, creamy yellow, and 2–4 in. (5–10 cm) in diameter with 9 glabrous tepals. The gynoecium is pubescent. The fruit is unknown.

This species is native to forests at about 5000 ft. (1524 m) in Keppel Province, Borneo. It was described by Dandy in 1928 based on a single meager specimen. Dandy reported that the material is "very meager, but merits description since it represents a very distinct species." He reported that it is most closely related to *Magnolia pachyphylla*. It differs from that species in having smaller, differently shaped leaves and a pubescent gynoecium. The epithet *persuaveolens* means "very fragrant." This species is not known to be in cultivation.

Magnolia poilanei Gagnepain.

In Humbert, *Supplément de la Flore Générale de l'Indo-Chine* 1:40. 1938. Poilane's Magnolia. Type specimen: *Poilane (Institut des Recherches) 12433.* Herbarium of the Musée National d'Histoire Naturelle, Paris.

Magnolia poilanei is a small evergreen tree about 30 ft. (9.1 m) tall. Branchlets are pubescent when young, becoming glabrous. The leaves are lanceolate to elliptic, 4–6 in. (10–15 cm) long, and about 2 in. (5 cm) wide. They are thinly coriaceous, and slightly pubescent, becoming less so with age. The leaf apex is obtuse to almost acuminate; the base is cuneate to obtuse. The petiole is pubescent, becoming glabrous, and about 0.5 in. (1.3 cm) long. The flowers are white with 9 tepals, the outer 3 pubescent at the base. The gynoecium is ellipsoid and densely pubescent. Fruit is unknown.

This species is native to forests of about 4000 ft. (1200 m) in Binh-thuan Province of Annam. It was described by Gagnepain in 1938 and named in honor of the collector of the type specimen. This species is not known to be in cultivation.

Magnolia pulgarensis (Elmer) Dandy.

[*Talauma pulgarensis* Elmer.] *Bulletin of Miscellaneous Information, Kew* 1928: 187. 1928. Pulgar Magnolia. Type specimen: *Elmer 13192.* Herbarium of the Royal Botanic Garden, Edinburgh.

Magnolia pulgarensis is an evergreen shrub 10–13 ft. (3–4 m) tall with glabrous branchlets. Leaves are oblanceolate or broadly lanceolate, 6–8 in. (15–20.3 cm) long, and 2–3 in. (5–7.6 cm) wide. They are glabrous and thinly coriaceous. The leaf apex is acuminate, the base cuneate to attenuate. The petiole is glabrous, about 1 in. (2.5 cm) long. The flowers are creamy white with 7–9 glabrous tepals. The stamens are about 0.5 in. (1.3 cm) long. The gynoecium is glabrous. Fruit is unknown.

This species is native to ridges at 2000 ft. (600 m) on Mt. Pulgar on the island of Palawan, Philippines. It was first collected by Elmer and named *Talauma pulgarensis* by him in 1913. Dandy transferred the species to *Magnolia* in 1928, although the fruit has not been seen and is required to determine the genus to which this species belongs. The epithet *pulgarensis* refers to the type locality, Mt. Pulgar. This species is not known to be in cultivation.

Magnolia talaumoides Dandy.

Journal of Botany 68:208. 1930. Type specimen: *Poilane (Institut des Sciences) 6370.* Herbarium of the Musée National d'Histoire Naturelle, Paris.

Magnolia talaumoides is an evergreen shrub or small tree 13–16 ft. (4–5 m) tall with glabrous branchlets. The leaves are oblong or obovate, 8–11 in. (20–29 cm) long, and 3–4 in. (7.6–10 cm) wide. They are slightly coriaceous and glabrous. Leaf apex is acuminate, the base cuneate to attenuate. The petiole is glabrous, about 1 in. (2.5 cm) long. The flowers are somewhat malodorous with 9–10 glabrous tepals. The outer 3 tepals are greenish, the inner 6 white. The stamens are about 0.5 in. (1.3 cm) long. The gynoecium is ellipsoid and pubescent. The follicetum is oblong and about 2.5 in. (6.4 cm) long.

This species is native to forests of 1600–4000 ft. (480–1200 m) in central and southern Annam. It was named by Dandy in 1930. The species is probably most closely related to *Magnolia fistulosa*, from which it differs in having smaller flowers borne on longer pedicels. The epithet *talaumoides* means "talauma-like." This species is not known to be in cultivation.

Magnolia thamnodes Dandy.

Journal of Botany 68:208. 1930. Shrubby Magnolia. Type specimen: *Poilane in Herbarium Chevalier P282.* Herbarium of the Musée National d'Histoire Naturelle, Paris.

Magnolia thamnodes is a small, evergreen shrub 3–6 ft. (1–2 m) in height with pubescent branchlets, becoming glabrous. The leaves are oblong to elliptic, 4–6 in. (10–15 cm) long, and 2–3 in. (5–7.6 cm) wide. They are thinly coriaceous and glabrous. Leaf apex is rounded to obtuse and usually slightly emarginate; the base is cuneate to obtuse. The petiole is glabrous and up to 0.5 in. (1.3 cm) long. The flowers are creamy white with 9 tepals, the outer 3 pubescent at the base. The stamens are about 0.5 in. (1.3 cm) long. The gynoecium is ellipsoid and glabrous. The fruit is unknown.

This species is native to forests of 3300 ft. (1990 m) in Kampot Province of Cambodia. It was named by Dandy in 1930 based on a single specimen. The species is most closely related to *Magnolia craibana*, from which it differs in being pubescent and lacking glaucous branchlets. Both *M. thamnodes* and *M. craibana* have been described without fruiting specimens. Once fruit is examined, Dandy reported, both may be attributed to *Talauma*. The epithet *thamnodes* means "shrubby." This species is not known to be in cultivation.

Magnolia subgenus *Magnolia* section *Gynopodium* Dandy.

[*Micheliopsis* King, *Parakmeria* Cheng.] *Curtis's Botanical Magazine* 165:sub t 16. 1948. TYPE: *Magnolia nitida*.

Species in section *Gynopodium* are evergreen trees or large shrubs. The plants are completely glabrous and have small, glossy leaves and free stipules. The flowers are white to creamy yellow with 9(12) tepals. The gynoecium has a short stalk separating it from the point of attachment of the anthers. Species are generally found to be diploid with a chromosome number of $2n = 38$. However, Rui-Yang et al. (1985) reported that *Magnolia lotungensis* is hexaploid ($2n = 114$). This seems unlikely and is considered suspect until verified by additional chromosome counts.

Section *Gynopodium* contains three or more species of temperate and tropical eastern Asia from southeastern Tibet and northeastern Upper Burma to southeastern China and Formosa. The most distinctive feature of the section is the short stalk between the gynoecium and the attachment of the anthers. This is most noticeable when the plant is in fruit. This short-stalked gynoecium is also found occasionally in section *Maingola*, which differs from section *Gynopodium* in flower and fruit characters. Species of the genus *Michelia* also have stalked gynoecia, leading some authors to incorrectly place these *Magnolia* species in *Michelia*. The leaves do resemble those of *Michelia*, but *Magnolia* species differ from that genus in having terminal, not axillary, flowers. In some cases, authors created new genera (*Parakmeria, Micheliopsis*) for the species in this section. The stalked gynoecium gives *Gynopodium* its name: "gyno" refers to the gynoecium, and "podium" literally means "foot," but is more loosely a reference to being elevated.

Three species are treated here. *Magnolia nitida,* occasionally cultivated, is native to Tibet, Upper Burma, and northwestern Yunnan Province of China. *Magnolia kachirachirai* is native to southern Formosa, and *M. lotungensis* is native to Hainan, Guangdong, and Hunan provinces of China. Dandy (unpublished) considered *Parakmeria omeinsis* and *P. yunnanensis* to be magnolias belonging to this section. Due to lack of information regarding these species, however, they are not included here.

Magnolia nitida W. W. Smith.

Notes of the Royal Botanic Garden, Edinburgh 12:212. 1920. Glossy Magnolia. Type specimen: *Forrest 15059.* Herbarium of the Royal Botanic Garden, Edinburgh.

Magnolia nitida is an evergreen tree or large shrub 20–50 ft. (6–15 m) tall. The bark is dark grayish brown. The glabrous branchlets are short with prominent petiole rings. New leaves are bronze-red, becoming bronze and later green. The leaves are elliptic or oblong, 3–5 in. (7.6–12.7 cm) long, and 1–2 in. (2.5–5 cm) wide. Leaves are coriaceous, bright green and glossy above, and shiny pale green beneath. Leaf apex is acute or obtuse, the base cuneate. The petioles are about 1 in. (2.5 cm) long, glabrous, and yellowish green. The fragrant flowers are creamy white to yellow and 2–3 in. (5–7.6 cm) across with 9(12) tepals. The outer 3 tepals are sometimes tinged with purple at the base. The tepals are about 2 in. (5 cm) long and obovate. The stamens are yellowish, about 0.5 in. (1.3 cm) long. The gynoecium is green and stipitate (shortly stalked), the stigmas crimson. Pedicels are glabrous. The follicetum is oblong, 2–3 in. (5–7.6 cm) long, and bright green, the stipitate gynoecium noticeable. The outer seed coats are bright orangish red; the seeds are aromatic. Chromosome number: $2n = 38$.

This species is native to forests and thickets at 5900–12,100 ft. (1770–3660 m) altitude in northwestern Yunnan Province of China and adjacent southeastern Tibet and northeastern Upper Burma. The distribution of this species is very similar to that of *Magnolia rostrata. Magnolia nitida* was named in 1920 by W. W. Smith based on Forrest's collection in 1917 in northwestern Yunnan at 9800–11,100 ft. (2940–3330 m) altitude. The epithet *nitida* means "shining" or "glossy," referring to the leaves.

Magnolia nitida was introduced into cultivation in England by way of seed from Forrest's collections. Cultivated plants are usually smaller than specimens found in the wild. The species is prized for its glossy, evergreen leaves and ornamental, apple-green follicetum with orangish red seeds. It is not hardy enough to be cultivated widely, yet is grown in mild areas of England and in sheltered locations in North America. It prefers a moist soil and humid environment and will grow in sun or shade. Cold winters seem to do the greatest damage and may cause leaf drop. Seeds of *M. nitida* germinate readily, but cuttings present difficulties, losing their leaves before they begin to root. This species takes about fifteen years to flower from seed. Flowers March–April; fruits August–September. Zones (8)9–10. No cultivars have been selected from nor hybrids made with this species.

Magnolia kachirachirai (Kaneh. & Yamom.) Dandy.

[*Michelia kachirachirai* Kaneh. & Yamom., *Michelopsis kachirachirai* (Kaneh. & Yamom.) Keng.] *Bulletin of Miscellaneous Information, Kew.* 1927: 264. 1927. Kachirachirai Magnolia. Type specimen: *Kanehira, November 1924.* Location unknown.

Magnolia kachirachirai is a large evergreen tree reaching 60 ft. (18 m) tall with a trunk diameter of 3 ft. (1 m). The leaves are lanceolate to oblong, 4–5 in. (10–12.7 cm) long, about 1 in. (2.5 cm) wide, and coriaceous. Leaves are dark glossy green above and pale green beneath. Leaf apex is acute, the base cuneate. The petioles are about 1 in. (2.5

cm) long and glabrous. The flowers are creamy yellow with 9(12) tepals and are 1–2 in. (2.5–5 cm) across. The outer 3 tepals are about 1 in. (2.5 cm) long. The stamens are about 0.5 in. (1.3 cm) long. The gynoecium is short-stipitate. The pedicel is about 0.5 in. (1.3 cm) long and glabrous. The follicetum is ovoid to ellipsoid and 1–2 in. (2.5–5 cm) long.

This species is native to forests and cliffs of southern Formosa at altitudes of 1600–5000 ft. (480–1500 m). It was first collected by Kanehira in 1924 in Taito Province, Formosa. Based on this fruiting specimen, Kanehira and Yamomoto in 1926 named the species *Michelia kachirachirai*, believing the stipitate gynoecium to be a characteristic only of the genus *Michelia*. The following year Dandy (1927b) transferred the species to *Magnolia* since fruit and flower characteristics other than the stipitate gynoecium show its affinity to this genus. Kanehira's type specimen is pictured with Kanehira and Yamomoto's original description. The Japanese vernacular name for this species is Kachirachirai-no-ki, from which the scientific name is derived. *Magnolia kachirachirai* is not known in cultivation. It is perhaps not as desirable as *M. nitida*, due to its smaller leaves and flowers and its low tolerance for cold.

Magnolia lotungensis Chun & Tsoong.

Acta Phytotaxonomica Sinica 8:285. 1963. Lotung Magnolia. Type specimen: *Chun & Tsoong 50122*. Herbarium of the Institute of Botany, Austro-Sinica Academy, Canton.

Magnolia lotungensis is an evergreen tree reaching 60 ft. (18 m) in height with a trunk diameter of 3 ft. (0.9 m). The bark is light gray, and branchlets are slender. The leaves are elliptic, 2–4 in. (5–10 cm) long, 1–2 in. (2.5–5 cm) wide, and coriaceous. Leaves are glossy green above and pale green beneath. The leaf apex is shortly acute; the base is cuneate. The petioles are about 0.5 in. (1.3 cm) long. The flowers are reported to be creamy yellow and fragrant, but further details are unknown. The follicetum is oblong and about 2 in. (5 cm) long. The outer seed coats are red. Rui-yang et al. (1985) report a chromosome count of $2n = 114$ for this species, but this count is considered suspect until verified by further work.

This species is native to forests of Hainan, Guangdong, and Hunan provinces of China. It was named by Chun and Tsoong in 1963. The vernacular name is Lotu Mulan (lotung magnolia). Little is known about this species, and the original description lacks information about the flowers. *Magnolia lotungensis* is not known to be in cultivation.

Magnolia subgenus *Magnolia* section *Lirianthe* (Spach) Reichenbach.

[*Lirianthe* Spach, *Sphenocarpus* Wall.] *Der Deutsch Botaniker* 1(1):192. 1841. TYPE: *Magnolia pterocarpa*.

The single species in section *Lirianthe*, *Magnolia pterocarpa*, is an evergreen tree with white flowers having 9 tepals. Flowers usually have two or more bracts, which leave ring scars when they fall. The follicetum is made up of numerous carpels ending in long, flattened beaks.

Magnolia pterocarpa is native to northeastern India and Burma. It is unique within *Magnolia* in having flattened beaks on the carpels. This unique feature led Wallich and Spach to create separate genera (*Sphenocarpus* and *Lirianthe*, respectively) for this species. Apart from the distingishing carpel beaks, this section is very similar to the closely related section *Gwillimia*. Occasionally the pedicel is branched, each branch terminating in a flower. This is an abnormal growth habit rather than a characteristic to be used to distinguish the section. *Magnolia pterocarpa* is diploid with a chromosome number of $2n = 38$.

Magnolia pterocarpa Roxb.

[*Liriodendron grandiflorum* Roxb., *Magnolia sphenocarpa* Wall., *Michelia macrophylla* D. Don, *Sphenocarpus pterocarpus* (Roxb.) C. Koch, *Talauma grandiflora* (Roxb.) Spach, *T. roxburghii* Don.] *Plants of the Coast of Coromandel* 3:62, t 266. 1820. Type specimen: not designated. **Fig. 7.7.**

Magnolia pterocarpa is a medium to large tree reaching 60 ft. (18 m) in height. The young branchlets are covered with tawny pubescence, becoming almost glabrous with age. The leaves are obovate to elliptic, 8–16 in. (20.3–40.7 cm) long, and 4–8 in. (10–20 cm) wide. The leaves are slightly pubescent at first, becoming glabrous above and

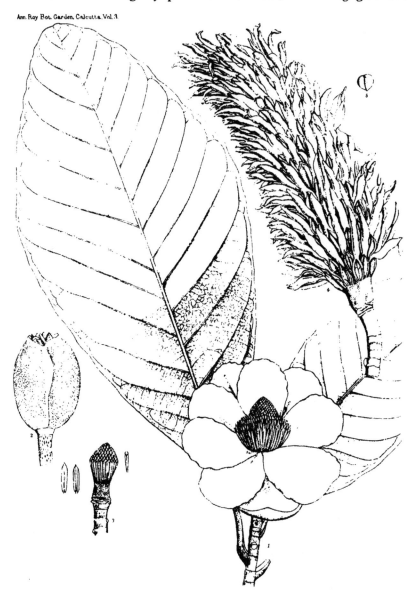

Ann. Roy Bot. Garden, Calcutta. Vol. 3.

Fig. 7.7. *Magnolia pterocarpa* from King 1891.

pubescent beneath. Leaf apex is rounded to obtuse; the base is cuneate. The petioles are 1–2 in. (2.5–5 cm) long and glabrous or pubescent. The fragrant flowers are white, about 3.5 in. (9 cm) across, with 9 tepals. The outer 3 tepals are greenish, obovate, and 1–2 in. (2.5–5 cm) long. The inner 6 tepals are white and slightly wider than the outer 3. The stamens are 0.5–1 in. (1.3–2.5 cm) long. The follicetum is oblong and 5–10 in. (12.7– 25.4 cm) long. The outer seed coats are orange. Chromosome number: $2n = 38$.

This species is native to forests at altitudes of up to about 3600 ft. (1080 m) in India, Assam, Burma, and eastern Pakistan. It was named by Roxburgh in 1820. In 1832 Roxburgh described another species, *Liriodendron grandiflorum*, using almost the same description as that of *Magnolia pterocarpa*. Roxburgh appears to have named the same plant twice, so the earlier name, *M. pterocarpa*, is the valid one. Roxburgh designated no type specimen, but it is presumed that his collection at the British Museum labeled *Liriodendron grandiflora* is the type material he used for both his descriptions. The epithet *pterocarpa* means "winged fruit," referring to the flattened beaks of the carpels.

The wood of *Magnolia pterocarpa* is white, soft, and used mostly for temporary boxes and crates, as it does not hold up well when wet. It is also a preferred fuel wood in Assam. The leaf buds are chewed by the Assamese to blacken their gums and teeth. There are many vernacular names including Baramphthuri-sopa, Chapite-jamja, Dieng-long-krop, Doloi-champa, Phapitemhaija, and Thoutha.

Magnolia pterocarpa is cultivated in the gardens in warmer parts of India and is especially prized for its fragrant flowers. It is not hardy enough to be grown in North America or Europe. No cultivars have been selected from nor hybrids made with this species.

Magnolia subgenus *Magnolia* section *Maingola* Dandy.

Curtis's Botanical Magazine 165:16. 1948. TYPE: *Magnolia maingayi*.

Species in section *Maingola* are evergreen trees or large shrubs. Leaves have short petioles and free stipules. Flowers have 9 tepals, and the gynoecium occasionally has a short stalk. The fruit are usually distorted due to incomplete fertilization or ovule abortion. The carpels may have a short beak. The chromosome number is known only for *Magnolia griffithii*, which is diploid with $2n = 38$ chromosomes.

Seven species are included in this section. These species are native to eastern Asia from Assam, Upper Burma, and Indochina to Sumatra and Java. Dandy (unpublished) apparently intended to name several more species within this section, but did not do so before his death.

Magnolia aequinoctialis Dandy.

Bulletin of Miscellaneous Information, Kew 1928: 185. 1928. Type specimen: *Houtvester Sumatra's Oostkust 25*. Buitenzorg Herbarium, Buitenzorg, Java.

This species is a large evergreen tree reaching 75 ft. (22.6 m) in height. The branchlets are densely pubescent at first, becoming glabrous or nearly so. The leaves are elliptic, 6–8 in. (15–20 cm) long, and 2–4 in. (5–10 cm) wide. Leaves are chartaceous and pubescent beneath when young, especially on the midrib. The leaf apex is acuminate; the base is cuneate to obtuse. Petiole is about 0.25 in. (0.6 cm) long, densely pubescent at first, becoming glabrous. The flowers are creamy yellow and fragrant with

9 tepals. The outer 3 tepals are narrowly oblong, greenish, about 1 in. (2.5 cm) long and sepaloid. The stamens are about 0.4 in. (1.2 cm) long. The gynoecium is sessile, cylindric, and pubescent or tomentose. The pedicel is about 1 in. (2.5 cm) long and pubescent. The follicetum is about 1 in. (2.5 cm) long.

Magnolia aequinoctialis is native to forests and jungles at 4000–5600 ft. (1200–1690 m) in northern and central Sumatra. It was named by Dandy in 1928. The epithet *aequinoctialis* pertains to the equinox and refers either to the proximity of this species to the equator, or to its flowering period during the vernal equinox. The species is closely related to *M. macklottii*, also of Sumatra. *Magnolia aequinoctialis* differs in having pubescent carpels and smaller, more elliptic leaves. This species is not known to be in cultivation.

Magnolia annamensis Dandy.

Journal of Botany 68:209. 1930. Annam Magnolia. Type specimen: *Chevalier 38877,* Herbarium of the Musée National d'Histoire Naturelle, Paris.

Plants of *Magnolia annamensis* are evergreen trees or shrubs with pubescent branchlets. The leaves are elliptic to narrowly oblong, 10–12 in. (25.4–30.5 cm) long, and 4–6 in. (10–15 cm) wide. They are coriaceous and mostly glabrous. The petiole is about 0.5 in. (1.3 cm) long and pubescent at first, becoming glabrous or nearly so. The gynoecium is sessile, ellipsoid, and pubescent. The flowers and fruit are unknown.

This species is known only from forests at 4900 ft. (1470 m) in the Hon-ba Mountains in Nha-trang Province of Annam. It was named by Dandy in 1930 based on a single specimen. The epithet refers to the plant's locality. Dandy called this a "remarkably distinct" species; however, it is very closely related to *Magnolia griffithii* and *M. maingayi*. It differs from the former in having a sessile gynoecium, and from the latter in having wider and more coriaceous leaves, longer petioles, and less densely pubescent carpels. This species is not known to be in cultivation.

Magnolia griffithii Hook f. & Thoms.

[*Michelia griffithii* (Hook f. & Thoms.) Finet & Gagnep.] In Hook f., *Flora of British India* 1:41. 1872. Griffith's Magnolia. Type specimen: *Griffith (Kew 66)*. Herbarium of the Royal Botanic Gardens, Kew. **Fig. 7.8.**

Magnolia griffithii is a large, evergreen tree. The branchlets are covered with yellowish tomentum at first, becoming glabrous or nearly so. The leaves are elliptic or elliptic-oblong, 9–16 in. (23–40.7 cm) long, and 4–5 in. (10–12.7 cm) wide. They are thinly coriaceous, glabrous above and pubescent to tomentose beneath. The leaf apex is acuminate; the base is obtuse to cuneate. The petiole is stout, about 0.25 in. (0.6 cm) long, and tomentose, becoming nearly glabrous. The flowers are slightly fragrant, 3–4 in. (7.6–10 cm) in diameter, and creamy white or pale yellow with 9 tepals. The outer 3 tepals are narrowly oblong, about 2 in. (5 cm) long, and glabrous. The inner 6 tepals are slightly smaller. The stamens are about 0.5 in. (1.3 cm) long. The gynoecium is short stipitate, cylindric, and glabrous. The pedicel is slender, about 2 in. (5 cm) long, and covered with white, silky hairs. The follicetum is 4–6 in. (10–15 cm) long. Chromosome number: $2n = 38$.

This species is native to forests and jungles at altitudes of about 1300 ft. (390 m) in Assam and northern Upper Burma. It was named by Sir Joseph Hooker in 1872. The species was thought by Finet and Gagnepain (1906) to belong to the genus *Michelia*. This species is not known to be in cultivation.

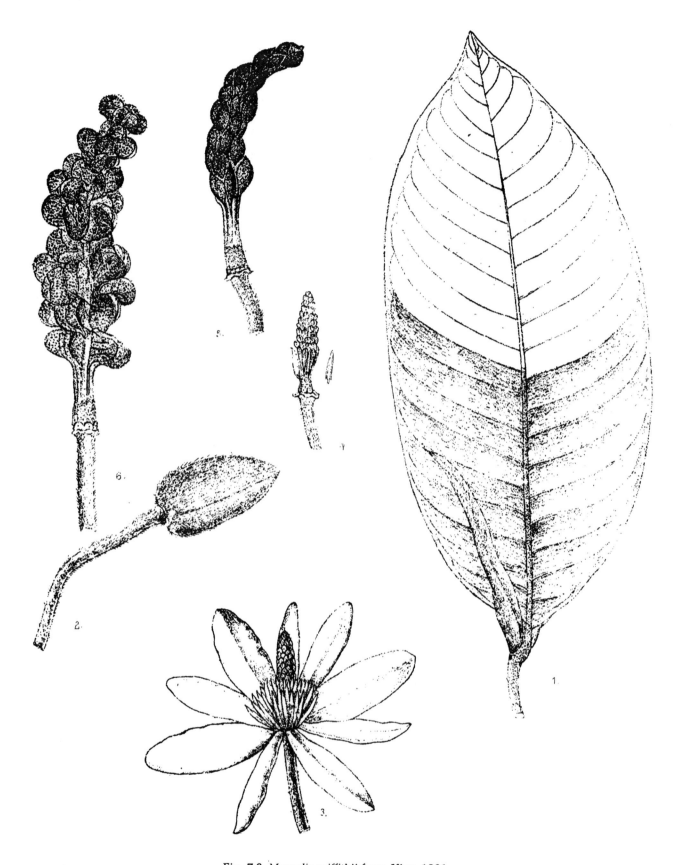

Fig. 7.8. *Magnolia griffithii* from King 1891.

Magnolia gustavii King.

Annals of the Royal Botanic Garden, Calcutta 3:209, t 61. 1891. Gustav's Magnolia. Type specimen: *Gustav Mann May 1890;* duplicates at the Gray Herbarium, Harvard University, Cambridge, Massachusetts, and the Royal Botanic Gardens at Kew and Edinburgh. **Fig. 7.9.**

Magnolia gustavii is a large evergreen tree with glabrous branchlets. The leaves are oblong to lanceolate, 5–7 in. (12.7–17.8 cm) long, and 1–2 in. (2.5–5 cm) wide. They are membranaceous, glossy above, and almost glaucous beneath. Leaf apex is acuminate, the base cuneate and unequal. The petiole is about 0.25 in. (0.6 cm) long. The flowers are 3–4 in. (7.6–10 cm) in diameter, with 9 tepals. The outer 3 tepals are about 2 in. (5 cm) long and coriaceous; the inner 6 are about the same length and membranaceous. The stamens are about 0.5 in. (1.3 cm) long. The gynoecium is sessile and glabrous. The follicetum is 2–4 in. (5–10 cm) long.

This species is native to the Makum Forest at 1000–2000 ft. (300–600 m) in Assam. It was named by King in 1891 in honor of Gustav Mann, the collector of the type specimen.

Fig. 7.9. *Magnolia gustavii* from King 1891.

Magnolia macklottii (Korth.) Dandy.

[*Magnolia javanica* Koord. & Valet, *Manglietia macklottii* Korth.] *Bulletin of Miscellaneous Information, Kew* 1927: 263. 1927. Type specimen: *Korthals* without number. Leiden Herbarium, the Netherlands.

Magnolia macklottii is a large tree reaching about 65 ft. (20 m) in height. The branchlets are pubescent, becoming glabrous. Leaves are oblanceolate, 8–10 in. (20.3–25.4 cm) long, and 2–3 in. (5–7.6 cm) wide. They are chartaceous, glabrous above, and pubescent beneath, becoming almost glabrous. Leaf apex is acuminate; the base is obtuse and slightly unequal. The petiole is about 0.25 in. (0.6 cm) long, pubescent at first, becoming glabrous. The flowers are creamy yellow with 9 glabrous tepals. The outer 3 tepals are narrowly oblong, about 1 in. (2.5 cm) long, and greenish. The inner 6 tepals are creamy yellow. The stamens are about 0.25–0.5 in. (0.6–1.3 cm) long. The gynoecium is sessile or slightly stipitate, cylindric, and glabrous. The follicetum is about 2 in. (5 cm) long.

This species is native to evergreen forests at altitudes of up to 5000 ft. (1500 m) in Sumatra and Java. It is the only *Magnolia* species native to Java and is very rare there. In Sumatra it has an allied species, *Magnolia aequinoctialis*. *Magnolia macklottii* differs in having a more glabrous gynoecium (sometimes slightly stipitate) and larger, more oblanceolate, leaves. The two species overlap in range. *Magnolia macklottii* was named *Manglietia macklottii* by Korthals in 1851 and was transferred to *Magnolia* by Dandy in 1927. The species is assumed to have been named in honor of Macklott, but it is not known who Macklott was. This species is not known to be in cultivation.

Magnolia maingayi King.

Journal of the Asiatic Society of Bengal 58, 2:369. 1889. Maingay's Magnolia. Type specimen: *Maingay (Kew Herbarium 17)*. Herbarium of the Royal Botanic Gardens, Kew. **Fig. 7.10.**

Magnolia maingayi is an evergreen tree or shrub 50–65 ft. (15–20 m) high. The branchlets are covered with white or yellowish tomentum. The leaves are obovate to oblanceolate and 8–10 in. (20.3–25.4 cm) long by 3–4 in. (7.6–10 cm) wide. The leaves are chartaceous to thinly coriaceous, glabrous above, and glaucous beneath. The petiole is about 0.3 in. (0.8 cm) long. The flowers are fragrant and yellowish white with 9 tepals. The outer 3 tepals are oblong, 1–2 in. (2.5–5 cm) long, and glabrous or pubescent outside at the base. The inner 6 tepals are smaller and glabrous. The stamens are about 0.5 in. (1.3 cm) long. The gynoecium is sessile, oblong, and densely tomentose. The follicetum is cylindric, 1–3 in. (2.5–7.6 cm) long, and pubescent.

This species is native to dense forests at altitudes of 2500–5000 ft. (750–1500 m) in the Malay Peninsula from Penang and Perak to Singapore, and in southwestern Sarawak. It was named by King in 1889. The species is not known to be in cultivation.

Magnolia pealiana King.

[*M. membranacea* var. *pealiana* (King) P. Parment.] *Annals of the Royal Botanic Garden, Calcutta* 3:210, t 59. 1891. Peal's Magnolia. Type specimen: *Gustav Mann, January 1890 and September 1890*. Herbarium of the British Museum, London. **Fig. 7.11.**

Magnolia pealiana is a large evergreen tree. Branchlets are pubescent at first, becoming glabrous. The leaves are oblong to elliptic, 7–8 in. (18–20 cm) long and 2–3 in. (5–7.6 cm) wide. They are coriaceous, glabrous above, and pubescent beneath, becoming glabrous. The leaf apex is acuminate, the base cuneate. The petiole is about 0.25 in. (0.6 cm) long, pubescent at first. The flowers are 3–4 in. (7.6–10 cm) in diameter with 9 tepals. The outer 3 tepals are oblong, about 2 in. (5 cm) long, while the inner

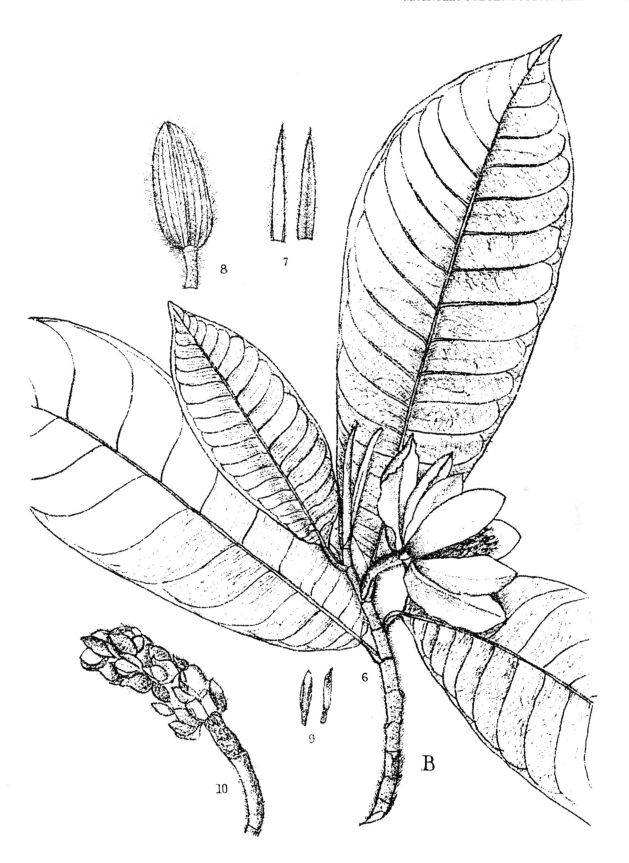

Fig. 7.10. *Magnolia maingayi* from King 1891.

6 are slightly smaller. The stamens are about 0.5 in. (1.3 cm) long. The gynoecium is about twice that length, cylindric, and glabrous, with a short stalk. The follicetum is 3–5 in. (7.6–12.7 cm) long.

This species is native only to the Makum Forest in the Lakhimpur District of Assam. It was named by King in 1891. It is closely related to *Magnolia griffithii* and probably should be included within that species; the differences between the two are small. *Magnolia pealiana* has smaller, less densely pubescent leaves than *M. griffithii*. *Magnolia pealiana* is named in honor of S. Peal who helped King with his work on the magnolias and their plant relatives in Assam. The vernacular name is Gahori-soppa. The soft, white wood of the species is used to make tea-boxes. This species is not known to be in cultivation.

Fig. 7.11. *Magnolia pealiana* from King 1891.

MAGNOLIA subgenus *YULANIA* (Spach) Reichenbach.
Deutsch Botaniker 1(1):192. 1841.

Species in subgenus *Yulania* are deciduous trees or shrubs. The flowers appear before the leaves (the species are said to be precocious) or with the leaves. The outer whorl of tepals is small and calyx-like, and the anther sacs open laterally or sublaterally. Subgenus *Yulania* includes three sections.

Yulania includes many of the cultivated Asiatic species, including *Magnolia denudata* and six others.

Buergeria also includes four widely cultivated Asiatic species, the most common of which is *Magnolia kobus* var. *stellata*.

Tulipastrum contains only two species, *Magnolia liliiflora* of Asia and *M. acuminata* of North America.

Magnolia subgenus *Yulania* section *Yulania* (Spach) Dandy.

[*Yulania* Spach] *Camellias and Magnolias Conference Report* 72. 1950. TYPE: *Magnolia denudata*.

Plants in this section are deciduous trees or shrubs. The flowers are precocious with 9 or more tepals of white to pink or purple. The tepals are all similar, not differentiated into sepaloid and petaloid whorls. These species are hexaploid with $2n = 114$ chromosomes.

Section *Yulania* comprises seven species native to eastern Asia. This section receives its name from the common name of *Magnolia denudata*, the Yulan (Lily-tree), long cultivated in China and Japan.

Magnolia denudata Desr.

[*Gwillimia yulan* (Desf.) C. deVos, *Magnolia conspicua* Salisb., *M. conspicua* var. *rosea* Veitch, *M. denudata* var. *purpurascens* Rehder & Wilson, *M. heptapeta* (Buc'hoz) Dandy, *M. obovata* Thunb. (in part), *M. obovata* var. *denudata* (Desr.) DC., *M. precia* Corr. ex Vent., *M. purpurea* var. *denudata* (Desr.) Loud., *M. yulan* Desf., *Michelia yulan* (Desf.) Kostel, *Yulania conspicua* (Salisb.) Spach. Excluding *Mokkwuren flore albo* Kaempfer.] *Lamarck, Encyclopédie Méthodique, Botanique* 3:675. 1791. Yulan, Lily-Tree, Naked Magnolia. Type specimen: Kaempfer's table 43, Mokkwuren 1. **Plates 18–20.**

Plants of *Magnolia denudata* are deciduous trees or large shrubs reaching 50–65 ft. (15–19.6 m) tall, with rough, gray-brown bark. The twigs are dark brown and pubescent at first, becoming glabrous with age. Terminal buds are covered with yellowish tomentum. The obovate, elliptic, or broadly elliptic leaves are 3–6 in. (7.6–15 cm) long and 2–4 in. (5–10 cm) wide. They are dark green and glabrous above and light green and pubescent beneath, becoming almost glabrous at maturity. Leaves are abruptly acuminate at the apex and cuneate or obtuse at the base. The petiole is about 1 in. (2.5 cm) long and tomentose, becoming almost glabrous. The fragrant flowers are vase-shaped with 9(10–12) white tepals, each occasionally tinged with pink or purple at the base or midrib. Tepals are obovate, 2–4 in. (5–10 cm) long, and 1–2 in. (2.5–5 cm) wide. The stamens are rose-purple and about 0.25 in. (0.6 cm) long. The pedicel is about 0.5 in.

(1.3 cm) long and pubescent when young. The reddish brown follicetum is cylindric, rarely ovoid, and 2–6 in. (5–15 cm) long; the outer seed coats are red. Chromosome number: $2n = 114$.

Magnolia denudata is native to woods and open areas at about 4000 ft. (1200 m) in the eastern provinces of China. It is found from southeastern Jiangsu and Zhejiang provinces, through southern Anhui and northern Jiangxi to southwestern Hunan provinces. It has long been cultivated in both China and Japan and may have naturalized in some areas where it is not native.

The nomenclatural history of *Magnolia denudata* is long and complex and linked with that of *M. liliiflora*. Although such history is discussed by Johnstone (1955; see "Magnolia Books," chapter references), Treseder (1978; see "Magnolia Books," chapter references), Ueda (1985), and Meyer and McClintock (1987), a concise but complete account is offered here.

In 1712, Englebert Kaempfer published an account of the plants now known as *Magnolia denudata, M. liliiflora,* and *M. kobus*. In 1791, Sir Joseph Banks published a volume of Kaempfer's drawings, including only the Japanese names of the plants. *Mokuren* (spelled *Mokkwuren* by Kaempfer) is the Japanese name for magnolia and was the name on Kaempfer's plates of plants now known as *Magnolia denudata* and *M. liliiflora*. Banks erroneously switched the text corresponding to the plates, so that *Mokkwuren flore albo* incorrectly referred to the purple-flowered species and *Mokkwuren frutex tulipifer* referred to the white-flowered species. The same year Kaempfer's engravings were published (1791), Desrousseaux named the two *Mokkwuren* plants *Magnolia denudata* and *M. liliiflora*, with Kaempfer's plates being used as the types. Unfortunately his descriptions compounded Banks' error in switching the text; therefore *M. denudata* is described as having red flowers and *M. liliiflora* as having white flowers. Otherwise, Desrousseaux's descriptions and typification are complete and accurate.

In 1776, years before the posthumous publication of Kaempfer's engravings in 1791, Pierre Joseph Buc'hoz, a French naturalist and physician, published two volumes of color plates. One volume was largely devoted to Chinese plants and made no mention of the origin of the plates. Captions of most were simply the Chinese name of the plant without any further description. Pictured in this volume were the Yulan, or Lily tree, and the Mulan, or Woody Orchid. Three years later Buc'hoz published another work adding names and descriptions to some of the plates published in 1776. The Yulan and Mulan plates were reproduced in this work and placed in the new genus *Lassonia*. The Yulan was named *Lassonia heptapeta* (now *Magnolia denudata*) and the Mulan was named *Lassonia quinquepeta* (now *Magnolia liliiflora*). The description, based on nothing other than these plates, given by Buc'hoz includes a "calyx of one piece, but fringed, a corolla of 7 petals in one species and 5 in the other." The figures and descriptions inaccurately represent magnolias in several ways: the fringed calyx does not exist in *Magnolia*; the numbers and arrangement of tepals are inaccurate, giving rise to the inaccurate epithets for these species; the floral structures of both the androecium and the gynoecium match no known plant and certainly not magnolias; and the bud and branch orientation in the figures are not magnolia-like. In short, the plants represented by the figures (see Fig. 7.12, 7.13) may resemble *Magnolia denudata* and *M. liliiflora*, but they are botanically inaccurate and cannot be considered as types on which to base a species. The drawings were obviously Chinese impressionist art and not intended for botanical use.

Due to all these inaccuracies, *Lassonia quinquepeta* and *L. heptapeta* clearly cannot refer to *Magnolia liliiflora* and *M. denudata*. Nonetheless, these names were transferred to

Fig. 7.12. *Magnolia quinquepeta* as figured by Buc'hoz.

Magnolia by Dandy in 1934. However, in his 1950 and later works, Dandy listed these names as synonyms of *M. liliiflora* and *M. denudata*. From 1950 to 1976 the latter names were most commonly used. However, Spongberg (1976) resurrected the old names of *Magnolia quinquepeta* and *M. heptapeta* and was followed by the authors of *Hortus Third* (1976), Treseder (1978; see "Magnolia Books," chapter references), and Ueda (1985).

With the resurrection of the Buc'hoz names, a great controversy ensued as to whether or not these names were valid. This controversy was, it is hoped, put to rest with the publication of an excellent study by Meyer and McClintock (1987) in which the Buc'hoz names are rejected for reasons of inaccuracy as outlined here. The correct

Fig. 7.13. *Magnolia heptapeta* as figured by Buc'hoz.

names are those of Desrousseaux (*Magnolia liliiflora* and *M. denudata*), based on Kaempfer's engravings as the types.

The epithet *denudata* refers to the precociousness of the species—it flowers before the leaves are produced and while the branches are still "nude." This characteristic also gives rise to the common name Naked Magnolia. The Chinese name Yulan (Lily-tree) refers to the resemblance of the flowers to lilies. P. M. Cibot (1778) described the species as resembling a "naked walnut tree with a lily at the end of every branch." *Magnolia denudata* has long been of religious significance, with its clear, white flowers symbolizing purity. Buddhist monks in China planted the Yulan at temples as long ago as the T'ang Dynasty (A.D. 618–906). It has appeared in Chinese art for centuries, and the flowers are used for food and the bark for medicine. It was introduced into Western gardens by Sir Joseph Banks about 1780 and has become one of the most popular species in cultivation. Banks introduced the species under the name *Magnolia conspicua*, under which name it was cultivated until it was discovered that *M. denudata* was the earliest published name.

When *Magnolia denudata* is grown where it escapes frost damage, it can be one of the most beautiful of magnolias. The white flowers presented on bare branches early in the spring are a welcome sight. However, frosted flowers become brown mush, which can be rather unsightly. Therefore it is best to grow the Yulan in a protected area of the garden to avoid disappointment in case of late frosts.

The Yulan is quite hardy and not particular in its requirements. It prefers moist soil and light shade but performs well in almost any situation. This species flowers at an early age, about three years from seed. Seed set is usually low, and layering is difficult; the most common methods of propagation are grafting and budding. *Magnolia kobus* and *M.* × *soulangiana* are commonly used for rootstock. Flowers March–April; fruits August–September. Zones (4)5–9.

CULTIVARS

'Caerhays Clone'   Selected for its overall superiority; listed by Treseder's Nurseries, Cornwall, England, ca. 1973.

'Elongata'   Flowers slightly larger than typical; white with pink tinge at base of tepals. Listed by Overlook Nurseries, Mobile, Alabama, 1957. (Possibly *Magnolia sprengeri* var. *elongata*.)

'Forrest's Pink'   Flowers pinker than any form of *M. denudata*, without a trace of purple. Selected by Treseder's Nurseries, Cornwall, England, from the original tree at Caerhays Castle, Cornwall, England, collected by Forrest. Date of selection unknown.

'Gere'   A late-flowering cultivar similar to 'Wada's Japanese Clone'. Selected from a tree in Urbana, Illinois, by J. C. McDaniel and described by him in 1980.

'Globosa'   Flowers small and globose. Described by Pampanini as a variety in 1915. Probably no longer grown.

'Late Clone'   Later flowering than typical. Listed by Louisiana Nursery of Opelousas, Louisiana.

'Moon Garden'   Pure white, fragrant flowers. Listed by Lousiana Nursery of Opelousas, Louisiana.

'Purple Eye'   This cultivar is sometimes listed as a form of *Magnolia denudata*, but is most probably a hybrid. See description in chapter 10.

'Sawada's Cream' ('Sawada')   Flowers small, creamy white. This form was named by Phil Savage, Bloomfield Hills, Michigan, and has been used as a parent in many of his crosses.

'Sawada's Pink'   Tepals have a slight pink tinge. Obtained by Phil Savage from nurseryman K. Sawada of Mobile, Alabama. 'Sawada's Cream' and 'Sawada's Pink' are not registered cultivars. They are included here because of their importance in the breeding program of Phil Savage, Bloomfield Hills, Michigan.

'Wada's Form'   See 'Wada's Japanese Clone'

'Wada's Japanese Clone' ('Wada's Form')   Flowers are pure white and later-flowering than typical. Vigorous. Introduced by K. Wada, Hakoneya Nurseries, Japan.

HYBRIDS

Crosses between *Magnolia denudata* and *M.* × *veitchii* were made in the 1960s by Frank Santamour of the U.S. National Arboretum. These hybrids are more vigorous than either parent and produce large white to pink flowers. They were distributed by the U.S. National Arboretum to various other arboreta for testing, but none has yet been named. Oswald Blumhardt of New Zealand repeated this cross and reported a few promising hybrids in 1982, but none of them has yet been named. 'Helen Fogg' resulted from Phil Savage's cross between these two species. The following named hybrids are available in the trade.

M. *acuminata* × M. *denudata* = 'Butterflies', 'Elizabeth', 'Ivory Chalice', 'Sundance', 'Yellow Fever', 'Yellow Garland'

M. *acuminata* var. *subcordata* 'Miss Honeybee' × M. *denudata* 'Sawada's Cream' = 'Goldfinch'

M. *campbellii* × M. *denudata* = M. × *veitchii*

M. *cylindrica* × M. *denudata* 'Sawada's Pink' = 'Fireglow'

M. *denudata* × M. *kobus* var. *stellata* 'Waterlily' = 'Emma Cook', 'Pristine'

M. *denudata* × M. *liliiflora* = M. × *soulangiana*

M. *denudata* × M. *salicifolia* = 'Wada's Snow White'

M. *denudata* × M. *sargentiana* var. *robusta* = 'Ann Rosse', 'Marj Gossler'

M. *denudata* × M. *sprengeri* 'Diva' = 'Legacy'

M. *denudata* 'Sawada's Pink' × (M. × *veitchii* 'Peter Veitch') = 'Helen Fogg'

M. *denudata* × unknown = 'Lacey', 'Purple Eye'

Magnolia campbellii Hook f. & Thoms.

Flora Indica 1:77. 1855. Type specimen: not designated.

Plants of *Magnolia campbellii* are deciduous trees (rarely large shrubs) to 115 ft. (35 m) tall with gray-tan bark. Twigs are glabrous and tan, sometimes becoming almost black. The terminal buds are covered with yellow hairs. The elliptic to obovate leaves are 6–12 in. (15–30.5 cm) long and 2–6 in. (5–15 cm) wide. Leaves are dark green and glabrous above and pale green and pubescent beneath, with an acuminate apex and rounded (rarely cuneate) and often unequal base. Petioles are about 1 in. (2.5 cm) long, glabrous or sometimes pubescent. The fragrant flowers are white to pale pink or crimson, and reach 10 in. (25.4 cm) in diameter. The inner tepals are erect, and the outer tepals open nearly horizontally, producing a cup-and-saucer shaped flower. The 12 (rarely 16) tepals are spatulate, concave, 3–5 in. (7.6–12.7 cm) long, and 1–2 in. (2.5–5 cm) wide. The stamens are about 1 in. (2.5 cm) long and rose-colored. The follicetum is cylindric, 4–8 in. (10–20 cm) long, and reddish brown. The outer seed coat is scarlet-red. Chromosome number: $2n = 114$.

Magnolia campbellii is native to forests and thickets at 7000–11,000 ft. (2100–3300 m) in the Himalayas from eastern Nepal eastward to western Yunnan Province of China. The species was first collected in 1838 by Dr. W. Griffith and described but not named. Griffith's description was published posthumously in 1854. Joseph Hooker also collected the species, in 1849, and he and Thomson named it in 1855. Hooker was not aware of Griffith's work and therefore did not refer to it. At the time of Hooker's collection, *M. campbellii* was one of the most common trees in the forests, and he described mountainsides becoming pink in the spring when it bloomed. Its harvest for firewood and timber for planking and tea-boxes has made the tree scarce. Those plants remaining in the wild are usually suckers from harvested trees. Trees are often 30–50 ft. (9–15 m) tall before they reach maturity (25–30 years of age). If cut before maturity, the trees produce no seeds to perpetuate the colony.

The epithet *campbellii* was given by Hooker in honor of Archibald Campbell, political resident at Darjeeling. Vernacular names include Lal-champ, Penre, Sagok, Sigumgrip, Patagari.

Magnolia campbellii was beautifully illustrated in *Illustrations of Himalayan Plants* by Hooker in 1855. This illustration aroused the interest of English gardeners, and several attempts were made to introduce the species. These attempts were not successful until the late 1870s when live plants were sent from Calcutta. The plants sent to England were killed in the winter, but a specimen in southern Ireland survived. This tree, in a garden at Lakelands, Cork, flowered in 1885 and was featured in *Curtis's Botanical Magazine* that year, together with a description by Hooker.

This species is often considered by writers and gardeners to be the most spectacular of all magnolias, and the most desirable. The plants usually reach heights of 30–60 ft. (9–18 m) in cultivation, beginning as single-stemmed trees when young and becoming more densely branched with age. The leaves are highly variable, especially in size, shape, and degree of pubescence. Flower color is also variable, from white to pink to almost crimson. *Magnolia campbellii* is quite tender and susceptible to frost damage so must be protected from frosts and high winds. The significant concern in growing this species is that it requires twenty to twenty-five years to flower from seed, and even then blooms are sparse the first few years. This interval before flowering may be shortened to ten years by growing plants that are grafted onto *M. campbellii* var. *mollicomata* or to twelve years by growing plants grafted onto *M. sargentiana* var. *robusta*. Nursery professionals at Golden Gate Park in San Francisco, California, reported flowering in five years after grafting onto *M.* × *soulangiana* and growing the trees in containers. Several similar methods for earlier flowering have been tried, such as allowing the plants to become rootbound, binding the trunk, and placing the tree under various other types of stress. While these methods may decrease the time before flowering, this induced stress is not good for the overall health of the tree and is not encouraged.

The species is thought to be self-infertile, as solitary trees do not produce seed. Propagation from seed, when available, is not difficult, but grafting and layering are more common methods of propagation. Flowers February–March (April); fruits August–September. Zones (5)6–7.

Crosses between typical *Magnolia campbellii* and *M. campbellii* var. *mollicomata* were made by Charles Raffill of the Royal Botanical Garden at Kew in 1946. Raffill's work, and some probable earlier crosses at Sidbury Manor, are thoroughly discussed by Treseder (*Magnolias,* 1978) under the Sidbury-Raffill crosses. (I discuss Raffill's work only briefly here; interested readers may refer to Treseder's discussion for details.) Rafill's purpose in crossing these two forms was to combine the flower color of *M. campbellii* with the flower poise of *M. campbellii* var. *mollicomata*. He also wanted to produce plants which bloomed at an early age and bloomed later in the spring, avoiding frost damage. Raffill proposed the name *M.* × *kewensis* for these crosses. However, this name had been assigned to plants resulting from a cross between *M. kobus* and *M. salicifolia*. Also, since Raffill's crosses were between different forms of a single species, a grex name is not applicable. (Grex names are used only for hybrids having two different species as parents.) As a consequence, the resulting plants are listed here as cultivars. These include 'Sidbury', 'Charles Raffill', and 'Kew's Surprise'.

CULTIVARS

'Betty Jessel' A late-flowering form (April–May). Flowers color is nearly crimson. Raised from seed from the original tree of 'Darjeeling'. Grown by Sir George Jessel in Kent and named in honor of his wife. Registered in 1967.

'Caerhays Clone' Flowers deep pink. Listed by Treseder's Nurseries, Cornwall, England, in 1973.

'Charles Raffill' The earliest of Raffill's crosses (made in 1946 between *Magnolia campbellii* and *M. campbellii* var. *mollicomata*) to bloom, at age thirteen. It is a vigorous grower with flowers 8–10 in. (20–25.4 cm) across. Flower buds deep rose, opening to pink with purple undertones. Inner surface of tepals white tinged with rose. A hardy form.

'Darjeeling' Flowers dark rose or wine-colored. Original tree in Darjeeling Botanic Garden, India. Named in 1967. Often confused with 'Betty Jessel'.

'Elisabeth Holman' A superior seedling raised by Michael Williams at Lanarth, Cornwall, England, 1951–52, and given to Nigel Holman, Cornwall; named by Holman in honor of his wife.

'Eric Walther' Fastigiate form with rose-pink flowers. Cross between *Magnolia campbellii* and *M. campbellii* var. *mollicomata*. Described by Elizabeth McClintock in 1965 and named after the first director of the Strybing Arboretum, San Francisco, California.

'Hendricks Park' Flowers deep rose, to 11 in. (28 cm) across. Original plant at Hendricks Park, Eugene, Oregon. Listed by Gossler Farms Nursery, Springfield, Oregon, 1971.

'Kew No. W4' Flowers deep rose-pink outside, pale pink inside. Original plant at the Royal Botanical Gardens, Kew, 1968.

'Kew No. W5' Flowers bright pink. Original plant at the Royal Botanical Gardens, Kew, 1968.

'Kew No. 40' Vigorous form with pale purple flowers. Original plant at the Royal Botanical Gardens, Kew, 1968.

'Kew's Surprise' A seedling of 'Charles Raffill', the product of Raffill's cross between *Magnolia campbellii* and *M. campbellii* var. *mollicomata*. Flower shape like that of var. *mollicomata*; flowers are white inside, crimson rose outside. Named in 1967. Original tree at the Royal Botanic Gardens, Kew.

'Landicla' Flowers large, deep pink-purple. Listed by Treseder's Nurseries, Cornwall, England, in 1973.

'Late Pink' Flowers pink; later-blooming than other forms. Registered ca. 1961 by Eric Walther of the Strybing Arboretum, San Francisco, California.

'Maharaja' Flowers reaching 11 in. (28 cm) across. Tepals with purple shading at the bases. Named by D. Todd Gresham, Santa Cruz, California, in 1964.

'Pale Pink Seedling' Exceptionally large, pale pink flowers, possibly resulting from a cross between a white-flowered form and a pink-flowered form. Listed by Treseder's Nurseries, Cornwall, England, ca. 1973.

'Queen Caroline' Red-purple flowers reaching 9 in. (23 cm) across. Registered by the Royal Botanic Gardens at Kew in 1977.

'Samson et Delilah' A pink-flowered form cultivated at Strybing Arboretum, San Francisco, California, possibly introduced from France.

'Sidbury' A vigorous form flowering at a younger age than typical. Result of a cross between *Magnolia campbellii* and *M. campbellii* var. *mollicomata*, with bright pink flowers. The cross was made at Sidbury Manor in Devon prior to Rafill's crosses in 1946. Listed by Hillier and Sons Nursery, Winchester, England in 1973.

'Trewithen Dark Form'   Flowers deep rose outside, pale pink inside. Illustrated by Johnstone (1955; see "Magnolia Books," chapter references).

'Trewithen Light Form'   Flowers light rose-pink outside, pale pink inside. Illustrated by Johnstone (1955; see "Magnolia Books," chapter references).

'Veitch Clone'   Pink-flowered form offered by Treseder's Nurseries, Cornwall, England, ca. 1973.

'Wakehurst'   Flowers dark rose, otherwise similar to 'Charles Raffill'. Offered by Hillier and Sons Nursery, Winchester, England, ca. 1973.

HYBRIDS

Treseder reported that hybrids between *Magnolia* × *soulangiana* and *M. campbellii* were made by Charles Raffill at Kew in 1943. A plant thought to be from these crosses is cultivated at Lanarth, Cornwall, England. The flowers are white inside and crimson outside with 8 tepals. Apparently no hybrids from this cross have been named.

> *M. campbellii* × *M. denudata* = *M.* × *veitchii*
>
> *M. campbellii* × *M. liliiflora* = 'Early Rose', 'Star Wars'
>
> *M. campbellii* × *M. sargentiana* var. *robusta* = 'Buzzard', 'Hawk'
>
> *M. campbellii* × (*M.* × *soulangiana* 'Amabilis') = 'First Flush'
>
> *M. campbellii* × unknown (pink Gresham Hybrid?) = 'Frank Gladney'

Magnolia campbellii var. *alba* Treseder.

Magnolias 90–93. 1978. **Plates 7–8.**

Magnolia campbellii var. *alba* (sometimes listed as *M. campbellii* 'Alba') differs from typical *M. campbellii* in its white flowers with a light pink streak down the center of the tepals. It reaches maturity at about fifteen years of age, in almost half the time of the typical form, and flowers several days after that form in cultivation. The leaves and flower buds are larger than typical.

This was the form first discovered by Griffith in 1838 at an altitude of about 8200 ft. (2460 m) in Sikkim. It was not cultivated until about 1926, when a seed from Darjeeling germinated at Caerhays Castle, Cornwall. Treseder's Nurseries sold this under the name 'Caerhays White Clone' in 1973. The Caerhays plants are fertile. Several light pink clones have been produced, assumed to be hybrids between typical *Magnolia campbellii* and *M. campbellii* var. *alba*. (See 'Pale Pink Seedling' under *M. campbellii*.) Flowers February–March (April); fruits August–September. Zones (6)7.

CULTIVARS

'Caerhays White Clone'   Flowers white, leaves larger than typical. Listed by Treseder's Nurseries, Cornwall, England, in 1973.

'Chyverton'   Seedling from the original tree of 'Caerhays White Clone'. Listed by Treseder's Nurseries, Cornwall, England, in 1965.

'Ethel Hillier'   Hardy form with large flowers. Tepals white with pale pink tinge. Introduced by Hillier and Sons Nursery, Winchester, England, in 1973.

'Maharanee'   Flowers smooth white, 8–10 in. (20.3–25.4 cm) across. Named by D. Todd Gresham, Santa Cruz, California, in 1964.

'Stark's White'   Similar to 'Strybing White', but the outer tepals do not droop, and the flowers are slightly smaller. Cultivated by John Stark of Oakland, California. Seed obtained by Stark from Clarke Nursery in San José, California. Clarke Nursery grew it from seed received from an unspecified nursery in India. Named in 1962. **Plate 9.**

'Strybing White'   Flowers white, about 12 in. (30.5 cm) across. The inner tepals enclose the gynandrophore, as in typical *Magnolia campbellii*, but the outer tepals are drooping rather than saucer-shaped. Grown from seed from India. Named ca. 1961 by Eric Walther of the Strybing Arboretum, San Francisco, California. **Plate 10.**

'White Form'   Flowers large, cream-colored. Original plant at Caerhays Castle, Cornwall, England. Named by J. C. Williams in 1951.

HYBRIDS

M. campbellii var. *alba* × *M. sargentiana* var. *robusta* = 'Michael Rosse', 'Moresk', 'Princess Margaret'

Magnolia campbellii var. *mollicomata* (W. W. Smith) F. Kingdon-Ward.

[*Magnolia campbellii* subsp. *mollicomata* (W. W. Smith) Johnstone, *M. campbellii* subsp. *mollicomata* convar. *williamsiana* Johnstone, *M. mollicomata* W. W. Smith.] *Gardeners Chronicle* 3, 137:238. 1955.

Plants of *Magnolia campbellii* native to the eastern end of the species' range (Burma, Tibet, and western Yunnan Province of China) are recognized as *M. campbellii* var. *mollicomata*. These plants differ from the typical in having pedicels covered with yellow tomentum, larger flowers of paler color, and more elongated floral buds. Variety *mollicomata* reaches flowering age in ten to twelve years, half the time of the typical form, and blooms later in the season.

Magnolia campbellii var. *mollicomata* was first discovered by George Forrest in Yunnan Province, China, in 1904 and described by W. W. Smith in 1920 as *M. mollicomata*. This new name arose from an error on the part of Smith and Forrest wherein they confused this species with *M. rostrata*. The error appears to have been straightened out by later botanists, but the history of this confusion remains in the literature.

In 1955, Kingdon-Ward proposed the name *Magnolia campbellii* var. *mollicomata* to indicate the close relationship of this plant to *M. campbellii*. The same year, Johnstone proposed the name *M. campbellii* subsp. *mollicomata* for the same reason. Either of these names may be used; however, I have chosen the rank of variety for consistency throughout the genus, as discussed in chapter 5.

Some botanists, including Dandy, have collapsed var. *mollicomata* within *Magnolia campbellii*, holding that there is no constant character which defines the variety. I am of the opinion, however, that the differences mentioned above warrant the designation var. *mollicomata*. The epithet *mollicomata* (*mollis* = soft, *comosus* = bearing hairs) refers to the soft hairs on the pedicels which, along with the other above-mentioned characters, differentiate this variety from the typical form.

Variety *mollicomata* was introduced into cultivation in England by Forrest in 1914, when he sent seed to the Royal Botanic Garden at Kew. Although its flowers do not display the brighter pink coloration of the typical form, plants reach flowering age in half the time. An additional advantage is that it blooms later in the season than the typical form, thereby avoiding some late frosts. A discussion of Raffill's crosses between this variety and the typical *Magnolia campbellii* is presented under the latter heading. Flowers February–March (April); fruits August–September. Zones (6)7.

A purple-flowered form of *Magnolia campbellii* var. *mollicomata* was discovered by Forrest in 1924 in northwestern Yunnan Province, China. A seed from this collection yielded an exceptional form grown at Lanarth in Cornwall, England. This plant is known by the cultivar name 'Lanarth'. It reproduces true to type and has been used as the parent for plants now referred to as the Lanarth Group. Seedlings of 'Lanarth' take sixteen to twenty years to reach flowering age. Plants are vigorous and develop a fastigiate form. The leaves appear almost puckered due to distinct venation. The flowers are a stunning violet-red or magenta, fading to purple, and are about 9 in. (23 cm) across. The plants are so vigorous that they are often grafted onto less vigorous root-stocks to reduce their typical 100 ft. (30 m) height so the beauty of the flowers may be enjoyed by others than the birds. 'Lanarth' seedlings are variable so cannot carry the name.

Other plants were raised from Forrest's seed at Werrington Park and Borde Hill in England, but neither these nor their offspring are cultivated elsewhere. This entire group of purple-flowered plants of var. *mollicomata* was referred to as *M. campbellii* subsp. *mollicomata* convar. *williamsiana* by Johnstone. These plants are not deserving of botanical status, and the cultivar name 'Lanarth' is to be used instead.

CULTIVARS

'Fastigiata'   Fastigiate form offered by Hillier and Sons Nursery, Winchester, England, in 1956.

'Lanarth'   The pyramidal, purple-flowered form grown at Lanarth in Cornwall, England, from seed collected by Forrest. It reproduces true to type and has given rise to the group of seedlings known as the Lanarth Group. Flowers March–April. See text for further discussion. **Plate 11.**

'Mary Williams'   Flowers purple, stamen filaments violet with creamy rose anthers. Named by J. C. Williams of Caerhays in 1954.

HYBRIDS

M. campbellii var. *mollicomata* × *M. liliiflora* 'Nigra' = 'Caerhays Surprise'

M. campbellii var. *mollicomata* 'Lanarth' × *M. liliiflora* = 'Apollo', 'Vulcan'

M. campbellii var. *mollicomata* × *M. sargentiana* var. *robusta* = 'Treve Holman', 'Mossman's Giant'

M. campbellii var. *mollicomata* (Lanarth Group) × *M. sargentiana* var. *robusta* = 'Mark Jury'

Magnolia sprengeri Pampan.

[*Magnolia conspicua* var. *purpurescens* Bean, *M. denudata* var. *purpurescens* (Maxim.) Rehder & Wilson, *M. diva* Stapf ex Millais, *M. sprengeri diva* Stapf, *M. sprengeri purpurescens* Stapf, *M. sprengeri* subsp. *diva* (Stapf) Hooper.] *Nuovo Giornale Botanico Italiano*, New Series 22:295. 1915. Sprenger's Magnolia. Type specimen: *Silvestri 4104*. Herbarium Universitatis Florentinae, Florence, Italy.

Plants of *Magnolia sprengeri* are deciduous trees or shrubs to 50 ft. (15 m) tall with grayish brown bark. Twigs are grayish brown and glabrous, sometimes developing pubescence toward the end of the year's growth. Terminal buds are covered with yellow hairs. The leaves are obovate, 4–7 in. (10–18 cm) long, and 2–4 in. (5–10 cm) wide. Leaves are glabrous above and pubescent beneath, especially when young. The leaf apex is abruptly acuminate, the base cuneate, sometimes slightly unequal. The petiole is about 1 in. (2.5 cm) long and brown. The fragrant flowers are 6–8 in. (15–20 cm) across with 12(14) tepals arranged in three whorls. Tepals are spatulate or obovate, 2–4 in. (5–10 cm) long and 1–3 in. (2.5–7.6 cm) wide. Flowers are pale pink or white inside, reddish purple outside. Stamens are about 0.5 in. (1.3 cm) long and reddish purple. The folicetum is cylindric, 2–6 in. (5–15 cm) long, and reddish brown. The outer seed coats are red. Chromosome number: $2n = 114$.

This species is native to woods and thickets of central China, western Henan, western Hubei and eastern Sichuan provinces, at 3000–6500 ft. (900–1950 m) elevation. It was named by Pampanini in 1915 based on specimens collected by Silvestri in 1912 and 1913 in northwestern Hubei Province. The species was named in honor of Carl Sprenger (1846–1917), a German nurseryman in Vomero, Italy.

The type specimen of *Magnolia sprengeri* was collected by Silvestri when this precocious species was in flower, so no leaves could be included. The lack of leaf information in the type specimen has led to problems in the naming and classification within the species. Silvestri's specimen on which Pampanini's description was based has no leaves, and the flower color is impossible to ascertain.

E. H. Wilson collected two magnolias on his expeditions to China in 1906 and 1910. He and Alfred Rehder classified these collections. They named Wilson's white-flowered specimen with leaves at least twice as long as wide, *Magnolia denudata* var. *elongata*. A similar pink-flowered form with leaves less than twice as long as wide they named *Magnolia denudata* var. *purpurescens*. These two names were published in Sargent (1913).

In the meantime, plants from Wilson's collections had been introduced into cultivation through Veitch and Sons in Exeter, England. When these trees reached maturity and flowered, most had white flowers and appeared to correspond to Wilson's *Magnolia denudata* var. *elongata*. A plant from Wilson's original collection purchased by J. C. Williams and grown at Caerhays, Cornwall, was quite different, having beautiful, large pink flowers. Williams' plant was assumed at the time to be *M. denudata* var. *purpurascens*. Mr. Williams sent a flower of his plant to Kew to be used by an artist for an illustration in *Curtis's Botanical Magazine*. The taxonomist, Otto Stapf, examined it and recognized that it was not a form of *M. denudata*, but a new plant. Among other things, it had twelve tepals, while *M. denudata* has nine, leading Stapf to conclude that the Caerhays plant was something other than *M. denudata*. Wilson eventually acknowledged this difference as well.

Dandy brought Silvestri's type specimens of *Magnolia sprengeri* to Stapf's attention. Meager though the specimens are, it was clear to Dandy and Stapf that Wilson's two "varieties" of *M. denudata* belonged with *M. sprengeri*. Stapf published a description of *M. sprengeri diva* (the pink form) in *Curtis's Botanical Magazine* in 1927, using the Caerhays flower as the basis for the description and the illustration. He also mentioned *M. sprengeri elongata* (Wilson's white-flowered *M. denudata* var. *elongata*) in the note to the plate.

With the publication of these two forms of *Magnolia sprengeri*, the problems of classification within the species would seem to have been solved. But which of the two forms is typical *M. sprengeri*? Silvestri's type specimen provides no clues, for it could represent either form. Similarly, Pampanini, in his original description, omits flower color and describes the leaves only as "deciduous, produced after the flowers." In view of this lack of information about the typical form, Johnstone (1955; see "Magnolia Books," chapter references) proposed the use of *M. sprengeri* var. *elongata* and *M.*

sprengeri var. *diva* to refer to the white and pink forms, respectively. This treatment was followed by Treseder (1978; see "Magnolia Books," chapter references). Yet this solution is not acceptable botanically since it leaves the species without a typical var. *sprengeri*. (The designation of a typical variety carrying the name of the species is discussed in chapter 5.)

In his 1976 treatment Spongberg considered the pink-flowered form to be typical, based on Wilson's notes (Sargent 1913) as to the relative abundance of the two forms. Wilson wrote that the white form was "rather rare" and referred to the pink form as "common" and "plentiful." It seems likely that Silvestri encountered the more plentiful form. A decision is needed as to which form is typical. Lacking further evidence, I agree with Spongberg (1976) the pink form was most likely collected by Silvestri. Based on this assumption, the pink form with leaves less than twice as long as wide is treated here as typical *Magnolia sprengeri* var. *sprengeri*. *Magnolia sprengeri* var. *diva* is then a synonym of this variety. The original tree at Caerhays, pictured in Stapf's description of *M. sprengeri* var. *diva*, should be designated as *M. sprengeri* 'Diva'. Only vegetatively propagated clones from this tree are included in this cultivar designation. Seedlings of the Caerhays plant are variable so cannot carry the cultivar name. The white form with leaves at least twice as long as wide is *M. sprengeri* var. *elongata* and is described in more detail in the following section.

Magnolia sprengeri was introduced into cultivation by E. H. Wilson after he turned seeds over to Veitch and Sons, Exeter, England for propagation. When the Veitch firm was dissolved in 1913, some of these seedling plants went to Kew Gardens, Bodnant, and Caerhays. The plants at Kew and Bodnant have been identified as *M. sprengeri* var. *elongata*. The plant at Caerhays is the one which has been named 'Diva'.

The original shipment of seeds to Veitch contained seeds of both plants in Wilson's collections, and Wilson had accidentally mixed them together. In a manuscript which remained incomplete at the time of his death, Wilson wrote that he "sent seeds (#668V) packed in earth but unfortunately mixed with them seeds of another magnolia having identical fruits but white flowers" (Howard 1980). Thus, it would appear that this shipment was a mixture of Wilson's var. *purpurescens* (pink-flowered) and var. *elongata* (white-flowered). This mix-up of seeds explains the difference in flower color of the Kew, Bodnant, and Caerhays plants.

Magnolia sprengeri seems to require no special conditions for cultivation and blooms at about twenty years from seed. The species is largely propagated by grafting. It usually flowers later than *M. campbellii* and thus often escapes frost injury. Flowers March–April; fruits August–September. Zones (6)7–9.

CULTIVARS

'Claret Cup' Flowers 8 in. (20.3 cm) across with 12 tepals. Tepals rosy purple outside, pale purple to white inside. Seedling from the original tree of 'Diva'. Registered by Lord Aberconway of Bodnant, North Wales, 1963.

'Copeland Court' Flowers uniformly crimson-pink, darker and brighter than 'Diva', with 12–14 tepals. A compact, symmetrical, and erect growing plant, densely branched. Seedling of 'Diva'. Listed by Treseder's Nurseries, Cornwall, England, 1973.

'Diva' Flowers 8 in. (20.3 cm) across, rosy pink outside, pale pink inside. The original tree at Caerhays Castle, Cornwall, England, is quite prolific, with flowers even on the lowermost branches. The flowers are erect, the tepals curling slightly at the tip.

The name 'Diva' is Italian for "goddess." Named by Stapf in 1927. Zones (7)8–9. See text for a historical discussion of this form. **Plate 74.**

'Eric Savill' Seedling of 'Diva'; flower color a deep red-purple. Originated at Savill Garden, England, and registered by John Bond in 1982.

'Wakehurst' Flowers pink inside, purple outside. Named by Sir Henry Price, Wakehurst Place, Sussex, England, in 1948.

HYBRIDS

M. denudata × *M. sprengeri* 'Diva' = 'Legacy'

M. liliiflora 'Nigra' × *M. sprengeri* 'Diva' = 'Galaxy', 'Northwest', 'Spectrum'

M. liliiflora 'Darkest Purple' × *M. sprengeri* 'Diva' = 'Raspberry Swirl'

M. sargentiana var. *robusta* × *M. sprengeri* 'Diva' = 'Caerhays Belle'

M. × *soulangiana* 'Lennei' seedling × *M. sprengeri* 'Diva' = 'Paul Cook'

M. × *soulangiana* 'Wada's Picture' × *M. sprengeri* 'Diva' = 'Big Dude'

Magnolia sprengeri var. *elongata* (Rehder & Wilson) Johnstone.

[*Magnolia conspicua* var. *elongata* (Rehder & Wilson) A. W. Hill, *M. denudata* var. *elongata* Rehder & Wilson, *M. elongata* Millais, *M. sprengeri elongata* (Rehder & Wilson) Stapf.] *Asiatic Magnolias in Cultivation* 87. 1955. Type specimen: not designated.

Magnolia sprengeri var. *elongata* differs from the typical species, *M. sprengeri* var. *sprengeri*, in having slightly smaller, white flowers with shorter, narrower tepals and a shrubby growth habit. The leaves are narrow, oblong, or obovate and more than twice as long as broad. This variety is also smaller and less vigorous. Wilson collected this plant in 1906 and 1910 and named it *M. denudata* var. *elongata* in 1913. In 1927 Stapf referred to it briefly as *M. sprengeri* var. *elongata* in a description of *M. sprengeri* var. *diva*, but he did not go on to describe var. *elongata*. In 1955 Johnstone published a complete description of the variety; thus he is credited with naming it. See the discussion under typical *M. sprengeri* for more historical information.

Magnolia sprengeri var. *elongata* is not grown as frequently as the pink-flowered forms. It may not be grown at all in North America, as it is not in great demand since it differs only slightly from other available forms. It does flower later than some Asian species, such as *M. kobus,* and therefore escapes some frost damage. Johnstone (1955; see "Magnolia Books," chapter references) and Treseder (1978; see "Magnolia Books," chapter references) note two distinct forms of var. *elongata* in cultivation, differing mainly in flower color, one being white, the other cream-colored. Both forms may have purple coloring at the base of the tepals. The white-flowered form has thick, stiff tepals which form a cup-shaped flower. The tepals of the cream-flowered form reflex when the flower opens. Variety *elongata* takes approximately twenty years to flower from seed. Flowers March–April; fruits August–September. Zones 5–9. No cultivars have been selected, nor hybrids made with this variety.

Magnolia dawsoniana Rehder & Wilson.

In Sargent, *Plantae Wilsonianae* 1:397. 1913. Dawson's Magnolia. Type specimen: *Wilson 1241,* 1907–1909 Arnold Arboretum expedition. Herbarium of the Arnold Arboretum, Harvard University, Cambridge, Massachusetts. **Plate 14.**

Plants of *Magnolia dawsoniana* are deciduous shrubs (rarely trees) to 50 ft. (15 m) tall with smooth, brownish gray bark. Branchlets are tan or yellowish. Leaves are obovate, 4–6 in. (10–15 cm) long and 2–3 in. (5–7.6 cm) wide with pronounced veins. They are dark green and glabrous above, lighter green and glabrous beneath. The leaf apex is rounded or acuminate, the base cuneate, often unequal. The petioles are about 1 in. (2.5 cm) long and glabrous. The fragrant flowers are about 10 in. (25.4 cm) across with 9–12 tepals arranged in three whorls. The tepals are oblong, 2–5 in. (5–12.7 cm) long and 1–2 in. (2.5–5 cm) wide. The inner surface of the tepals is white, the outer surface tinged pink. The stamens are rose, about 1 in. (2.5 cm) long. The follicetum is cylindric, 2–5 in. (6–12.7 cm) long, and reddish brown. The outer seed coats are red. Chromosome number: $2n = 114$.

Magnolia dawsoniana is native to a small area in eastern Sikang Province, China, at altitudes of 6500–7500 ft. (1950–2250 m). It was described by Rehder and Wilson in 1913 and named in honor of Jackson T. Dawson, first superintendent of the Arnold Arboretum.

Rehder and Wilson thought *Magnolia dawsoniana* to be closely related to *M. denudata*. Superficially they do look similar. But the species is very similar to *M. sargentiana*, which is probably its closest relative. *Magnolia dawsoniana* differs from the latter species in having fewer tepals (9–12 compared with 12–14) and leaves glabrous on both sides.

Magnolia dawsoniana was introduced to North America by Wilson in 1908. It reached Europe in 1919, sent by Wilson along with other magnolias to Léon Chenault (a nurseryman in Orléans, France) for propagation and distribution. *Magnolia dawsoniana* takes about twenty years to flower from seed, and few seeds are even produced. Grafted plants bloom at about ten years of age. This plant prefers a sunny site for maximum flower production but is otherwise not very particular about growing conditions. The species can set a prodigious number of flowers and create quite a display. It often grows as a large, spreading shrub, especially under favorable conditions. For this reason, the plant needs plenty of room to grow. It does not often set seed and is usually grafted to decrease the time before flowering. Flowers March–April; fruits August–September. Zones (5)6. No hybrids are currently available.

CULTIVARS

'Chyverton Red' ('Chyverton')   A hardy form with crimson-colored flowers. Stamens are also crimson. The coloring varies from year to year depending on temperature: flowers appear pale crimson in warm years, deep crimson in cooler years. The original tree is at Chyverton, Cornwall. Listed by Treseder's Nurseries, Cornwall, England, in 1965.

'Clarke'   Flowers rich, deep pink, about 8 in. (20 cm) in diameter. Hardier than *Magnolia sargentiana* and *M. campbellii*. Originated at Clarke Nursery, San José, California.

'Lanarth'   Flowers white with lavender tinge; buds purple. Grown by M. P. Williams at Lanarth, England, and listed by Treseder's Nurseries, Cornwall, England, in 1965.

'Ruby Rose'   Large, red-purple flowers reaching 12 in. (30.5 cm) across, with 12 tepals. Possibly a hybrid. Seedling raised by Rose Del Grasso and registered by Eugene R. German, Fort Bragg, California, ca. 1980.

'Strybing Clone'   Flowers larger than typical, pink. Listed in 1990 by Louisiana Nursery, Opelousas, Louisiana.

'Washington Park Clone'   Flowers deep red with broad tepals. Originated at the Strybing Arboretum, San Francisco, California.

Magnolia sargentiana Rehder & Wilson

[*M. conspicua* var. *emarginata* Finet & Gagnep., *M. denudata* var. *emarginata* (Finet & Gagnep.) Pamp., *M. emarginata* (Finet & Gagnep.) Cheng.] In Sargent, *Plantae Wilsonianae* 1:398. 1913. Sargent's Magnolia. Type specimen: *Wilson 914A*, Arnold Arboretum Expedition of 1907–1909. Herbarium of the Arnold Arboretum, Harvard University, Cambridge Massachusetts.

Magnolia sargentiana is a large, deciduous tree (rarely a shrub) to 65 ft. (19.6 m) tall. Bark is brownish gray, branchlets are yellowish brown; terminal buds are covered with yellowish hairs. The leaves are obovate or oblanceolate, 4–8 in. (10–20 cm) long and 2–4 in. (5–10 cm) wide. They are green and glossy above and grayish green and pubescent beneath. Leaf apex is rounded, acuminate or sometimes emarginate; the base is cuneate, often unequal. The petiole is 1–2 in. (2.5–5 cm) long and dark brown. The fragrant flowers are 8–10 in. (20.3–25.4 cm) across with 12(14) tepals. The tepals are oblong, 2–5 in. (5–12.7 cm) long, and about 1 in. (2.5 cm) wide. The inner surface of the tepals is light pink, the outer surface purplish pink. The pedicel is pubescent, about 1 in. (2.5 cm) long. Stamens are about 0.5 in. (1.3 cm) long and rose-colored. The follicetum is cylindric, 4–6 in. (10–15 cm) long, and reddish brown. The outer seed coats are reddish. Chromosome number: $2n = 114$.

Magnolia sargentiana is a woodland plant native to northern Yunnan, western Sichuan, and eastern Sikang provinces of China, at altitudes of about 5200–8500 ft. (1560–2550 m). This species was found growing in eastern Sikang by Armand David, a French missionary, in 1869. It was not named until 1906, when Finet and Gagnepain named it *Magnolia conspicua* [*M. denudata*] var. *emarginata*. Wilson collected fruiting specimens in 1908, and in 1913 he described it as a new species, *M. sargentiana*. The name honors Charles Sprague Sargent, the first director of the Arnold Arboretum. The vernacular name for this species is Yin-chin-hwa. Its bark (called "wu-p'i" or "hsin-p'i") is used in Sichuan for medicinal purposes.

This species is easily confused with *Magnolia dawsoniana*. Dandy was not satisfied that *M. dawsoniana* and *M. sargentiana* constitute separate species. Indeed, they both inhabit the same regions of western China and may be a single species, yet for now the existing taxonomy is still appropriate. The flowers of *M. sargentiana* resemble those of *M. dawsoniana* in shape, size, and poise, but the flowers of *M. dawsoniana* have only 9–10 tepals compared to 12 in *M. sargentiana*. *Magnolia dawsoniana* also lacks the pubescent pedicel of *M. sargentiana*; on *M. dawsoniana* the pedicel is glabrous or nearly so. Leaves of *M. dawsoniana* are also thinner and of a lighter green color.

Magnolia sargentiana was introduced into cultivation in North America by Wilson in 1908. Plants from Wilson's introduction were then sent to the Royal Botanic Garden, Kew, in 1913 and 1918. They were sent by Léon Chenault, the French propagator of the species for the Arnold Arboretum. *Magnolia sargentiana* has since become one of the most desirable magnolias, perhaps second only to *M. campbellii* in beauty. It is a tall, upright tree with spindly growth, producing large pink flowers at about twenty-five years from seed. Reportedly this plant sometimes flowers on almost a biennial cycle, blooming so profusely on alternating years that it "rests" the following year. The flower buds are perhaps more susceptible to frost damage at the pre-bloom stage than buds of most of the cultivated Asian species. *Magnolia sargentiana* likes a moist, well-drained soil and filtered sunlight. This species does not produce seed heavily. Flowers March–April; fruits August–September. Zones (6)7–9. No cultivars have been selected nor hybrids made with this species.

Magnolia sargentiana var. *robusta* Rehder & Wilson.

In Sargent, *Plantae Wilsonianae* 1:399. 1913. Type specimen: *Wilson 923a*. Herbarium of the Arnold Arboretum, Harvard University, Cambridge Massachusetts. **Plates 65–66.**

Plants of *Magnolia sargentiana* with a more compact and bushy growth habit, foliage that is longer (6–8 in./15–20.3 cm) and narrower (2–3 in./5–7.6 cm), larger flowers (8–12 in./20.3–30.5 cm across), and larger folliceta (6–8 in./15–20.3 cm long) belong to *M. sargentiana* var. *robusta*. This variety flowers at about thirteen years of age. Flowers are produced on upper and lower branches with up to 16 pink tepals, lighter pink than those of the typical *M. sargentiana* var. *sargentiana*. Flower buds are sickle-shaped, the perule splitting along two sutures. Flowers on the typical species occur only on the upper branches, have 12 dark pink tepals, and have ovoid flower buds, the perule splitting along a single suture. *Magnolia sargentiana* var. *robusta* is known only from the type locality in western Sichuan Province, China, where it grows in forests to 7600 ft. (2280 m) altitude.

Johnstone and Treseder (see "Magnolia Books," chapter references) each comment extensively on the differences between this variety and the typical species. Both authors maintain that this form is sufficiently different from *Magnolia sargentiana* to warrant specific status. On the other hand, Dandy included this variety within typical *M. sargentiana*. He reported that it was based only on a single specimen from within the range of typical *M. sargentiana* and wrote that he "cannot accept that this single plant represents a botanical taxon." This is an unusual remark by Dandy in light of the fact that he named at least six species based on single herbarium specimens (see section *Gwillimia*). Until more is known about all the plants involved, over the full range of the species, this form is best left as a variety of *M. sargentiana*.

Magnolia sargentiana var. *robusta* was introduced into cultivation in North America in 1908. Its history of introduction and distribution is the same as that of typical *M. sargentiana*. In 1913 seedlings were sent by the Arnold Arboretum to Léon Chenault in France for propagation and distribution. Plants first flowered in France in 1923 and at Caerhays Castle, England, in 1931.

Variety *robusta* is more commonly cultivated than typical *Magnolia sargentiana*. Its shrubby growth habit, its larger flowers spread throughout the tree, and its early flowering age all combine to make it more generally appealing. Its cultural requirements are the same as for the typical form with the exception that this plant needs greater shelter from high winds due to its larger flowers. The larger flowers catch more wind and increase branch weight, making branches more susceptible to breakage, and the large tepals soon take on a tattered appearance in strong wind.

CULTIVARS

'Blood Moon' A form with darker flowers than are typical of *M. sargentiana* var. *robusta*; tree to 50 ft. (15 m) tall. Hardy to 0°F (−18°C). Original tree at Strybing Arboretum, San Francisco, California. **Plate 67.**

'Chyverton Dark' Flowers fuschia in bud, mauve when open. Listed by Treseder's Nurseries, Cornwall, England, ca. 1973; cultivated at Chyverton, Cornwall, England.

'Multipetal' Form with double flowers (19–27 tepals). Original plant in Waterford, Ireland. Named by Sir Peter Smithers in 1983.

'White' Flowers pink when opening, fading to pure white. Listed by Treseder's Nurseries, Cornwall, England, ca. 1965.

'White Clone' Flowers pearly white. Listed by Treseder's Nurseries, Cornwall, England, ca. 1973.

HYBRIDS

M. campbellii × *M. sargentiana* var. *robusta* = 'Buzzard', 'Hawk'

M. campbellii var. *alba* × *M. sargentiana* var. *robusta* = 'Michael Rosse', Moresk', 'Princess Margaret'

M. campbellii var. *mollicomata* × *M. sargentiana* var. *robusta* = 'Treve Holman', 'Mossman's Giant'

M. campbellii var. *mollicomata* Lanarth Group × *M. sargentiana* var. *robusta* = 'Mark Jury'

M. denudata × *M. sargentiana* var. *robusta* = 'Ann Rosse', 'Marj Gossler'

M. sargentiana var. *robusta* × *M. sprengeri* 'Diva' = 'Caerhay's Belle'

Magnolia amoena Cheng.

Contributions from the Biological Laboratory of the Science Society of China, Botanical Series 9:280, fig. 28. 1934. Beautiful Magnolia, Tienmu Magnolia. Type specimen: description based on three specimens—*Cheng 4444A*, a fruiting specimen; *S. Chen 2692*, a flowering specimen; and *M. Chen 654*. All three specimens were collected in northwestern Zhejiang Province and are now in the Herbarium, University of Nanjing, China. **Fig. 7.14.**

Magnolia amoena is a deciduous tree 25–40 ft. (7.6–12 m) tall with gray or grayish white bark; branchlets are purplish. Flower buds are covered with white pubescence.

Fig. 7.14. *Magnolia amoena* from Chien and Cheng 1934.

The membranaceous leaves are oblong, 4–6 in. (10–15 cm) long, and 1–2 in. (2.5–5 cm) wide. Leaf apex is gradually acuminate, the base cuneate. The petiole is less than 1 in. (2.5 cm) long. The fragrant flowers are about 2 in. (5 cm) across with 9 pink tepals. The tepals are spatulate, about 2 in. (5 cm) long. Stamens are purplish red, about 0.5 in. (1.3 cm) long. The follicetum is cylindric, about 2 in. (5 cm) long.

Magnolia amoena is native to forests of the Tienmu Mountains in northwestern Zhejiang Province of China at altitudes of 2300–3300 ft. (690–990 m). It was described by W. C. Cheng in 1933. The epithet *amoena* means "beautiful" or "pleasing." The vernacular name Tienmu Mulan means Tienmu Magnolia, referring to the mountains in which it grows. *Magnolia amoena* shares its range with *M. cylindrica* and *M. denudata*. It seems to be closely related to those species and may, in fact, prove to be only a pink form of one of the two.

Several attempts have been made to introduce this species into cultivation. The Arnold Arboretum of Harvard University acquired seed (quantity not known) in 1982 which did not germinate. In 1986, the arboretum again received seed (only ten seeds), six of which germinated and grew to 3.5 in. (9 cm) the first year. This success in germination and development was conducted by Rob Nicholson and the arboretum staff. The arboretum plans to make some plants of this species available if propagation efforts continue to be successful. Nicholson (1987) estimated hardiness for this species to be Zones (6)7–9 based on the fact that *Magnolia amoena* shares a native range with *M. cylindrica*, *M. denudata*, and *M. officinalis*. No cultivars have been selected from nor hybrids made with this species.

Magnolia zenii Cheng.

In P'ei, *Contributions from the Biological Laboratory of the Science Society of China, Botanical Series* 8:291, fig. 20. 1933. Zen's Magnolia. Type specimen: description based on three specimens—*W. C. Cheng 4233* (in flower), *C. P'ei 2417* (in flower), and *C. P'ei 3123* (in fruit). Dr. T. R. Dudley, research botanist at the U.S. National Arboretum, has chosen *Cheng 4233* as the lectotype; the two remaining specimens are paratypes. These specimens are located at the Herbarium of the Jiangsu Botanical Institute, Nanjing, China. **Plates 82–83.**

Magnolia zenii is a small, deciduous tree 16–23 ft. (5–7 m) tall with gray bark and purplish branchlets. Winter buds are covered with silky hairs. The leaves are oblong, 3–6 in. (7.6–15 cm) long, and 1–3 in. (2.5–7.6 cm) wide. They are membranaceous, dark green and glabrous above, and mostly glabrous and pale green on the underside. Midrib and leaf veins are pubescent. Leaf apex is acute, the base cuneate. The petiole is about 0.5 in. (1.3 cm) long. The fragrant flowers are about 5 in. (12.7 cm) across with 9 tepals. The white tepals are about 3 in. (7.6 cm) long, with purple bases and veins. Stamen filaments are red. The pedicel is less than 0.5 in. (1.3 cm) long and densely pubescent. The follicetum is cylindric and 2–3 in. (5–7.6 cm) long. The outer seed coats are scarlet.

This species is native to forests of Jiangsu Province of China at 800–1000 ft. (240–300 m). It is a rare endemic, restricted to a small population of only a few dozen plants at the type locality. It was originally described by W. C. Cheng in 1933 and named in honor of H. C. Zen, director of the China Foundation for the Promotion of Education and Culture, for his encouragement of biological research. The species is endemic to Mount Bao-hua (also known as Mount Pao-hua). The vernacular name, Bao-hua Yulan, refers to its type locality and its similarity to the Yulan (*M. denudata*), which Cheng considered to be a closely related species.

Although overall *Magnolia zenii* is similar to *M. denudata*, it differs in having more oblong leaves with abrupt apices and smaller flowers with purple shading. Cheng also considers *M. zenii* to be closely related to *M. cylindrica*, which has oblanceolate leaves with obtuse or acute apices. The leaves, in fact, most closely resemble those of *M. kobus*, though the flowers of *M. kobus* have an outer whorl of sepaloid tepals.

Magnolia zenii was introduced into cultivation in the United States in 1980. The same year, members of the Sino-American Botanical Expedition to China observed the plant in cultivation in the Botanical Garden Memorial Sun Yat-Sen in Nanjing, China. Two members of the expedition, Dr. T. R. Dudley of the U.S. National Arboretum and Dr. Stephen Spongberg of the Arnold Arboretum, received seeds of the species. The seeds were germinated by them at their respective institutions, and plants are now established at these two gardens. Del Tredici and Spongberg (1989) described the propagation of the species by seed and cuttings.

One of the four plants at the Arnold Arboretum bloomed at about seven years of age. Flowers had only 6–7 tepals, due to lack of expansion of 2–3 of the inner tepals. A plant at the U.S. National Arboretum first bloomed in 1990 at the age of nine years. There is a strong fragrance to these flowers. The species may be marginally hardy in Zone 5, yet further observation is necessary to confirm this. Flowers March–April; fruits June–July. No cultivars have been selected from nor hybrids made with this species.

Magnolia subgenus *Yulania* section *Buergeria* (Siebold & Zuccarini) Baillon

[*Buergeria* Sieb. & Zucc.] *Addisonia* 7:2. 1866. TYPE: *Magnolia kobus* var. *stellata.*

Species in section *Buergeria* are deciduous, precocious trees or shrubs. Flowers are usually white, occasionally pink; the outer whorl of tepals is small and sepaloid. The follicetum is cylindric and often distorted due to ovule abortion. The species in this section are diploid with a chromosome number of $2n = 38$.

This section contains four species native to China and Japan: *Magnolia kobus* and its varieties, native to Japan and South Korea; *M. salicifolia*, native to Honshu, Kyushu, and Shikoku islands of Japan; *M. biondii*, native to Shaanxi, Henan, Hubei, and Sichuan provinces of China; and *M. cylindrica*, native to northern Fukien Province, China.

The treatment in this book of *Magnolia stellata* (Star Magnolia) and *M. × loebneri* as varieties of *M. kobus* has been arrived at after careful consideration of both the plants and the literature. While a taxonomist, unfortunately, never knows his or her degree of correctness as to the status or the relationships between plants, after considering the evidence a judgment must be made based on that evidence. The decision regarding the status of these three taxa was, by far, the most difficult taxonomic decision I made in this book. This was especially true since the taxa are extensively cultivated. After studying all the available information, I agree with Spongberg's (1976) recognition of *M. stellata* and *M. × loebneri* as varieties of *M. kobus*. My conclusion is based on the following.

Prior to 1954 the Star Magnolia had long been considered a species, *Magnolia stellata*. In three papers published between 1954 and 1957, Dr. Benjamin Blackburn questioned its status as a species, observing that flowers of *M. stellata* produce a reduced outer whorl of tepals, as do the flowers of *M. kobus*. Prior to Blackburn's writings, *M. stellata* was believed to lack these sepaloid tepals, based on the original description by Maximowicz. Blackburn noted that the sepaloid tepals fall before the flower is mature, concluding that Maximowicz must have examined mature flowers. Blackburn also commented on the tendency of *M. stellata* seedlings to resemble *M. kobus* or represent an intermediate form between the two species. He reported that only a small fraction of *M. stellata* seedlings resembled *M. stellata*, while about half appeared to be true *M. kobus*. The remaining seedlings were intermediate between the two, therefore more properly classified as *M.* × *loebneri*.

These findings led Blackburn to propose that *Magnolia stellata* be reduced to a form (1954) or a variety (1955, 1957) of *M. kobus*. "Hybrids," or intermediates, between typical *M. kobus* and *M. kobus* var. *stellata* were then considered by Blackburn to be only variants of typical *M. kobus*. Blackburn also reported that *M. stellata* was not native to Japan, as had previously been thought by many botanists. He regarded it as a variety of garden origin. Inami (1959) refuted Blackburn's claims that *M. stellata* was not truly native to Japan, reporting its natural occurrence in the Aichi prefecture of Honshu, Japan.

Blackburn's taxonomy seemed to go unheeded until the publication of Spongberg's work in 1976. Spongberg followed Blackburn's treatment of the Star Magnolia as a variety of *Magnolia kobus*. The Star Magnolia therefore was identified as *M. kobus* var. *stellata*. But rather than including plants formerly known as *M.* × *loebneri* within the type *M. kobus*, as was suggested by Blackburn, Spongberg created a new var. *loebneri* to encompass the intermediate forms.

Spongberg also reported that the few records of *Magnolia kobus* var. *stellata* growing in the wild are "questionably valid," and that origin was still in question. In response to this, Ueda (1988) presented the distribution of the Star Magnolia in Japan, based on herbarium specimens and observations. He reports its occurrence in the Gifu, Aichi, and Mie prefectures of Honshu, Japan. Ueda's distribution map of *M. kobus* and *M. kobus* var. *stellata* is shown in Figure 7.15 and corresponds to that presented by Inami (1959).

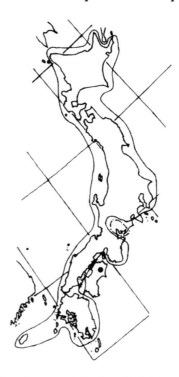

Fig. 7.15. Distribution of *Magnolia kobus* (solid line) and *M. kobus* var. *stellata* (broken line) in Japan. Reprinted from Ueda 1988 with permission from the Journal of the Arnold Arboretum.

Ueda also reported that, in his experience, *M. stellata* seed have always produced *M. stellata* seedlings, not seedlings resembling *M. kobus* as Blackburn had claimed.

Ueda used the information he presented to support maintaining species status for *Magnolia stellata*. However, the information he provided does not contradict the treatment of the Star Magnolia as a variety of *M. kobus*. In fact, the geographic distribution is supportive of such a treatment. Varieties need not have overlapping distributions, as Ueda implies. Nor does the fact that Star Magnolias, in Ueda's experience, reproduce true from seed disqualify its treatment as a variety. Therefore, the Star Magnolia is recognized here as *M. kobus* var. *stellata*. Intermediates and hybrids, both intentional and spontaneous, between typical *M. kobus* and the Star Magnolia belong to *M. kobus* var. *loebneri*.

Magnolia kobus DC.

[*Buergeria obovata* Siebold & Zuccarini, *Magnolia borealis* (Sargent) Kudo, *M. glauca* var. α Thunberg, *M. kobus* var. *borealis* Sargent, *M. praecocissima* Koidz., *M. praecocissima* var. *borealis* (Sargent) Koidz., *M. thurberi* Parsons ex W. Robinson, *M. tomentosa* Thunberg, *Yulania kobus* (DC.) Spach.] *Regni Vegetabilis Systema Naturale* 1:456. 1817. Type specimen: not designated. **Plates 41–42.**

Plants of *Magnolia kobus* are deciduous trees or shrubs to 80 ft. (24 m) tall with rough, brownish or silvery gray bark. Twigs are yellow-green becoming brown and pubescent. The terminal buds are covered with yellowish or silvery hairs. Leaves are obovate to oblanceolate, 4–8 in. (10–20 cm) long, 2–4 in. (5–10 cm) wide, dark green and glabrous above, and pale green beneath. The leaf apex is abruptly acuminate; the base is attenuate to cuneate. Petioles are about 1 in. (2.5 cm) long, yellow-green and pubescent. The white flowers are fragrant, with 9–12 tepals. The inner 6–9 petaloid tepals are about 2–3 in. (5–7.6 cm) long. The 3 outer sepaloid tepals are about 1 in. (2.5 cm) long and white to greenish. Stamens are about 0.5 in. (1.3 cm) long, usually purple or rose-colored at the base. The pedicel is about 0.5 in. (1.3 cm) long and more or less pubescent. The follicetum is cylindric in shape or distorted, 2–5 in. (5–12.7 cm) long, and reddish brown. The outer seed coat is red. Chromosome number: $2n = 38$.

Magnolia kobus is native to forests and thickets of Japan and is also found on Quelpaert Island, South Korea. The wood of *M. kobus* is used by the Japanese for utensils, engravings, matches, and fuel. This species was first described by de Candolle in 1817. His description was based on several earlier accounts, including a figure by E. Kaempfer (1712), a description by R. A. Salisbury (1806) of *M. gracilis* (now considered a synonym for *M. liliiflora*), and his own specimens.

de Candolle gave this species the epithet *kobus* because this was given by Kaempfer as the Japanese vernacular name. The correct vernacular name, however, is *Kobushi*. Ueda (1986b) has attempted to show that de Candolle's name *M. kobus* is invalid by virtue of his inclusion of a reference to *M. gracilis* (*M. liliiflora*). But considering the nature of this inclusion, and the inclusion of Kaempfer's plate, it is my opinion that Ueda does not present a sufficiently strong case to warrant changing the long-standing epithet of this species.

In 1908 Sargent described *Magnolia kobus* var. *borealis*, a more northerly form of the species, reputedly more vigorous and attaining a greater size than its southern counterpart. Johnstone (1955; see "Magnolia Books," chapter references) was the first to question the distinction of this variety, followed by Spongberg (1976) and Gardiner (1989; see "Magnolia Books," chapter references). The characters defining the varieties are not consistent, and single plants showing characters of both typical *M. kobus* and *M. kobus* var. *borealis* are not uncommon. Therefore, recognition of a separate botanical variety does not appear justified. Variety *borealis* is included here within typical *M. kobus*. It may be important in the horticultural trade to differentiate between typical *M. kobus* var. *kobus* and the larger, more vigorous forms previously known as *M. kobus* var. *borealis*. In this case the use of the cultivar name 'Borealis' is suggested to differentiate.

Magnolia stellata and *M. × loebneri* are here recognized as *M. kobus* var. *stellata* and *M. kobus* var. *loebneri*, respectively. A discussion of the evidence leading to this decision is presented in the introduction to this section.

Magnolia kobus is closely related and similar in appearance to *M. salicifolia*, yet differs in having pubescent buds and pedicels, and in having leaves which are broadest in shape above the middle. It differs from its variety *stellata* in having a larger, treelike habit, flowers with fewer tepals (usually 9, the outer 3 sepaloid), and larger leaves.

This species was collected, together with var. *stellata*, by Dr. George Rogers Hall during his trip to Japan about 1861. Hall then gave plants of *Magnolia kobus* to Parsons Brothers Nursery of Long Island, New York, in 1862. Parsons named it *M. thurberi* and distributed it under that name until it was realized that it was actually *M. kobus*. It was probably introduced into England about 1879 by Charles Maries, a collector for Veitch Nursery.

Magnolia kobus is adaptable to many situations and is also quite cold hardy. Seed germinates readily, but seedlings will require a variable ten to thirty years to flower. Plants propagated from grafts or cuttings take approximately ten years to flower. The species is often used as a hardy, vigorous rootstock in grafting many species of *Magnolia*. Flowers March–April; fruits August–September. Zones 4–7(8).

CULTIVARS

All cultivars are white-flowered forms.

'Borealis'   A hardy form which is more vigorous and reaches larger sizes than typical *Magnolia kobus*. Formerly var. *borealis*; see text for discussion. **Plate 43.**

'Fastigiata'   A dwarf fastigiate form introduced in 1925 by Wada of Hakoneya Nursery, Japan.

'Janaki Ammal'   A colchicine-induced polyploid form originating at the Royal Horticultural Society Garden at Wisley. Listed in some accounts as a cultivar of *Magnolia kobus* var. *stellata*.

'Nana Compacta'   A slow-growing, compact form originating at the Kohankie Nursery in Painesville, Ohio, before 1950.

'Norman Gould'   A very compact, colchicine-induced polyploid form. Flowers larger with large tepals and puckered leaves. Grown at the Royal Horticultural Society garden at Wisley. Listed in some accounts as a cultivar of *Magnolia kobus* var. *stellata*, but determined by Kehr (1991) to belong to *M. kobus* var. *kobus*.

'Pickard's Stardust'   Thought to be a seedling from *M. kobus*. Flowers white with upright tepals; very fragrant. Registered by Pickard, Magnolia Gardens, Canterbury, England, ca. 1980.

HYBRIDS

"Crosses" between *Magnolia kobus* and *M. kobus* var. *stellata* are common, and many named cultivars exist. These, formerly included in *M. × loebneri*, are listed under *M. kobus* var. *loebneri*. *Magnolia kobus* has also been crossed with several magnolias in section *Yulania*, but none of these have produced commercial cultivars. Named hybrids include

M. kobus × *M. liliiflora* 'Nigra' = 'Marillyn'
M. kobus × *M. salicifolia* = *M.* × *kewensis*

Magnolia kobus var. *stellata* (Siebold & Zuccarini) Blackburn.

[*Buergeria stellata* Siebold & Zuccarini, *Gwillimia halleana stellata* C. deVos, *Magnolia halleana* W. Robinson, *M. kobus* f. *stellata* (Siebold & Zuccarini) Blackburn, *M. stellata* Maxim., *M. stellata* var. *keiskei* Makino, *M. stellata* var. *rosea* Veitch, *M. tomentosa* Thunb., *Talauma stellata* (Siebold & Zuccarini) Miquel.] *Amatores Herbarii* 17(1–2):1. 1955. Star Magnolia. Type specimen: collections from Japan by Siebold, now in the Leyden Herbarium, the Netherlands. **Plates 46–47.**

Magnolia kobus var. *stellata* differs from the species in having a smaller, more shrubby growth habit, reaching only 5–15 ft. (1.5–4.6 m). Often its leaves are narrower, and its flowers have a greater number of petaloid tepals (12–33). The sepaloid tepals often fall before anthesis. It blooms at a very early age, one to two years, and is one of the earliest magnolias to bloom in the spring. The flowers of *M. kobus* var. *stellata* are typically white, sometimes tinged with pink; several very pink forms exist.

The native range of this plant has been questioned for many years. Until the 1950s wild specimens were unknown, or were thought to be plants escaped from cultivation. This led to the theory that the Star Magnolia originated in cultivation. Some authors, however, have considered it native to China or Japan. In an attempt to solve the problem of the nativity of the plant, Ueda (1988) undertook a study of the natural distribution of the Star Magnolia and found it to be endemic to Mie, Gifu, and Aichi prefectures of central Honshu, Japan (see Figure 7.15). It is found growing in sunny spots on slopes, usually in damp, swampy areas. Under these conditions it suckers freely, producing dense thickets.

This plant was first described by Thunberg in 1784 as *Magnolia tomentosa*. Unfortunately, the specimen on which Thunberg based his description is a species of *Edgeworthia* rather than a magnolia. Because of this misidentification of the type specimen, Thunberg's name for this magnolia (*M. tomentosa*) has been disregarded. Ueda (1986b) made a case for adopting *M. tomentosa* as the name for this plant, inasmuch as it is apparent that Thunberg meant to name the magnolia and not the Edgeworthia. While Ueda's argument is sound, *M. tomentosa* is not used here due to the continuing question about its validity.

If Thunberg's *Magnolia tomentosa* is to be disregarded, the earliest name for this plant is *Buergeria stellata*, given by Siebold and Zuccarini in 1846. The plant was transferred from *Buergeria* to *Magnolia* by Maximowicz in 1872, creating *Magnolia stellata*. Blackburn (1955) considered this plant to be a variety of *M. kobus*, and Spongberg (1976) followed him. For reasons discussed at the beginning of section *Buergeria*, the treatment of the Star Magnolia as a variety of *M. kobus*, as suggested by Blackburn and Spongberg, is followed here.

The epithet *stellata* means "star-like" and refers to the flowers, in which the tepals radiate out from the center like the rays of a star. The Japanese vernacular name for *M. kobus* var. *stellata* is Shide-kobushi. Blackburn (1954) translates this as "kobus of homes," referring to its use as a garden plant. Ueda (1988) translates Shide-kobushi as "zigzag petaled kobus," referring to the flower petals resembling the *shide* or zigzag-folded papers used in Japanese ceremony.

Siebold attempted to introduce this magnolia into Western gardens on several occasions. His lack of success in this introduction was due to political difficulties hampering foreign trade. The plant was introduced in 1862 by Dr. George Rogers Hall of Bristol, Rhode Island. Hall collected it in Japan and gave his specimens to the nursery of Samuel Parsons, Long Island, New York, for propagation and distribution. Parsons reported that propagation efforts were few until about 1869, when his interest inspired greater effort. In 1875 William Robinson named the plant *Magnolia halleana* after Dr. Hall, and this name is still sometimes used, although it is not valid as it was published after *M. stellata*.

The Star Magnolia is one of the most popular magnolias today, and many forms are available. It is quite adaptable to many situations, is wind tolerant, and tolerates shade although it is more vigorous and blooms more profusely in sunny locations. Because of its small stature it has a place in almost any landscape. It blooms early in the spring and is a welcome sight after a long winter. Because of this early flowering time, however, it is subject to damage from frost.

As noted earlier, Blackburn (1955, 1957) reported that *Magnolia kobus* var. *stellata* did not reproduce true from seed. (Blackburn is familiar with cultivated plants in the United States.) Ueda (1988) stated that in his experience this was not the case. (Ueda is familiar with wild plants in Asia.) These seemingly conflicting accounts may not be conflicting at all. To explain, it is likely that plants cultivated in North America and Europe are widely removed, both geographically and genetically, from their wild Asian forebears. Plants in cultivation in the Occident may have undergone spontaneous hybridization and some purposeful selection, now behaving differently from plants found in the wild. Treseder (1978; see "Magnolia Books," chapter references) offers a similar explanation in ascribing the origin of the Western cultivated *M. kobus* var. *stellata* as possibly a cross between *M. kobus* and *M. salicifolia*. This also explains the exceptionally high degree of variability among seedlings.

Variety *stellata* is easily grown from seed as it germinates readily; but in order to avoid any unwanted variation, cuttings, layering, or grafting offer useful alternatives. Grafted plants seem to develop faster than those propagated from cuttings. Flowers March–April; fruits August–September. Zones 5–9.

CULTIVARS

'Centennial'   Flowers to 5 in. (12.7 cm) across with about 30 tepals, white with a pink tinge. Selected from 'Rosea' seedling and named by the Arnold Arboretum of Harvard University in 1972 in honor of its centennial. Hardy to Zone 5. **Plate 48.**

'Chrysanthemumiflora'   Flowers with over 40 tepals of a clear pink color. A seedling of 'Rubra' selected by Wada of Hakoneya Nurseries, Yokohama, Japan.

'Dawn'   Flowers pale pink, 40 or more tepals. Selected by Harold Hopkins in 1975 from a tree in Bethesda, Maryland.

'Dr. Massey'   Flower buds pink, opening to white.

'Green Star'   Flowers pure white without a hint of pink. When first opening, flowers have a thin line of yellowish green down the middle of some of the tepals. Found growing at the University of Tennessee campus in Knoxville, Tennessee, by Frank Galyon and named by him in 1962. Seldom seen in trade. Used by Dr. Galyon for breeding purposes.

'Jane Platt' ('Rosea Jane Platt')   Flower color lighter pink than that of 'Rosea', with 20–30 tepals. Selected by Gossler Farms Nursery, Springfield, Oregon, from the garden of Jane Platt, Portland, Oregon, and named in her honor. Hardy to Zone 6. **Plate 49.**

'Keiskei'   A dense, shrubby form with small, purplish flowers, long cultivated in Japan. Similar to 'Rubra' in having dark flower color.

'Kikuzaki'   Flowers small, 2 in. (5 cm) across, light pink, with 20–30 tepals. Blooms profusely, even at an early age.

'King Rose'   Pink in bud, becoming lighter pink when open; inner surface of tepals almost white. Flowers with 20–25 tepals. Hardy to Zone 5.

'Nobilis'   Plant with treelike form and pure white flowers.

'Pink'   See 'Rosea'

'Pink Stardust'   Flowers light pink with 40–50 tepals. Offered by Louisiana Nursery, Opelousas, Louisiana.

'Red'   Flowers 3–5 in. (7.6–10 cm) across with 10–16 tepals; outside dark purple-red, inside white. Registered in 1946 by Sawada of Overlook Nurseries, Mobile, Alabama.

'Rohrbach'   Buds pink, flowers remaining pink at anthesis. Introduced by the U.S. Department of Agriculture, Beltsville, Maryland.

'Rosea' ('Pink')   Buds pink, flowers somewhat smaller than typical, blooms pink fading to white. Develops as a small shrub. Cultivated since about 1899 and still a favorite. Hardy to Zone 5.

'Rosea Jane Platt'   See 'Jane Platt'

'Royal Star'   Pink buds open to large, white flowers 6 in. (15 cm) across with 25–30 tepals. Blooms 7–10 days later than typical, avoiding some frosts. Robust, fast-growing, roots easily. Seedling of Vermeulen's 'Waterlily'. Selected in 1955 and registered in 1960 by J. Vermeulen and Sons, New Jersey. Hardy to Zone 4. **Plate 50.**

'Rubra'   Flowers deep rose to almost red throughout flowering time. The flower color is darker than that of 'Rosea'. Forms a small tree. Janaki Ammal's (1953) chromosome count for this cultivar is $2n = 51$ chromosomes; *Magnolia kobus* var. *stellata* is diploid ($2n = 38$). Janaki Ammal suggested that the clone she tested might be *M. kobus* var. *stellata* × *M. liliiflora*, explaining the darker flower color and unusual chromosome number. However, the cultivar 'Rubra' does not resemble any of the Kosar/deVos hybrids (see chapters 9 and 10 for discussion) between these species. 'Rubra' was introduced from Japan by K. Wada, Hakoneya Nurseries, Yokohama, Japan, ca. 1925. Hardy to Zone 5. **Plate 51.**

'Scented Silver'   A form with pure white flowers that are much more fragrant than typical. Habit is upright, growing with a single trunk. Selected by Frank Galyon, Knoxville, Tennessee, and registered in 1990.

'Super Star'   See *Mangolia kobus* var. *loebneri*

'Two Stones'   Polyploid form induced by colchicine treatment. Similar to typical *Magnolia kobus* var. *stellata*, but larger in all characters, including flowers. Created and selected by August Kehr, Hendersonville, North Carolina. Registered in 1991.

'Water Lily'   Distinct from the 'Waterlily' clones (description following) in having large, white flowers and numerous tepals. First listed by Hillier Nurseries, Winchester, England, in 1972.

'Waterlily'   Treseder (1978; see "Magnolia Books," chapter references) reported that three clones with this name are available. They are as follows:

A clone originating at Greenbriar Farms in Norfolk, Virginia. The habit is more upright than typical, flowers are slightly larger and quite fragrant. Tepal number averages about 20. This is thought to be the form most common in the United States.

A clone received as 'Rosea' by John Vermeulen, Neshanic Station, New Jersey, in the mid 1920s. It was given to Vermuelen by a gardener in Long Island, New York, Mr. Brouwer. When later comparing this clone with other 'Rosea' plants, Vermeulen realized it was superior. The flowers are clear pink and have up to 32 tepals.

A clone located at the Arnold Arboretum, Cambridge, Massachusetts, and distributed from there by E. H. Wilson ca. 1930. The flowers are pink fading to white and have 34–40 tepals.

HYBRIDS

"Crosses" with typical *Magnolia kobus* are listed under *M. kobus* var. *loebneri*.
M. denudata × *M. kobus* var. *stellata* 'Waterlily' = 'Emma Cook', 'Pristine'
M. kobus var. *stellata* × *M. liliiflora* = 'Ann', 'Betty', 'George Henry Kern', 'Jane', 'Judy', 'Orchid', 'Pinkie', 'Randy', Ricki', 'Susan'
M. kobus var. *stellata* × *M. salicifolia* = *M.* × *proctoriana*

Magnolia kobus var. *loebneri* (Kache) Spongberg.
   [*M.* × *loebneri* Kache.] *Journal of the Arnold Arboretum* 57(3):287. 1976. Loebner's Magnolia. **Plate 44.**
   Magnolia kobus var. *loebneri* includes plants resulting from crosses between *M. kobus* var. *kobus* and *M. kobus* var. *stellata*. The "hybrids" represented by this variety are intermediate between the parents in size, habit, and tepal number and shape. Plants develop as large shrubs or small trees with white or pink flowers having 11–16 narrow tepals.
   Magnolia kobus var. *loebneri* is not known in the wild. This is not surprising since the varieties of *M. kobus, kobus* and *stellata,* are not native to areas near enough to allow natural hybridization. When these varieties are grown together in cultivation, however, spontaneous "hybrids" do occur.
   The first intentional cross between *Magnolia kobus* var. *kobus* and *M. kobus* var. *stellata* was made by M. Loebner, Pillnitz, Germany, in the early 1900s. The resulting plant was originally named *M.* × *loebneri* by Paul Kache, in 1920. This name was used until 1976, when Spongberg proposed to make the "hybrids" a variety of *M. kobus*, creating *M. kobus* var. *loebneri*. For reasons discussed at the beginning of section *Buergeria*, this varietal status is used here.
   Magnolia kobus var. *loebneri* was first introduced into cultivation by Loebner in 1923, when he sold several of his plants to the German nurseryman Wilhelm Kordes in Sparrieshoop, Germany. Since then many additional crosses, intentional and spontaneous, have resulted in a great many diverse forms. Flower color varies from white to pink, and flower size ranges from 4 in. (10 cm) to 6 in. (15 cm) across. Habit varies from multi-stemmed shrubs to single-stemmed trees of 25 ft. (7.8 m) tall. All these plants flower at an early age and are easily propagated from cuttings Flowers March–May; fruits August–September. Zones 5–9.

   CULTIVARS

'Ballerina'   Flowers fragrant, tepals 30 or more, pale pink at the base. Blooms later than other *Magnolia kobus* var. *loebneri* clones. A cross between 'Spring Snow' and *M. kobus* var. *stellata* 'Waterlily' made by J. C. McDaniel of Urbana, Illinois, and registered by him in 1969. Slightly smaller and slower growing than 'Merrill', because of backcross with *M. kobus* var. *stellata*. Zone 5.

'Donna'   Flowers white, 8 in. (20.3 cm) across, opening flat, blooming in May. This form may have the largest flowers of any *Magnolia kobus* var. *loebneri* cultivar. Selected by Harry Heineman, Scituate, Massachusetts.

'Dr. Merrill'   See 'Merrill'

'Dr. Van Fleet'   Pink flowers on an upright tree. Registered by Phil Savage, Bloomfield Hills, Michigan, in 1967.

'Encore'   White flowers with a pink tinge at the base of the 18–25 tepals. Produces multiple buds at branch terminals and sometimes along the branches. Since these buds mature at different times, the flowering period is extended and persistent, and the plant produces an "encore" of flowers. This form has a compact to nearly dwarf habit. Easily propagated from cuttings. Selected from open-pollinated 'Ballerina' seedlings by August Kehr, Hendersonville, North Carolina. Registered 1991.

'Leonard Messel'   Flower buds dark purple. Flowers have 12 linear tepals 3 in. (7.6 cm) long, opening to dark pink to purple outside, white inside. The flowers are quite frost-resistant. Thought to be a cross between typical *Magnolia kobus* and *M. kobus* var. *stellata* 'Rosea'. Registered in 1955 by Messel. Zone 5.

'Merrill' ('Dr. Merrill')   Flowers large, white, with about 15 tepals. Blooms at an early age. Growth to 30 ft. (9.1 m), with low branches and nice form. Named after Dr. E. D. Merrill, former director of the Arnold Arboretum of Harvard University, Cambridge, Massachusetts. Registered in 1952 by Dr. Karl Sax of the arboretum. Zone 5.

'Neil McEachern'   Grown from seed of *Magnolia kobus* var. *stellata* 'Rosea', collected from the garden of Neil McEachern at Lake Como, Italy, around 1953. Flowers nearly identical to those of 'Rosea'. Develops as a very large shrub or small tree. Registered about 1968 by John Bond of the Great Park in Windsor, England.

'Powder Puff'   Upright tepals give flowers the appearance of a brush or powder puff. Flowers white with 18–25 tepals. Selected by August Kehr, Hendersonville, North Carolina, from an open-pollinated seedling of 'Ballerina'. Registered 1990.

'Snowdrift'   Flowers larger than those of *Magnolia kobus* var. *stellata*, with 12 tepals. Leaves also slightly larger. This plant is one of the original seedlings grown at Hillier and Sons Nursery, Winchester, England. Described by Treseder in 1978.

'Spring Snow'   Flowers pure white, fragrant, 15–20 tepals. Flowers appear later than those of some *Magnolia kobus* cultivars, and flower buds appear to tolerate cold weather better than those of most other *M. kobus* var. *loebneri* cultivars. Rounded growth habit, to 30 ft. (9 m). Produces seed readily. Selected by J. C. McDaniel from a plant on the University of Illinois campus at Urbana and registered by him in 1970. Zone 5. **Plate 45.**

'Star Bright'   A vigorous form with flowers similar to var. *stellata*, pure white. Introduced by Tom Dodd Nurseries, Semmes, Alabama, ca. 1965.

'Super Star'   Flowers white, 4–6 in (10.2–15.2 cm) across, with 11–16 narrow tepals. A fast-growing tree rather than the typical shrub. Grows an average of 3–6 ft. (1–2 m) per year, producing a rounded canopy. A chance seedling discovered and named by Ray Bracken of Piedmont, South Carolina. Sold by Bracken as *Magnolia stellata*, but its habit is clearly that of *M. kobus* var. *loebneri*. Mr. Bracken reports that the habit and flowers are similar to 'Merrill', but 'Super Star' grows two or three times faster.

'White Rose'   Open-pollinated seedling of 'Ballerina'. Flowers white with about 22 tepals. The tepals are firm and droop only well after blooms have faded. Registered ca. 1988 by W. J. Siedl of Manitowoc, Wisconsin.

'Willowwood'   Abundant blooms to 7 in. (17.8 cm) across, with 11–14 tepals. Selected by Ben Blackburn in 1952 from a tree at Willowwood Arboretum of Rutgers University, Gladstone, New Jersey.

HYBRIDS

No hybrids have been made using this variety.

Magnolia salicifolia (Siebold & Zuccarini) Maxim.

[*Buergeria salicifolia* Siebold & Zuccarini, *Magnolia famasiha* P. Parment., *M. salicifolia* f. *fasciata* (Millais) Rehder, *M. salicifolia* var. *concolor* (Miq.) Maxim., *M. salicifolia* var. *fasciata* Millais, *M.* × *slavinii* Harkness.] *Bulletin de l'Académie Impériale des Sciences de St. Pétersbourg* 17:419. 1872. Japanese Willow-Leaf Magnolia. Type specimen: collection by Siebold, Leyden Herbarium, the Netherlands.

Magnolia salicifolia is a deciduous tree or large shrub, 30–50 ft. (10–15 m) tall, with rough, silver-gray bark. Twigs are yellowish brown. Terminal buds are covered with long, yellow or silvery hairs. The leaves are lanceolate, rarely elliptic, usually broadest at or below the middle, 2–5 in. (5–15.2 cm) long, and 1–3 in. (2.5–7.6 cm) wide. Leaves are dark green and glabrous above and pale green and glaucescent beneath. Leaf apex is acute to acuminate; base is cuneate, rarely rounded. Petioles are glabrous, yellowish brown, and about 1 in. (2.5 cm) long. The fragrant flowers are 3–4 in. (7.6–10 cm) across with 6–9 (sometimes to 12) petaloid tepals and 3 outer sepaloid tepals. The petaloid tepals are spatulate, 2–4 in. (5–10 cm) long, and 1–2 in. (2.5–5 cm) wide, often with a pink tinge at the base. Stamens are white- or cream-colored and about 0.5 in. (1.3 cm) long. Pedicels are glabrous. The follicetum is cylindric, 2–3 in. (5–7.6 cm) long, and reddish brown. The outer seed coat is red. Chromosome number: $2n = 38$.

Magnolia salicifolia is native to montane woods at about 1650–4400 ft. (495–1320 m) altitude on Honshu, Kyushu, and Shikoku islands of Japan. It was first collected in the wild by Siebold and named *Buergeria salicifolia* by Siebold and Zuccarini in 1846. In 1866 the species was transferred to *Talauma* by Miquel. In 1872 it was transferred to *Magnolia* by Maximowicz. The epithet *salicifolia* refers to the resemblance of the leaves to those of willows (*Salix* spp.), as does the common name Willow-leaf Magnolia. The Japanese vernacular name is Tamusiba.

This species is closely related to *Magnolia kobus*, though it differs in having glabrous pedicels and leaf buds, which are pubescent in *M. kobus*. *Magnolia salicifolia* can be distinguished from other species in this section in that its leaves give off an anise scent when bruised. It is a variable species, and several forms have previously been recognized. A fastigiate form, once considered a variety and now known as Fasciata', was discussed by Millais (1927; see "Magnolia Books," chapter references). Specimens of this form show a wide, upright habit. A more common form is that which once had the name *M. salicifolia* var. *concolor*, named by Miquel in 1866. This form has a spreading habit, broader leaves, and long, strap-shaped tepals. It blooms later than the typical species. This form is found in Western cultivation and was most likely introduced in the early 1900s. In addition there are intermediates between typical *M. salicifolia* and the broader leaved form. Many other variations of *M. salicifolia* can be observed, so a complete understanding of the species requires more extensive field research. Variations have been noted in growth habit, flowering time, and degree of fragrance in leaves.

The species was introduced into cultivation in North America by Sargent from the collection of seed made by Sargent and Veitch on Mt. Hakkoda on the island of Honshu,

Japan, in 1892. Sargent apparently sent Veitch germinated seeds or seedlings from his collection, and the species was brought into cultivation in England in 1906.

Magnolia salicifolia is not as showy as some other, similarly precocious species and is not as commonly grown as *M. kobus,* but it is hardy. In the wild it is found growing along streams or in other moist areas. This kind of environment is apparently ideal for cultivation, but the soil needs to be well drained. *Magnolia salicifolia* readily sets fruit and is easily grown from seed. Flowers March–May; fruits (August) September–October. Zones (5)6–9.

CULTIVARS

'Else Frye'   Upright form with large flowers. The 9 tepals are white with a pink tinge at the bases. Selected by Joseph Witt in the University of Washington Arboretum, Seattle, Washington, in 1961.

'Fasciata' ('Fastigiata')   Fastigiate form listed by Millais in 1927. Original tree at Tilgate, Sussex, England.

'Fastigiata'   See 'Fasciata'

'Iufer'   This is a small, pyramidal form. Flowers are white; stamens have red tips. Introduced by Iufer Nursery in Salem, Oregon. Registered in 1986.

'Jermyns'   A slow-growing, shrubby form with larger flowers and leaves than the species. Later blooming than typical. Listed by Hillier and Sons Nursery, Winchester, England, in 1973.

'Kochanakee'   Vigorous pyramidal form with large fragrant flowers. Origin unknown.

'Miss Jack'   Vigorous, fast-growing form with narrow leaves. Flowers white, tinged with pink. Zone 6.

'Treseder'   Compact, conical form originating at Treseder's Nurseries, Cornwall, England.

'Van Veen'   Flowers exceptionally fragrant.

'W. B. Clarke'   Flowers larger than typical. Heavily foliaged. Originated at W. B. Clarke Nursery, San José, California.

HYBRIDS

Plants formerly known under the synonym *Magnolia* × *slavinii* are listed under *M.* × *proctoriana.* Other named hybrids include

M. denudata × *M. salicifolia* = 'Wada's Snow White'
M. kobus × *M. salicifolia* = *M.* × *kewensis*
M. kobus var. *stellata* × *M. salicifolia* = *M.* × *proctoriana*

Magnolia biondii Pampan.

[*Magnolia aulacosperma* Rehder & Wilson., *M. conspicua* var. *fargesii* Finet & Gagnep., *M. denudata* var. *fargesii* (Finet & Gagnep.) Pamp., *M. fargesii* (Finet & Gagnep.) Cheng, *M. obovata* Pavol (not Thunb.).] *Nuovo Giornale Botanica Italiano, New Series* 17:275. 1910. Type specimen: *Silvestri 734.* Herbarium Universitatis Florentinae, Florence, Italy. **Plates 4–6.**

Magnolia biondii is a small, deciduous tree to 55 ft. (16.5 m) tall with smooth, gray bark. Branchlets are densely pubescent, becoming almost glabrous with age. Leaves are oblong to elliptic, 5–7 in. (12.7–17.8 cm) long, and 2–4 in. (5–10 cm) wide. They are dark green and glabrous above and brighter green and pubescent beneath, becoming glabrous. Leaf apex is acuminate, base cuneate. The petiole is about 1 in. (2.5 cm) long and glabrous. The fragrant, white flowers are about 4 in. (10 cm) across, with 9 tepals.

The outer 3 tepals are lanceolate, about 0.5 in. (1.3 cm) long. The inner 6 tepals are obovate, about 2 in. (5 cm) long. The stamens are about 0.5 in. (1.3 cm) long. The gynoecium is glabrous. The follicetum is irregularly cylindric, about 5 in. (12.7 cm) long. The outer seed coat is red.

Magnolia biondii is native to woods and open country at about 2000–11,000 ft. (600–3300 m) altitude in Shaanxi, western Henan, western Hubei, and eastern Sichuan provinces, China. *Magnolia biondii* is the northernmost Chinese magnolia. It was first mentioned in the literature in 1900 when Diels suggested that Giraldi's collection of this plant might be a new species of *Magnolia*. However, Diels did not give the plant a name. It was named *Magnolia biondii* by Pampanini in 1910, in honor of A. Biondi.

Wilson also collected this species in 1907. Wilson's specimens lacked flowers, and, thinking he had collected a new species, he named it *Magnolia aulacosperma*. The term *aulacosperma* means "furrowed seeds," referring to ridges in the seed coats.

In 1928, it became apparent that *Magnolia biondii* (Pamp.) and *M. aulacosperma* were in fact the same species; thus the plant should be known by the older name, *Magnolia biondii*. The Chinese vernacular name translates as "hope of spring," as the tree flowers in mid to late winter. The common name, Chinese Willow-leaf Magnolia, refers to its similarity to *M. salicifolia*, the Japanese Willow-leaf Magnolia, both having willow-like leaves. *M. biondii* is distinguished from *M. salicifolia* in its pubescent pedicels and in that the leaves do not carry an anise scent. (If these are in fact the only genuine differences between the two species, they should most likely be considered as two varieties of the same species.)

Seed of this species was included in the collection sent by Wilson to the Arnold Arboretum of Harvard University in 1908. The seed did not germinate, however, and various attempts have been made since that time to introduce the species into cultivation. Some of these attempts resulted in a few plants being cultivated, but there is no verified report of the species in cultivation before the 1970s.

In 1976, Dr. Yu Chen Ting, then associated with Boston College, visited his family in Henan Province, China, which is one of the native localities of *Magnolia biondii*. With the encouragement of Dr. August Kehr of the Magnolia Society and a contribution from the members of the society, Dr. Ting agreed to collect *M. biondii* seeds during his visit. He was frustrated in his collecting mission, interrupted by a serious earthquake. He returned to China in 1977 with two colleagues and was able to obtain about 100 seed of the species. Dr. Ting brought these to the United States and distributed half to the Arnold Arboretum at Harvard University and half to Dr. J. C. McDaniel, a horticulturist and researcher at the University of Illinois, Urbana, for propagation and distribution.

The Arnold Arboretum successfully germinated about twenty-five seedlings from that lot; the results from Dr. McDaniel's lot are unknown. Ten to fifteen seedlings from the Arnold Arboretum plants were distributed to nurseries, and rooted cuttings were later distributed to a handful of Magnolia Society members. August Kehr received one of these seedlings and in 1982 grafted a twig of *Magnolia biondii* onto *M. kobus*. He reports that the graft took easily and grew nicely. This grafted plant bloomed on 21 February 1986, possibly the first of this species to bloom in cultivation in North America. Currently this species remains rare in cultivation. Flowers February–March; fruits August–September. No cultivars have been selected from nor hybrids made with this species.

Magnolia cylindrica Wilson.

Journal of the Arnold Arboretum 8:109. 1927. Type specimen: *Ching 2949*. Herbarium of the Arnold Arboretum, Harvard University, Cambridge, Massachusetts. **Plate 13.**

Magnolia cylindrica is a small tree or large shrub to 30 ft. (9 m) tall with smooth, gray bark. Branchlets are brown and pubescent, becoming glabrous with age. Leaves are obovate to elliptic, 3–6 in. (7.6–15 cm) long, 1–2 in. (2.5–5 cm) wide, membranaceous, dark glossy green and glabrous above, and pubescent to nearly glabrous below. The leaf apex is acute to rounded, the base cuneate to attenuate. The petiole is pubescent, up to 1 in. (2.5 cm) long. The flowers are white, fragrant at night, and have 9 tepals. The outer 3 tepals are small, membranaceous and sepaloid, a little over 0.5 in (1.5 cm) long. The inner 6 tepals are 3–4 in. (7.6–10 cm) long, 1–2 in. (2.5–5 cm) wide, and spatulate. Stamens are about 0.25 in. (0.6 cm) long with pinkish filaments. The gynoecium is glabrous. Pedicels are densely pubescent. The follicetum is cylindric, 2–3 in. (5–7.6 cm) long, with a greenish color tinged with red. The outer seed coat is red.

This species is native to woods and shaded ravines at 2000–4600 ft. (600–1380 m) altitude in northern Fukien Province of China. It was first discovered in 1925 by R. C. Ching, collector for E. H. Wilson, and was named by Wilson in 1927. Since the specimens studied by Wilson bore only immature fruit, the epithet *cylindrica* probably refers to the shape of the fruit. The fruits in this species are usually not distorted in shape as they are in other species in this section.

Magnolia cylindrica is thought by some to be most closely related to *M. kobus* and *M. salicifolia*. There are conflicting reports as to whether *M. cylindrica* has the anise-scented wood of *M. salicifolia*. Wilson's description of the type specimen includes anise-scented bark, but many specimens in cultivation lack this fragrance.

McDaniel (1974) raised the question troubling many: is the *Magnolia cylindrica* of Western gardens the true species as Wilson described it? The plants grown as this species, while fine horticultural specimens, do not conform to Wilson's description. They differ in leaf size and texture, position and poise of flowers on the branches, follicetum size and shape, flower coloration, and other aspects. McDaniel suggests that what is grown in gardens is possibly a hybrid between *M. cylindrica* and a species of section *Yulania*, probably *M. denudata*. The *M. cylindrica* in Western cultivation resembles *M. denudata* as much as it resembles any species in section *Buergeria*. McDaniel's discussion is not only convincing but well founded in that the two species grow together in the wild and bloom at the same time. The fact that *M. cylindrica* and *M. denudata* easily cross also supports this theory. A determination will not be resolved until more specimens of *M. cylindrica* are collected from natural populations in China. Further hybridization studies would also be helpful.

Magnolia cylindrica was introduced into cultivation in 1936 with seed sent to Mrs. J. Norman Henry of Gladwyne, Pennsylvania, from the Lu-Shan Botanical Garden in China. Seeds were also sent to England at this time, but these failed to germinate. In 1950 scions from Mrs. Henry's plants were sent to Hillier Sons Nursery to introduce the species into England. Many cultivated specimens have proven surprisingly hardy, to Zones (4)5–9 in North America. Flowers April–May; fruits September–October. No cultivars have been selected for this species.

HYBRIDS

August Kehr, Hendersonville, North Carolina, has crossed *Magnolia cylindrica* with *M. liliiflora*, using *M. cylindrica* as the pollen parent. None of these hybrids has been named. The following *M. cylindrica* hybrids are in cultivation:

M. cylindrica × *M. denudata* 'Sawada's Pink' = 'Fireglow'
M. cylindrica × *M. veitchii?* = 'Albatross'

Magnolia subgenus *Yulania* section *Tulipastrum* (Spach) Dandy.

[*Tulipastrum* (Spach) Reichenbach, *Tulipastrum* Spach.] *Camellias and Magnolias*, Royal Horticultural Society Conference Report 74. 1950. TYPE: *Magnolia acuminata*.

The two species in section *Tulipastrum* are deciduous trees or shrubs. In *Magnolia acuminata* flowers expand with or at about the same time as the leaves. In *M. liliiflora* flowers emerge before the leaves and continue as leaves emerge. The flowers are yellow, greenish yellow, or purple, with 9 (sometimes to 18) tepals. The outer 3 tepals are much smaller and narrower than the others. Fruit is usually more or less distorted due to ovule abortion. These species are tetraploid with a chromosome number of $2n = 76$.

There are two species in this section. *Magnolia acuminata* is native to eastern North America and is represented in the southern portion of its range by var. *subcordata*. Both forms are cultivated. The second species, *M. liliiflora*, has long been cultivated in China and Japan. It is not known in the wild but is thought to be native to China. Although these species do not appear at first glance to be closely related, the vegetative and floral structure of the two plants are quite similar. This and section *Rhytidospermum* are the only sections containing both Asian and American species.

Magnolia acuminata L.

[*Kobus acuminata* Nieuwland, *Magnolia acuminata* var. *maxima* Lodd. ex Loud., *M. acuminata* var. *ludoviciana* Sarg., *M. acuminata* var. *candolii* DC., *M. decandollii* Savi, *M. candolii* (DC.) Link, *M. virginiana* ε *acuminata* L., *Tulipastrum acuminatum* (L.) Small, *T. acuminatum* var. *ludovicianum* (Sarg.) Ashe, *T. americanum* var. *vulgare* Spach.] *Systema Naturae* 10.2:1082. 1759. Cucumber Tree, Cucumber Magnolia, Blue Magnolia. Type specimen: *Clayton 404*. Herbarium of the British Museum, London. **Plate 1.**

Plants of *Magnolia acuminata* are deciduous trees or shrubs to 100 ft. (30 m) tall and 20 in. (51 cm) in diameter at breast height, with a pyramidal habit. The trunk is straight, the bark light brown to gray and furrowed. Branchlets are light brown to gray or reddish brown and glabrous to slightly pubescent. Buds are covered with silky, yellowish hairs. Leaves are elliptic, oblong, or ovate to obovate, 2–8 in. (5–20 cm) long, and 1–4 in. (2.5–10 cm) wide. They are glabrous to slightly pubescent above, pubescent to almost tomentose beneath, especially when young. Leaf apex is rounded to acute or acuminate; the base is rounded to cuneate, rarely almost cordate. Petioles are 0.5–2 in. (1.3–5 cm) long, greenish brown, and pubescent, becoming almost glabrous. The flowers are slightly fragrant, cup-shaped, and 2–4 in. (5–10 cm) across, with 9 (sometimes to 12) tepals. The inner tepals are glaucous, obovate to oblanceolate, 1–4 in. (2.5–10 cm) long, and 1–2 in. (2.5–5 cm) wide. Coloration is yellow or yellow-green, often with a purplish blue tinge. The outer 3 tepals are membranaceous, about 1 in. (2.5 cm) long, and reflexed. Stamens are about 0.5 in. (1.3 cm) long. The pedicel is brown, glabrous to slightly pubescent, and about 1 in. (2.5 cm) long. The follicetum is cylindric, oblong, 1–4 in. (2.5–10 cm) long, and about 1 in. (2.5 cm) wide, and glabrous. The outer seed coat is reddish orange. Chromosome number: $2n = 76$.

Magnolia acuminata is native to forests below 4000 ft. (1200 m) in eastern North America from New York State and southern Ontario south to the Florida panhandle, west to Arkansas and Louisiana. The species was first discovered by John Clayton, a county clerk and dedicated plant enthusiast, in Virginia in the 1700s. It was described by Catesby in 1739, and in 1759 Linnaeus named the species *Magnolia acuminata*, based on Catesby's description and plate. Earlier, in 1753, he had named the plant *M. virginiana* var. *acuminata*, also based on Catesby; therefore, he moved this plant from varietal to specific status.

The epithet *acuminata* refers to the acuminate leaf apex. The common names Cucumber Tree and Cucumber Magnolia refer to the fruit aggregates, which resemble small cucumbers, though fruit is often distorted because of incomplete pollination or ovule abortion. Blue Magnolia refers to the bluish cast of the glaucous tepals. This species is not common enough in the wild to be of much commercial importance. The straight trunks have been cut for use as troughs, and the light, soft wood is occasionally used in cabinetmaking.

Magnolia acuminata is a variable species, and there has been disagreement as to botanical varieties. Typical *M. acuminata* produces yellowish green flowers and glabrous twigs. The following five additional varieties appear in the literature.

> *Magnolia acuminata* var. *ludoviciana,* named by Sargent in 1919 to include plants in West Feliciana Parish of Louisiana having very large flowers and tomentose leaves. These characters do not appear to be consistent, so I have included these plants within typical *M. acuminata.*
>
> Variety *ozarkensis* was named by W. W. Ashe in 1926 to include those plants, especially in the Ozark Mountains, having leaves which are more glabrous than other plants of this species. This is a minor distinction, so I believe this group of plants is also best referred to typical *Magnolia acuminata.*
>
> Variety (or forma) *aurea* was named by Ashe in 1931 (as a variety) and by Hardin in 1954 (as a form) to include glabrous plants with flower color more yellow than typical. These are also referred to typical *Magnolia acuminata,* as the flower color is not a consistent character.
>
> Variety *alabamensis* was named by W. W. Ashe in 1931 to contain plants having pubescent twigs and greenish yellow flowers. This group, collected in Alabama by Ashe, is merely a western extension of *Magnolia acuminata* var. *subcordata* and is here included with that variety.
>
> Variety *subcordata* was named by Spach in 1839 to include the plants of *Magnolia acuminata* at the southern extreme of its distribution. Plants of *M. acuminata* var. *subcordata* have pubescent to tomentose twigs and yellow rather than greenish yellow flowers. This is the most commonly recognized variety and is the only one recognized here.

Magnolia acuminata was first introduced into cultivation around 1740 by John Bartram. He discovered the plant shortly after Catesby did, and several years later he sent seed to his friend Peter Collinson in London. It is cultivated mostly for its quality as a shade tree, bright pink-red fruit, and golden brown fall color. Its shaggy, furrowed bark is also interesting and unique among magnolias. The flowers resemble those of the tulip tree (*Liriodendron*) and are not particularly showy. It is one of the hardiest species of *Magnolia* so is often used as a rootstock for grafting less hardy species, or in breeding to produce hardy hybrids. Propagation from seed is not difficult, but seed set is usually low due to self-incompatibility and ovule abortion. Cuttings may be used for propagation, yet rooting percentage is low unless very juvenile material is used. Few cultivars have been selected for this species, and most of those are old clones. The range of variation of the species has not been explored to full advantage by plants breeders, so a greater selection effort would be rewarding. Flowers (April) May–June; fruits July–October. Zones 4–9.

CULTIVARS

'Busey' Hardier than the species, early maturing with yellow fall color. Named by J. C. McDaniel, Urbana, Illinois, in the early 1970s.

'Candollei' Flowers greenish. First named by de Candolle in 1824. It is doubtful that this form is still in cultivation.

'Dunlap' A hardy form first listed by Treseder's Nurseries, Cornwall, England, in 1965. Selected from a tree in Savoy, Illinois, probably by J. C. McDaniel of Urbana, Illinois.

'Excelsa' Larger in all ways than the species, with pale yellow flowers. Grown in France in 1857. Probably no longer in cultivation.

'Fertile Myrtle' Name given by Phil Savage, Bloomfield Hills, Michigan, to a particularly fertile form of the species. Not registered. Important in various magnolia breeding programs in North America.

'Gigantea' (*Magnolia gigantea* Hort. ex C. deVos) A vigorous form distributed by Ellwanger and Barry of Rochester, New York, in 1855. Possibly no longer in cultivation.

'Golden Glow' Golden yellow flowers on an upright tree. Selected from a tree in the Smoky Mountains, Sevier County, Tennessee, and registered by Dr. Frank Galyon in 1975.

'Klassen' Vigorous, large-flowered clone, self-fertile, with dark yellow fall color. Selected by J. C. McDaniel, Urbana, Illinois, in the early 1970s.

'Lanhydrock Clone' Free-flowering form first listed by Treseder's Nurseries, Cornwall, England, in 1965.

'Laser' A colchicine-induced 16x form (304 chromosomes) created and named by August Kehr, Hendersonville, North Carolina. Leaves larger and twigs thicker than typical. Produced from seedling of 'Fertile Myrtle'. Registered in 1991.

'Maxima' Growth more vigorous, leaves larger than typical; first listed by Loudon in 1838. Probably no longer in cultivation.

'Moyer Clone' Vigorous tree with large leaves, listed by Treseder's Nurseries, Cornwall, England, in 1965.

'Nelson' A "witches'-broom" form which originated before 1950 as a bud mutation on a large tree in Princeton, New Jersey, owned by William Nelson. The mutation was propagated by J. C. McDaniel in Urbana, Illinois, and the grafted plants maintain a bushy growth habit without flowering or reverting to normal *Magnolia acuminata*.

'Patriot' A colchicine-induced octaploid (8x = 152 chromosomes) created and named by August Kehr, Hendersonville, North Carolina. Leaves larger and twigs thicker than typical. Produced from seedling of 'Fertile Myrtle'. Registered in 1991.

'Philo' Highly self-compatible and produces a large crop of seeds. Also develops nice, yellow fall color. Selected from a tree in Philo, Illinois, by J. C. McDaniel, Urbana, Illinois. Registered in 1964.

'Rustica' Leaves narrower than those of species. Cultivated in Europe as early as 1777. Probably no longer in cultivation.

'Variegata'   Silvery white variegation on leaves. First mentioned by Bean (1951), possibly the same as plants sometimes known as *Magnolia acuminata* var. *aurea*.

HYBRIDS

The number of hybrids involving *Magnolia acuminata* has increased rapidly in the past few years, as breeders work toward producing a precocious magnolia with truly yellow flowers. *Magnolia acuminata* var. *subcordata* has been widely used in these breeding programs since the flowers are yellower; however, the offspring of such crosses are not as hardy as those in which the more northerly var. *acuminata* is the parent.

Working toward the goal of producing a yellow-flowering precocious form, Santamour (1976) published the results of experimental crosses using *Magnolia acuminata* as the seed parent and thirteen species, hybrids, and cultivars as the pollen parents. Hybrids resulted from at least six of the crosses.

Santamour also crossed *Magnolia acuminata* with *M. grandiflora* in hope of producing an evergreen magnolia with yellow flowers. His hybridizations were the first successful crosses between members of the two subgenera. The resulting seedlings had white flowers and were not significantly different from *M. grandiflora*. Plants were, however, hardier than that species and flowered at an early age. The crosses were only successful when *M. grandiflora* was used as the seed parent; reciprocal crosses failed. Santamour thought that possible maternal inheritance of yellow flower color, coupled with the lack of fruit set when *M. acuminata* is used as the maternal parent, frustrates making a cross yielding yellow flowers on an evergreen tree. Backcrossing these hybrids with *M. acuminata* might increase the amount of yellow pigmentation, but could inhibit the evergreenness expression as well.

Magnolia acuminata has been crossed with the following species and their cultivars: *M. campbellii, M. denudata, M. kobus, M. liliiflora, M. sargentiana, M. sprengeri, M.* × *kewensis, M.* × *loebneri, M.* × *soulangiana, M.* × *veitchii*. The following are hybrids available in the trade:

M. acuminata × (M. × brooklynensis 'Evamaria') = 'Yellow Bird'
M. acuminata × M. denudata = 'Butterflies', 'Elizabeth', 'Ivory Chalice', 'Sundance', 'Yellow Fever', and 'Yellow Garland'
M. acuminata × M. liliiflora = M. × brooklynensis
M. acuminata × (M. × veitchii 'Peter Veitch') = 'Curly Head'

Magnolia acuminata var. *subcordata* (Spach) Dandy.

[*Magnolia acuminata* subsp. *cordata* (Michx.) E. Murray, *M. acuminata* var. *alabamensis* Ashe, *M. acuminata* var. *cordata* (Michx.) Seringe, *M. cordata* Michx., *M. cordifolia* Page, *Tulipastrum americanum* var. *subcordatum* Spach, *T. cordatum* (Michx.) Small.] *American Journal of Botany* 51:1056. 1964. Yellow Cucumber Tree. Type specimen: *Michaux, Magnolia lutea fol acuminatis florib. luteis*. Herbarium of the Musée National d'Histoire Naturelle, Paris. **Plate 2.**

Magnolia acuminata var. *subcordata* is native to the southeastern United States, the piedmont and coastal plain of North Carolina, Georgia, and Alabama. It differs from typical *M. acuminata* in its shorter height at maturity, pubescent twigs, and leaves and flower color. Leaves of var. *subcordata* are smaller and more broadly ovate, with a more rounded apex and a more or less cordate base. Its flowers are canary-yellow, not yellow-green as in the typical variety. Variety *subcordata* is usually a shrub or small tree, rarely reaching 30 ft. (10 m) tall.

This variety was initially discovered by Michaux near Augusta, Georgia, about 1790. Michaux introduced the plant into cultivation in 1803. So few plants were found in the wild after this initial discovery that this variety was thought to be extinct except for cultivated specimens. In 1910, however, it was rediscovered by Louis A. Berckmans near the original sites of Michaux, and it has since been found to be rather widespread.

When Michaux discovered this plant, he named it *Magnolia cordata*. Thirty-five years later, Loudon was the first to suggest that it might be a variety of *M. acuminata*, although he did not propose a varietal name. The following year, Spach also suggested varietal status, naming the plant var. *subcordatum* of *Tulipastrum americanum*, his name for what is now *M. acuminata*. Since the synonym *Tulipastrum americanum* is no longer correct, Dandy published the correct name, *Magnolia acuminata* var. *subcordata*, in 1962. Although this plant was known as var. *cordata* for some time, the correct varietal name is derived from Spach's var. *subcordatum*. The -*um* ending reflects the change in gender from Tulipastr*um* to Magnoli*a*. This varietal term refers to the occasionally subcordate leaf base, also found in typical *M. acuminata*.

Grafted plants of *Magnolia acuminata* var. *subcordata* were introduced into cultivation in France and England by Michaux in 1803. Treseder (1978) reports that all the older European specimens are believed to have been propagated from Michaux's original introduction. Seeds were introduced to England by John Lyon and John Fraser separately in 1801, and seedlings were grown at their respective nurseries, although their subsequent development is unrecorded. Variety *subcordata* is cultivated principally for its yellow flowers, but also for its smaller form and shrubby habit. Being the more southerly form, it is not as hardy as typical *M. acuminata*. Flowers May–June; fruits August–September. Zones 6–8.

CULTIVARS

'Ellen' Leaves spotted with yellow, habit upright. Selected from seedlings by A. J. Fordham of the Arnold Arboretum, Cambridge, Massachusetts. Registered by Fordham about 1980.

'Miss Honeybee' Larger, light yellow, fragrant flowers opening earlier in the season. A more vigorous form. Registered about 1970 by James Merrill Nurseries of Painesville, Ohio.

'Skyland's Best' Flowers yellow, about 6 in. (15 cm) across. Blooms early spring, about six weeks after *Magnolia denudata*, and again in late summer. Foliage is dark green. Selected by Richard Figlar of Pomona, New York, and registered about 1980. **Plate 3.**

'Striata' Tepals striated with rose coloration. This form was in cultivation about 1915, though it is doubtful that the cultivar is still available.

HYBRIDS

Like typical *Magnolia acuminata*, var. *subcordata* has been used with increasing frequency in breeding programs in the search for a precocious yellow magnolia. This variety is not as hardy as typical *M. acuminata*, but its flowers have a deeper yellow color. Variety *subcordata* has been crossed with cultivars of *M. denudata*, *M. × soulangiana*, and *M. × brooklynensis*. Phil Savage of Bloomfield Hills, Michigan, crossed variety *subcordata* 'Miss Honeybee' with *Michelia figo*, a close magnolia relative, to produce the intergeneric hybrid named × *Yuchelia* (for *Yulania* and *Michelia*) in 1989. Other named hybrids include

M. acuminata var. *subcordata* 'Miss Honeybee' × *M. denudata* 'Sawada's Cream' = 'Goldfinch'

M. acuminata var. *subcordata* × (*M.* × *soulangiana* 'Alexandrina') = 'Yellow Lantern'

Magnolia liliiflora Desr.

[*Buergeria obovata* Sieb. & Zucc., *Gwillimia purpurea* (Curtis) C. deVos, *Lassonia quinquepeta* Buc'hoz, *Magnolia denudata* var. *purpurea* (Curtis) C. K. Schneider, *M. denudata* var. *typica* C. K. Schneider, *M. discolor* Vent., *M. glauca* var. *β* Thunb., *M. obovata* Thunb., *M. obovata* var. *discolor* DC., *M. obovata* var. *liliiflora* (Desr.) DC., *M. obovata* var. *purpurea* (Curtis) Sweet, *M. purpurea* Curtis, *M. quinquepeta* (Buc'hoz) Dandy, *M. soulangiana* var. *nigra* W. Robinson, *Mokkwuren* Kampf., *M. flore albo* Banks, *Talauma obovata* Hance, *T. sieboldii* Miguel, *Yulania japonica* var. *purpurea* (Curtis) Spach.] In Lamarck, *Encyclopédie Méthodique, Botanique* 3: 675. 1791. Purple Magnolia, Lily-flowered Magnolia, Mulan. Type specimen: table 44, *Mokkwuren 2* in Kaempfer 1791. **Plate 52.**

Magnolia liliiflora is a deciduous shrub or small tree to 15 ft. (4.6 m) tall with a rounded habit. The twigs are light gray to brown and glabrous. Leaves are elliptic to obovate, 4–8 in. (10–20 cm) long, and 2–4 in. (5–10 cm) wide. They are dark green, glabrous on both sides, but occasionally pubescent when young. Leaf apex is acuminate, the base cuneate or attenuate. The petiole is about 0.5 in. (1.3 cm) long, dark brown, and glabrous. Flowers are slightly fragrant, vase shaped, 4–5 in. (10–12.7 cm) in diameter, 9 (sometimes to 18) tepals. Inner tepals are obovate to spatulate, 3–5 in. (7.6–12.7 cm) long, 1–2 in. (2.5–5 cm) wide, purple on the outside and white inside. The outer 3 tepals are 1–2 in. (2.5–5 cm) long, about 0.5 in. (1.3 cm) wide, and purplish green. Stamens are about 0.5 in. (1.3 cm) long and reddish purple. The gynoecium is glabrous. The pedicel is less than 0.5 in. (1.3 cm) long, covered with yellowish pubescence or glabrous. The follicetum is cylindric, 1–2 in. (2.5–5 cm) long, and deep purple to brown. Chromosome number: $2n = 76$.

Magnolia liliiflora is widespread in cultivation, especially in Asia, but is not known in the wild. Accounts of wild plants in the literature usually refer to escaped or abandoned cultivated plants. The best estimate of its true native range is eastern China.

The nomenclatural history of *Magnolia liliiflora* is long and complex. It is tied with that of *M. denudata*, and the reader is referred to the discussion under *M. denudata* for a full account. The synonym *M. quinquepeta* is rejected for reasons given in that account, and the correct name of this species is *M. liliiflora*, named by Desrousseaux based on Kaempfer's engravings.

Magnolia liliiflora has been cultivated in Asia for centuries. It was introduced from Japan into English gardens in 1790 by the Duke of Portland and quickly became popular. It is especially prized for its deep purple, precocious blooms and its shrubby habit. It is a small tree and therefore suitable for many home landscapes. Some pruning may be required for shaping. The typical species is not as hardy as others such as *M. denudata*, *M. kobus*, or *M.* × *soulangiana*; however, the flowers are borne later and therefore often escape spring frost damage. 'Nigra' and 'O'Neill' are two cultivars which are hardier than typical. *Magnolia liliiflora* is sometimes used as a dwarfing rootstock for other species, especially *M. denudata*; and although it seldom sets seed in cultivation, it is easily propagated by cuttings and layering. Flowers April–May; fruits July–August. Zones 7–9; 'Nigra' and 'O'Neill' are hardy to Zone 6.

CULTIVARS

'Arborea'   More robust than the species; leaves slightly smaller with rusty reddish tinge. Cultivated as early as 1891.

'Darkest Purple'   Narrow tepals of rich purple color. Listed by Overlook Nurseries of Mobile, Alabama, in 1949 as a hybrid but reassigned as a selected cultivar of this species by William Kosar of the U.S. National Arboretum.

'Gracilis'   A small shrub with small, deep purple flowers and narrow tepals. The exterior of the tepals is uniformly colored. It was originally named *Magnolia gracilis* by Salisbury in 1807, later *M. liliiflora* var. *gracilis* by Rehder, and still later simply the cultivar 'Gracilis' by Rehder (1968).

'Holland Red' ('Norway Red')   Flowers deep reddish purple. Blooms later than most cultivars. Origin unknown. **Plate 53.**

'Inodora'   Flowers large, pale purple, without fragrance. Originally named *Magnolia inodora* by de Candolle (1817) and *M. liliiflora* var. *inodora* by Pampanini (1916). Varietal status is not warranted, so 'Inodora' is considered a cultivar.

'Lyons'   Upright, reddish purple flowers similar to those of 'O'Neill', but tepals twisted. Originated at Hendricks Park, Eugene, Oregon, and offered by Gossler Farms Nursery, 1989.

'Mini Mouse'   A dwarf form with leaves 2 in. (5 cm) long. Flowers 3 in. (8 cm) across and dull purple. Bred by Oswald Blumhardt of New Zealand in the late 1970s by crossing 'Nigra' and typical *Magnolia liliiflora.*

'Nigra'   This is the most popular cultivar of *Magnolia liliiflora*. It has deep purple flowers and is probably the hardiest cultivar of this species (Zone 6). It was named by Vietch in 1861. 'Nigra' has often been incorrectly attributed to *M.* × *soulangiana.* Several different forms are sold as this cultivar, but the original clone was described as having tepals which are dark purple both inside and outside. Few plants sold in North America match that description.

'Norway Red'   see 'Holland Red'

'O'Neill'   This is a hardy form selected by J. C. McDaniel in 1973 from a tree owned by Mabel O'Neill in Champaign, Illinois. It is very much like 'Nigra', but the tepals are slightly larger and darker, and it blooms slightly later in the season. The flowers are dark wine-red outside, lighter inside with dark veins. Zone 5. **Plate 54.**

'Purpurea'   This cultivar is probably the same as typical *Magnolia liliiflora*. It was first published in *Curtis's Botanical Magazine* in 1797 as *M. purpurea,* now considered a synonym for *M. liliiflora*. The original description of *M. purpurea* includes ovate tepals; specimens of the typical species as grown today generally show narrower, obovate tepals. Bean (1951) lists the cultivar as having larger flowers with deeper purple coloration at the base of the tepals than at the tip (compare with 'O'Neill'). Treseder (1978; see "Magnolia Books," chapter references) also notes this fading of color from the base to the tip of the tepals.

'Reflorescens'   Large, dark purple flowers in spring and again in August or September. This flowering habit is not uncommon in other forms of the species. 'Reflorescens' may be the same as typical *Magnolia liliiflora*. Introduced ca. 1850.

'Trewithen'   Large flowers, purple outside, pink inside. This form was first illustrated by Johnstone (1955; see "Magnolia Books," chapter references) then described by Krüssmann in 1961. It is probably not available in North America.

HYBRIDS

In 1981 Dr. Frank Santamour of the U.S. National Arboretum published results of his hybridization of *Magnolia grandiflora* with *M. liliiflora.* This cross was made in an effort to produce an evergreen magnolia with purple flowers. Santamour used *M. liliiflora* 'Darkest Purple' as the pollen parent. The resulting hybrids were evergreen but had white flowers and were difficult to distinguish from *M. grandiflora*. The hybrids are, however, more vigorous than either parent and probably hardier than *M. grandiflora*.

Several horticulturally important hybrids have been produced with *Magnolia liliiflora* as one of the parents, for it often brings hardiness and/or purple flower color to the resulting hybrid. An unnamed hybrid between *M. cylindrica* and *M. liliiflora* has been created by August Kehr, Hendersonville, North Carolina. The following are named hybrids:

> *M. acuminata* × *M. liliiflora* = *M.* × *brooklynensis*
> *M. campbelli* × *M. liliiflora* = 'Early Rose', 'Star Wars'
> *M. campbelli* var. *mollicomata* × *M. liliiflora* 'Nigra' = 'Caerhays Surprise'
> *M. campbelli* var. *mollicomata* 'Lanarth' × *M. liliiflora* = 'Apollo', 'Vulcan'
> *M. denudata* × *M. liliiflora* = *M.* × *soulangiana*
> *M. kobus* × *M. liliiflora* 'Nigra' = 'Marillyn'
> *M. kobus* var. *stellata* × *M. liliiflora* = 'Ann', 'Betty', 'George Henry Kern', 'Jane', 'Judy', 'Orchid', 'Pinkie', 'Randy', 'Ricki', 'Susan'
> *M. liliiflora* × 'Mark Jury' = 'Serene'
> *M. liliiflora* 'Darkest Purple' × (*M.* × *soulangiana* 'Lennei') = 'Purple Prince', 'Purple Princess'
> *M. liliiflora* 'Nigra' × (*M.* × *soulangiana* 'Lennei') = 'Fenicchia Hybrid'
> *M. liliiflora* 'Nigra' × (*M.* × *soulangiana* 'Rustica Rubra') = 'Dark Splendor', 'Orchid Beauty', 'Red Beauty'
> *M. liliiflora* 'Darkest Purple' × *M. sprengeri* 'Diva' = 'Raspberry Swirl'
> *M. liliiflora* 'Nigra' × *M. sprengeri* 'Diva' = 'Galaxy', 'Northwest', 'Spectrum'
> *M. liliiflora* × (*M.* × *veitchii*) = Gresham Hybrids; see chapter 9 for a listing.
> *M. liliiflora* × ? = 'Osaka'

REFERENCES AND ADDITIONAL READING

The Magnolia Society Journals

A primary source of information on magnolias is the collection of newsletters/journals put out by the Magnolia Society. Founded in 1963, the society has published two issues of the newsletter/journal per year (four issues in 1973), each filled with information for the magnolia grower. From 1964 to 1979 (volumes 1–15) the issues were published as *Newsletter of the American Magnolia Society*; from 1980 to 1984 (volumes 16–20) as *Magnolia, Journal of the American Magnolia Society*; from 1985 (volume 21) to present as *Magnolia, Journal of the Magnolia Society*. In this book, references to all issues are given the title *Journal of the Magnolia Society*. Only a few particularly pertinent articles, or those cited in the text, are listed in the references in this book, but all the issues are very informative. For information on availability of back issues or membership in the society, contact Phelan A. Bright, Secretary-Treasurer, the Magnolia Society, 907 S. Chestnut Street, Hammond, LA, 70403-5102, U.S.A.

Special Magnolia Issues

Several journals have published special issues on magnolias. These include the following:

Arboretum Bulletin, Winter 1947. University of Washington Arboretum Foundation, Seattle, WA. Vol. 10, No. 4.

Arnoldia, March/April 1981. Arnold Arboretum, Harvard University, Jamaica Plain, MA. Vol. 41, No. 2.

Journal of the California Horticultural Society, January 1962. California Horticultural Society, California Academy of Sciences, Golden Gate Park, San Francisco, CA. Vol. 23, No. 1.

Magnolia Books

To date, five excellent books on magnolias have been published. All but the first are currently out-of-print.

Gardiner, James M. 1989. *Magnolias.* Chester, CT: Globe Pequot Press. A book for general readership; beautiful photographs, limited scope.

Johnstone, George H. 1955. *Asiatic Magnolias in Cultivation.* London: The Royal Horticultural Society. A complete and beautifully illustrated book on the Asian species in cultivation.

Millais, John G. 1927. *Magnolias.* London: Longmans, Green and Company. The first book to bring together information on magnolias, both cultivated and uncultivated species.

Treseder, Neil G. 1978. *Magnolias.* London: Faber and Faber. A thorough and excellent source of information and history for the cultivated species.

Treseder, Neil G., and Marjorie Blamey. 1981. *The Book of Magnolias.* London: Collins Sons and Company. Contains about thirty beautiful full-page (some life-size) illustrations of magnolias with general information on those species and on cultivation.

Manuscript of Dr. James E. Dandy

Dr. James E. Dandy, whose name will be immediately recognized by many magnoliaphiles, worked on a revision of the magnolia family for almost fifty years. Much of this work remained unpublished at the time of his death in 1976. His original manuscript and notes are under the care of the Botany Department, British Museum of Natural History, London. A photocopy is available at the library of the United States National Arboretum, Washington, D.C.

Additional References

In addition to the above, the following are recommended reading on various subjects discussed in this chapter.

Agababian, V. S. 1972. Pollen morphology of the family Magnoliaceae. *Grana* 12:166–176.

Andrews, Frank Marion. 1901. Karyokinesis in *Magnolia* and *Liriodendron* with special reference to the behavior of chromosomes. *Botanisches Centralblatt—Beiheft* 2, 11:134–142.

Andrews, H. 1802. *Magnolia purpurea. The Botanist's Repository* 4:t 226. [*M. coco.*]

Ashe, W. W. 1917. Notes on trees and shrubs. *Bulletin of the Charleston Museum* 13(4):28. [*M. acuminata* var. *subcordata.*]

_____. 1926. Notes on woody plants. *Journal of the Elisha Mitchell Scientific Society* 41:267–269. [*M. acuminata.*]

_____. 1927. *Magnolia cordata* and other woody plants. *Bulletin of the Torrey Botanical Club* 54:579–582. [*M. acuminata* var. *subcordata.*]

_____. 1928. Notes on southeastern woody plants. *Bulletin of the Torrey Botanical Club* 55:463–466. [*M. virginiana.*]

_____. 1931. Notes on *Magnolia* and other woody plants. *Torreya* 31:37–41. [*M. acuminata, M. grandiflora,* and *M. virginiana* var. *australis.*]

Badger, Kemuel, and Marion T. Jackson. 1983. A study of the natural history of naturally occurring populations of *Magnolia tripetala* in Indiana. *Proceedings of the Indiana Academy of Science* 93:309–312.

Bailey, Liberty Hyde. 1921. *Magnolia stellata. Addisonia* 6(3):37–38, plate 211.

Baranova, Margarita. 1972. Systematic anatomy of the leaf epidermis in the Magnoliaceae and some related families. *Taxon* 21(4):447–469.

Bartrum, Douglas. 1957. Magnolias. In *Rhododendrons and Magnolias.* England: John Gifford Ltd. Pp. 111–171.

Bean, W. J. 1932a. Magnolias, Part 1. In *New Flora and Sylva* 4:232–240.

_____. 1932b. Magnolias, Part 2. In *New Flora and Sylva* 5:11–19.

_____. 1951. *Magnolia.* In *Trees and Shrubs Hardy in the British Isles,* vol. 2. London: John Murray. Pp. 272–289.

Blackburn, Benjamin C. 1954. *Magnolia kobus* forma *stellata. Popular Gardening* 5(3):73.

_____. 1955. A question about Shidekobushi: a reexamination of *Magnolia stellata. Amatores Herbarii* 17:1–2.

_____. 1957. The early-flowering magnolias of Japan. *Baileya* 5(1):3–13. [*M. kobus,* var. *stellata,* and var. *loebneri.*]

Brush, Warren D. 1946. Southern Magnolia. *American Forests* 52:32–33. [*M. grandiflora.*]

_____. 1956. Our native magnolias. *American Forests* 62(4):31–32, 58–59. [North American species.]

Canright, James E. 1952. The comparative morphology and relationships of the Magnoliaceae. I. Trends of specialization in the stamens. *American Journal of Botany* 39(7):484–497.

_____. 1953. The comparative morphology and relationships of the Magnoliaceae. II. Significance of the pollen. *Phytomorphology* 3(3):355–365.

_____. 1955. The comparative morphology and relationships of the Magnoliaceae. IV. Wood and nodal anatomy. *Journal of the Arnold Arboretum* 36(2/3):120–140.

_____. 1960. The comparative morphology and relationships of the Magnoliaceae. III. Carpels. *American Journal of Botany* 47(2):145–155.

_____. 1963. Contributions of pollen morphology to the phylogeny of some Ranalean families. *Grana Palynologica* 4(1):64–72.

Chapman, Douglas J. 1981. Magnolia cultivars flower from April through summer. *Weeds, Trees and Turf* 20(5):46, 48.

Chien, S. S., and W. C. Cheng. 1934. An enumeration of vascular plants from Chekiang, III. *Contributions from the Biological Laboratory of the Science Society of China* 9(3):279–285. [*M. amoena.*]

Chun, W. Y. 1963. Genus speciesque novae Magnoliacearum Sinensium. *Acta Phytotaxonomica Sinica* 8(4):281–286. [*M. lotungensis,* p. 285.]

_____ . 1964. A new Hainanese magnolia. *Acta Phytotaxonomica Sinica* 9(2):117–118. [*M. albosericea.*]

Cibot, P. M. 1778. Le Yu-lan. In *Mémoires concernant l'histoire, les sciences . . . de Chinois, par les missionaires de Pe-kin,* vol. 3. Pp. 441–443.

Coker, William Chambers. 1943. *Magnolia cordata. Journal of the Elisha Mitchell Scientific Society* 59(1):81–88. [*M. acuminata* var. *subcordata.*]

Cowan, John MacQueen. 1938. *Magnolia globosa. New Flora and Silva* 10:272–274.

Craib, W. G. 1929. Contributions to the flora of Siam. *Kew Bulletin* 1929:105–106. [*M. craibiana.*]

Dallimore, W. 1925. The magnolias. *Quarterly Journal of Forestry* 19:139–148.

Dandy, James E. 1927. Key to the species of *Magnolia. Journal of the Royal Horticultural Society* 52:260–264.

_____ . 1927b. The genera of Magnolieae. *Kew Bulletin.* 1927(7):257–264.

_____ . 1928a. *Magnolia sinensis* and *M. nicholsoniana. Journal of the Royal Horticultural Society* 53:115.

_____ . 1928b. Malayan Magnolieae. *Kew Bulletin* 1928:183–193. [*M. aequinoctialis, M. pachyphylla, M. persuaveolens, M. pulgarensis.*]

_____ . 1928c. New or noteworthy Chinese Magnolieae. *Notes of the Royal Botanic Garden, Edinburgh* 16:123–132. [*M. biondii, M. campbellii, M. championii, M. henryi, M. sargentiana, M. sinensis, M. wilsonii.*]

_____ . 1930a. A new magnolia from Honduras. *Journal of Botany,* British and Foreign 68:146–147. [*M. yoroconte.*]

_____ . 1930b. New Magnolieae from China and Indo-China. *Journal of Botany,* British and Foreign 68:204–214. [*M. annamensis, M. clemensiorum, M. nana, M. paenetalauma, M. talaumoides, M. thamnodes.*]

_____ . 1934. The identity of *Lassonia* Buc'hoz. *Journal of Botany,* British and Foreign 72:101–103. [*M. liliiflora* and *M. denudata* nomenclature.]

_____ . 1936. *Magnolia globosa. Curtis's Botanical Magazine* 159:9467.

_____ . 1948a. *Magnolia dawsoniana. Curtis's Botanical Magazine* 164:9678,9679.

_____ . 1948b. *Magnolia nitida. Curtis's Botanical Magazine* 165:16.

_____ . 1950a. A survey of the genus *Magnolia* together with *Manglietia* and *Michelia.* In *Camellias and Magnolias: Report of the Royal Horticultural Society Conference.* Pp. 64–81.

_____ . 1950b. The Highdown Magnolia. *Journal of the Royal Horticultural Society* 75:159–161. [*M. wilsonii.*]

_____ . 1962. Magnoliaceae. In Flora of Panama. *Annals of the Missouri Botanical Garden* 49:485–488. [*M. sororum.*]

_____ . 1964. *Magnolia acuminata* var. *subcordata.* In S. Tucker, Terminal ideoblasts in Magnoliaceous leaves. *American Journal of Botany* 51(10):1051–1062.

_____ . 1965. *Magnolia virginiana. Curtis's Botanical Magazine* 175:457.

_____ . 1973. *Magnolia hypoleuca. Baileya* 19(1):44.

_____ . (unpublished). Manuscript and notes on the Magnoliaceae. See information at the beginning of these references.

Daubenmire, Rexford. 1990. The *Magnolia grandiflora–Quercus virginiana* forest of Florida. *American Midland Naturalist* 123:331–347.

de Spoelberch, Philippe. 1988. *Magnolia campbellii* var. *alba* in Bhutan. *Journal of the Magnolia Society* 24(1):13–17.

The deciduous magnolias. *Gardeners Chronicle* 9 May 1891:590–591.

Del Tredici, Peter. 1981. *Magnolia virginiana* in Massachusetts. *Arnoldia* 41(2):36–49.

_____ . 1983. *Magnolia zenii. Journal of the Magnolia Society* 19(1):19.

_____. 1987. Cultivar registrations, 1984–1986. *Journal of the Magnolia Society* 22(2):11–12.

_____. 1989a. Magnolia cultivar registrations: Supplement III, 1987–1989. *Journal of the Magnolia Society* 25(1):19–20.

_____. 1989b. A world of magnolias. *Horticulture* 67(3):44–52.

Del Tredici, P., and Stephen A. Spongberg. 1989. A new magnolia blooms in Boston. *Arnoldia* 49(2):25–27. [*M. zenii.*]

Dirr, Michael A. 1985. *Magnolia grandiflora. American Nurseryman* 162(9):138.

Dodd, Tom. 1980. Paying a call on *dealbata. Journal of the Magnolia Society* 16(1):29–32.

Dudley, Theodore R. 1983. *Magnolia zenii:* a rare magnolia recently introduced into cultivation. *Journal of the Magnolia Society* 19(1):20–22.

Dunn, S. Troyte. 1903. New Chinese plants. *Journal of the Linnean Society of London (Botany)* 35:484. [*M. henryi.*]

Earle, T. T. 1938. Origin of the seed coats in *Magnolia. American Journal of Botany* 25:221–222.

Egolf, Robert L. 1971. *Magnolia coco. Journal of the Magnolia Society* 8(1):10–11.

Eiland, Robert. 1988. The story of 'Monland'—the Timeless Beauty™ Magnolia. *Journal of the Magnolia Society* 23(2):7–8.

Elwes, Henry John, and Augustine Henry. 1912. *Magnolia.* In *Trees of Great Britain and Ireland,* vol. 6. Edinburgh: Privately Printed. Pp. 1581–1599.

Felger, Richard S. 1971. The distribution of *Magnolia* in northwestern Mexico. *Journal of the Arizona Academy of Science* 6(4):251–253. [*M. schiedeana.*]

Figlar, Richard B. 1982. *Magnolia splendens:* Puerto Rico's lustrous magnolia. *Journal of the Magnolia Society* 18(1):13–16.

_____. 1984a. *Magnolia portoricensis:* Puerto Rico's other magnolia. *Journal of the Magnolia Society* 19(2):1–3.

_____. 1984b. *Magnolia splendens. Journal of the Magnolia Society* 20(1):23–24.

_____. 1986. Magnolias as houseplants. *Journal of the Magnolia Society* 21(2):1–5.

_____. 1988. New cultivars of *Magnolia grandiflora. Journal of the Magnolia Society* 23(2):1–5.

Findlay, T. H. 1952. Notes on certain magnolias planted in Windsor Great Park. *Journal of the Royal Horticultural Society* 77:43–46.

Fogg, John M. 1961. The temperate American magnolias. *Bulletin of the Morris Arbortum* 12(4):51–58.

_____. 1976. 1976 Registrations. *Journal of the Magnolia Society* 12(2):3.

Fogg, John M., and Peter Del Tredici. 1984. Recently registered magnolia cultivars. *Journal of the Magnolia Society* 20(1):15–20.

Fogg, John M., and Joseph C. McDaniel. 1975. *Check List of the Cultivated Magnolias.* Mt. Vernon, VA: The American Horticultural Society Plant Sciences Data Center. [A list of the cultivars in cultivation up to 1975, including descriptions, when and where selected, and by whom. Extremely informative and thorough.]

Freeman, Oliver M. 1937. A new hybrid magnolia. *National Horticulture Magazine* 16(3):132–135. [*M. grandiflora* × *M. virginiana.*]

Furmanowa, Miroslawa, and Joanna Jozefowicz. 1980. Alkaloids as taxonomic markers in some species of *Magnolia* and *Liriodendron. Acta Societatis Botanicorum Poloniae* 49(4):527–535.

Galyon, Frank B. 1966. Sharp's Magnolia: a new discovery. *Bulletin of the University of Tennessee Arboretum Society* Winter 1965–66. [*M. sharpii.*]

German, Eugene R. 1979. *Magnolia sharpii. Journal of the Magnolia Society* 15(2):21–22.

Gossler, James. 1975. *Magnolia sprengeri* 'Diva'. *The American Horticulturist* 54(1):14.

Grier, N. M. 1917. Note on fruit of Mountain Magnolia. *Rhodora* 19(226):256. [*M. acuminata.*]

Gresham, D. Todd. 1964 An appreciation of *Magnolia campbellii* subspecies *mollicomata*. *Bulletin of the Morris Arboretum* 15: 29–31.

_____. 1966. *Magnolia wilsonii* × *M. globosa:* a new hybrid. *Bulletin of the Morris Arboretum* 17(4):70–73.

Griesel, Wesley O., and Jacob B. Biale. 1958. Respiratory trends in perianth segments of *Magnolia grandiflora*. *American Journal of Botany* 45(9):660–663.

Hadfield, M. 1966. The Yulan and its offspring. *Gardeners Chronicle* 3, 159:277–278. [*M. denudata.*]

Hara, Hiroshi. 1977. Nomenclatural notes on some Asiatic plants, with special reference to Kaempfer's *Amoenitatum Exoticarum*. *Taxon* 26(5/6):584–587. [Background information for *M. liliiflora* and *M. denudata* nomenclature.]

Hardin, James W. 1954. An analysis of variation within *Magnolia acuminata*. *Journal of the Elisha Mitchell Scientific Society* 70:298–312.

_____. 1972. Studies of the southeastern United States flora III. Magnoliaceae and Illiciaceae. *Journal of the Elisha Mitchell Scientific Society* 88(1):30.

Hardin, James W., and Kimberly A. Jones. 1989. Atlas of foliar surface features in woody plants, X. Magnoliaceae of the United States. *Bulletin of the Torrey Botanical Club* 116(2):164–173.

Harvill, A. M. 1964. The magnolias of Virginia. *Castanea* 29:186–188.

_____. 1964. *M. grandiflora* in Gray's Manual range. *Rhodora* 66:159.

Hernandez-Cerda, Maria E. 1980. Magnoliaceae. In *Flora de Veracruz*. Fasciculo 14. [*M. macrophylla* var. *dealbata, M. grandiflora.*]

Hillier, Harold G., and C. R. Lancaster. 1975. Magnolias in the Hillier Gardens and Arboretum. In *Rhododendrons 1975 with Magnolias and Camellias*. London: The Royal Horticultural Society. Pp. 61–69.

Holman, Nigel. 1973. Asiatic magnolias in a Cornish garden. In *Rhododendrons 1973 with Magnolias and Camellias*. London: The Royal Horticultural Society. Pp. 67–75.

_____. 1979. *Magnolia heptapeta* et alia? *The Plantsman* 1(1):56–61. [*M. liliiflora* and *M. denudata* nomenclature.]

Hooker, Joseph D. 1885. *Magnolia campbellii*. *Curtis's Botanical Magazine* 111:6793.

_____. 1895. *Magnolia parviflora*. *Curtis's Botanical Magazine* 121:7411. [*M. sieboldii*.]

Hopkins, Harold C. 1975. Viva *dealbata*. *Journal of the Magnolia Society* 11(2):8–9. [*M. macrophylla* var. *dealbata.*]

_____. 1989. *Magnolia sinensis:* the pretty little bush. *Journal of the Magnolia Society* 24(2):13–17.

Howard, Richard A. 1948. The morphology and systematics of the West Indian Magnoliaceae. *Bulletin of the Torrey Botanical Club* 75(4):335–357. [West Indian species in section *Theorhodon.*]

_____. ed. 1980. Wilson's Magnolias. *Journal of the Magnolia Society* 16(2):3–26. [A collection of Wilson's notes annotated by Howard.]

Hume, Harold H. 1961. Variations in *Magnolia grandiflora*. *Bulletin of the Morris Arboretum* 12(2):15–16.

Inami, Kazuo. 1959. Distribution of *Magnolia stellata*. *Amatores Herbarii* 20(1):10–14.

Introduce new magnolia. *American Nurseryman* 118(6)(1963):13. [*M. grandiflora* 'Majestic Beauty'.]

Janaki Ammal, E. K. 1953. The race history of magnolias. *Indian Journal of Genetics and*

Plant Breeding 12:82–92. [Chromosome numbers.]

Johnson, A. T. 1948. Magnolias: their variety and garden merit. In *Rhododendrons, Azaleas, Magnolias, Camellias and Cherries.* London: My Garden Publishers. Pp. 39–54.

Johnson [Callaway], Dorothy L. 1987. *Magnolia sieboldii. American Nurseryman* 166(8):150.

————. 1989a. Species and cultivars of the genus *Magnolia* (Magnoliaceae) cultivated in the United States. Unpublished Master's Thesis, Cornell University, Ithaca, New York.

————. 1989b. Nomenclatural changes in *Magnolia. Baileya* 23(1):55–56. [*M. macrophylla* var. *ashei* and var. *dealbata*.]

Johnson, M. A. 1953. Relationship in the Magnoliaceae as determined by the precipitin reaction. *Bulletin of the Torrey Botanical Club* 80(4):349–350.

Johnson, M. A., and David E. Fairbrothers. 1965. Comparison and interpretation of serological data in the Magnoliaceae. *Botanical Gazette* 126(4):260–269.

Johnstone, George H. 1948. Magnolias. In W. Arnold-Forster, *Shrubs for the Milder Counties.* New York and London: Charles Scribner's Sons. Pp. 228–248.

————. 1950a. Chinese magnolias in cultivation. In *Camellias and Magnolias: Report of the Royal Horticultural Society Conference.* London: The Royal Horticultural Society. Pp. 53–63.

————. 1950b. The eastern magnolias. In *Camellias and Magnolias: Report of Royal Horticultural Society Conference.* London: The Royal Horticultural Society. Pp. 44–52.

Kaempfer, Englebert. Joseph Banks, ed. 1791. *Icones selectae plantarum quas in japonica collegit et delineavit Engelbertus Kaempfer; ex archetypis in museo brittannico asservatis.* Plates 43, 44. Londini.

Kapil, R. N., and N. N. Bhandari. 1964. Morphology and embryology of *Magnolia. Proceedings of the National Institute of Sciences of India* 30:245–262.

Kehr, August E. 1986. *Magnolia biondii,* the "Hope of Spring." *Journal of the Magnolia Society* 22(1):7–10.

————. 1991. Magnolia improvement by polyploidy. *Journal of the Magnolia Society* 26(2):26–28.

Kennedy, George G. 1916. Some historical data regarding the Sweet Bay and its station on Cape Ann. *Rhodora* 18(214):205–212. [*M. virginiana*.]

Kikuzawa, Kihachiro. 1987. Development and survival of leaves of *Magnolia obovata* in a deciduous broad-leaved forest in Hokkaido, northern Japan. *Canadian Journal of Botany* 65(2):412–417.

King, George. 1891. The Magnoliaceae of British India. *Annals of the Royal Botanic Garden, Calcutta* 3:197–223. [*M. campbellii, M. globosa, M. griffithii, M. gustavi, M. maingayi, M. pealiana, M. pterocarpa*.]

Kingdon-Ward, Frank. 1930. Three Indo-Himalayan Magnolias. *Gardeners Chronicle* 7 June 1930:451–452. [*M. campbellii* var. *mollicomata, M. globosa, M. rostrata*.]

Kosar, William F. 1962. Magnolias native to North America. *Journal of the California Horticultural Society* 23(1):2–12.

Krüssmann, Gerd. 1985. *Magnolia.* In *Manual of Cultivated Broadleaved Trees and Shrubs,* vol. 2. Portland, OR: Timber Press. Pp 265–77.

Kurz, Herman, and Robert K. Godfrey. 1962. *Magnolia.* In *Trees of Northern Florida.* Gainesville, FL: University of Florida Press. Pp. 123–131. [*M. acuminata* var. *subcordata, M. fraseri* var. *pyramidata, M. grandiflora, M. macrophylla* var. *ashei, M. virginiana*.]

Langford, Larry. 1990. Check list revision. *Journal of the Magnolia Society* 26(1):22–28.

Law, Yuh-Wu. 1984. A preliminary study on the taxonomy of the family Magnoliaceae. *Acta Phytotaxonomica Sinica* 22(2):89–109.

Layritz, Richard. 1947. Magnolias in Victoria, British Columbia. *Washington Arboretum Bulletin* 10(4):10.

Li, Shun-Ching. 1935. *Magnolia.* In *Forest Botany of China.* Shanghai: The Commercial Press, Ltd. Pp. 457–475.

_____ . 1973. *Magnolia.* In *Forest Botany of China Supplement.* Taiwan: Chinese Forestry Association. Pp. 156–159.

Liberty Hyde Bailey Hortorium. 1976. *Magnolia* In *Hortus Third,* 3rd ed. New York: Macmillan. Pp. 694–697.

Little, Elbert L. 1969. New tree species. *Phytologia* 18(3):198–199. [*M. striatifolia.*]

Loudon, J. C. 1838. *Magnolia. Arboretum and Fruticetum* 1:260–284, 5:1–12.

Lloyd, J. U., and C. G. Lloyd. 1884. *Magnolia.* In *Drugs and Medicines of North America,* vol. 2. Cincinnati: Clarke. Pp. 21–46.

Magnolia acuminata. Curtis's Botanical Magazine 50(1823):2427.

Magnolia auriculata. Curtis's Botanical Magazine 30:1206. [*M. fraseri.*]

Magnolia conspicua. Curtis's Botanical Magazine 39:1621. [*M. denudata.*]

Magnolia delavayi. Curtis's Botanical Magazine 135(1909):8282.

Magnolia grandiflora var. *lanceolata. Curtis's Botanical Magazine* 45:1952. [*M. grandiflora* 'Lanceolata'.]

Magnolia kobus. Curtis's Botanical Magazine 8(138)(1912):8428.

Magnolia liliiflora and *Magnolia sargentiana. Icones Plantarus Omeiensium* 1(2)(1944): plate 67–68.

Magnolia macrophylla. Curtis's Botanical Magazine 48:2189.

Magnolia pumila: Dwarf Magnolia. *Curtis's Botanical Magazine* 25:977. [*M. coco.*]

Magnolia purpurea. Curtis's Botanical Magazine 2:390. [*M. liliiflora.*]

Magnolia salicifolia. Curtis's Botanical Magazine 139(1913):8483.

Magnolia wilsonii. In The gardener's pocketbook. *The National Horticultural Magazine* 29(2)(1950):100–101.

McClintock, Elizabeth. 1962. *Magnolia campbellii* and its variants. *Journal of the California Horticultural Society* 23(1):30–36.

_____ . 1973. Magnolias and their relatives in the Strybing Arboretum, Golden Gate Park, San Francisco. *International Dendrology Society Yearbook* 1973: 44–51.

McCracken, F. I. 1985. Observations on the decline and death of Southern Magnolia. *Journal of Arboriculture* 11(9):253–256.

McDaniel, Joseph C. 1966a. 'Cairo'—an Illinois cultivar of *Magnolia grandiflora. Bulletin of the Morris Arboretum* 17(4):61–62.

_____ . 1966b. Variations in the Sweet Bay Magnolias. *Bulletin of the Morris Arboretum* 17(1):7–12. [*M. virginiana.*]

_____ . 1967. *Magnolia virginiana* var. *australis* 'Henry Hicks', a new evergreen magnolia. *The American Horticultural Magazine* 46(4):230–235.

_____ . 1968. Magnolias from middle America. *Journal of the Magnolia Society* 5(1):2–3. [*M. guatemalensis.*]

_____ . 1969. Modern Magnolias. *American Nurseryman* 150(10):56–67.

_____ . 1970. Two cultivars for upgrading *Magnolia virginiana* seedling production. *Proceedings of the International Plant Propagators Society* 20:199–202. ['Mayer' and 'Havener'.]

_____ . 1971. Some selections of *Celtis laevigata* and *Magnolia acuminata. Proceedings of* the *International Plant Propagators Society* 21:477–479. [*M. acuminata* 'Philo'.]

———. 1973. Sharpening our sights on *Magnolia acuminata*. *Journal of the Magnolia Society* 9(2):9–14.

———. 1974. *Magnolia cylindrica,* a Chinese puzzle. *Journal of the Magnolia Society* 10(1):3–7.

———. 1975. Hybridizing section *Rhytidospermum*. *Journal of the Magnolia Society* 11(1):9–11.

———. 1976a. Some Latin American magnolias. *Journal of the Magnolia Society* 12(2):24–26.

———. 1976b. *Magnolia ashei*. *American Nurseryman* 144:9, 44, 46, 48.

———. 1976c. The big leaf clan. *American Horticulturist* 55(6):18–21. [Section *Rhytidospermum*.]

———. 1976d. *Magnolia biondii* at last. *Journal of the Magnolia Society* 12(2):3–6.

———. 1984. Variations in sweet bay magnolias. *Journal of the Magnolia Society* 19(2):24–29. [*M. virginiana*.]

McLaughlin, Robert P. 1928. Some woods of the Magnolia family: a key to their structural identification. *Journal of Forestry* 26:665–667.

———. 1933. Systematic anatomy of the woods of the Magnoliales. *Tropical Woods* 34:3–39.

McVaugh, Rogers. 1936. The Cucumber Tree in eastern New York. *Journal of the Southern Appalachian Botanical Club* 1(4):39–41. [*M. acuminata*.]

Menzies, Arthur L. 1962. Some deciduous Magnolias in the Strybing Arboretum. *Journal of the California Horticultural Society* 23(1):37–41.

Meyer, Frederick G., and Elizabeth McClintock. 1987. Rejection of the names *Magnolia heptapeta* and *M. quinquepeta*. *Taxon* 36(3):590–600. [*M. liliiflora* and *M. denudata* nomenclature.]

Michaux, André. 1803. *Magnolia*. In *Flora Boreali-Americana*. Reprint. New York: Hafner Press, 1974. Pp. 327–328.

Miller, Ronald F. 1975. The deciduous magnolias of west Florida. *Rhodora* 77:64–75. [*M. acuminata, M. fraseri* var. *pyramidata, M. macrophylla* var. *ashei, M. tripetala*.]

Molina-R., Antonio. 1974. Una contribucion de varias plantas nuevas de America Central. *Ceiba* 18(1–2):95–106. [*M. hondurensis*.]

Mulligan, Brian O. 1987. *Magnolia salicifolia* 'Wada's Memory'. *Bulletin of the University of Washington Arboretum* 50(4):14–15.

Nash, George V. 1918. *Magnolia kobus*. *Addisonia* 3(3):55–56, plate 108.

Nehrling, Henry. 1933. Magnolias. In *The Plant World in Florida*. Eds. A. Kay and E. Kay. New York: Macmillan. Pp. 30–36.

Nicholson, George. 1895. The Magnolias. *Gardeners Chronicle* 3, 17:515–516.

———. 1903. *Magnolia*. *Flora and Sylva* 1(1):14–22.

Nicholson, Rob. 1987. The propagation of *Magnolia amoena*—A rare Chinese endemic. *Journal of the Magnolia Society* 22(2):17–18.

———. 1988. Magnolias past and future. *American Horticulturist* 67(6):14–17, 36.

Pattison, Graham. 1986. *Magnolia dealbata*. *Journal of the Magnolia Society* 21(2):17–18.

Pearce, S. A. 1959. Magnolias at Kew. *Journal of the Royal Horticultural Society* 84:418–426.

Pe'i, C. 1933. The vascular plants of Nanking II. *Contributions from the Biological Laboratory of the Science Society of China* 8(3):291–293. [*M. zenii*.]

Pfaffman, George A. 1975. A trip to see the rare Mexican magnolia tree species *M. dealbata*. *Journal of the Magnolia Society* 11(2):9–15.

Pickering, Jerry L., and David E. Fairbrothers. 1967. A serological and disc electrophoretic investigation of *Magnolia* taxa. *Bulletin of the Torrey Botanical Club* 94(6):468–479.

Pitkin, William. 1947. Magnolias in the parks of Rochester, New York. *Washington Arboretum Bulletin* 10(4):14.

Pittier, Henry. 1910. New or noteworthy plants from Colombia and Central America— 2. *Contributions to the [U.S.] National Herbarium* 13:93–94. [*M. poasana.*]

Postek, Michael T., and Shirley C. Tucker. 1983. Ontogeny and ultrastructure of secretory oil cells in *Magnolia grandiflora*. *Botanical Gazette* 144(4):501–512.

Pounders, Cecil. 1985. Southern Magnolia. *Alabama Nurseryman* April 1985: 10. [*M. grandiflora* and cultivars.]

Praglowski, Joseph. 1974. Magnoliaceae. *World Pollen and Spore Flora 3*.

Primack, Richard B., Edward Hendry, and Peter Del Tredici. 1986. Current status of *Magnolia virginiana* in Massachusetts. *Rhodora* 88:357–365.

Raulston, J.C. 1988. Trees for tomorrow. *American Nurseryman* 167(3):36–37. [*M. grandiflora* 'Little Gem' and *M. macrophylla* var. *ashei*.]

Rehder, Alfred. 1939. New species, varieties, and combinations from the Arnold Arboretum. *Journal of the Arnold Arboretum* 20:85–101. [*M. wilsonii, M. × proctoriana*.]

———. 1940. Magnoliaceae. In *Manual of Cultivated Trees and Shrubs Hardy in North America*, 2nd ed. New York: Macmillan. Pp. 252–261.

Rehder, A., and E. H. Wilson. 1913. Magnoliaceae. In Sargent, *Plantae Wilsonianae*, vol. 1. Pp. 391–410.

Reifner, Richard E. Jr., and Joanne Tremper. 1980. *Magnolia macrophylla* naturalized in Maryland. *Phytologia* 46(5):283–284.

Rockwell, Herbert C. Jr. 1966. The genus *Magnolia* in the United States. Unpublished Master's Thesis, West Virginia University, Morgantown, West Virginia.

Rui-Yang, Chen, Chen Zu-geng, Li Xiu-lan, and Song Wen-qin. 1985. Chromosome numbers of some species in the family Magnoliaceae in China. *Acta Phytotaxonomica Sinica* 23(2):103–105

Russell, James. 1984. *Magnolia dealbata* in Vera Cruz. *Journal of the Magnolia Society* 20(1):11–12.

Santamour, Frank S. 1976. Recent hybridizations with *Magnolia acuminata* at the National Arboretum. *Journal of the Magnolia Society* 12(1):3–9.

———. 1979. Intersubgeneric hybridization between *M. grandiflora* and *M. acuminata*. *Journal of the Magnolia Society* 15(2):11–13.

———. 1981. Intersubgeneric hybrids between *M. grandiflora* and *M. liliiflora*. *Journal of the Magnolia Society* 17(2):9–11.

Sargent, Charles S. 1889. *Magnolia glauca* in its most northern home. *Garden and Forest* 2:363–364. [*M. virginiana*.]

———. 1891. Magnolia. In *Silva of North America*. Boston: Houghton Mifflin. 1:1–16, pl.1–12.

———. 1905a. *Magnolia kobus*. In *Trees and Shrubs*. Boston: Houghton Mifflin. 2:57–58.

———. 1905b. *Magnolia pyramidata*. In *Trees and Shrubs*. Boston: Houghton Mifflin. 1:101. [*M. fraseri* var. *pyramidata*.]

———. ed. 1913. Magnolia. *Plantae Wilsonianae* 1:391–410.

———. 1919. Notes on North American trees IV. *Botanical Gazette* 67:208–242. [*M. acuminata, M. virginiana*.]

———. 1933. Magnolia. In *Manual of the Trees of North America*, 2nd ed. Boston: Houghton Mifflin. Pp. 342–352.

Savage, Philip J. 1969. The goddess of Changyang Hsien. *Journal of the Magnolia Society*

6(2):1–5. [*M. sprengeri* 'Diva'.]

———. 1974. The beautiful ivory nude. *Journal of the Magnolia Society* 10(2):3–9. [*M. denudata.*]

———. 1976. Sights and scents among the hardy umbrella trees. *Journal of the Magnolia Society* 12(1):14–17. [Section *Rhytidospermum.*]

Sawada, K. 1950. Oriental magnolias in the south. *The National Horticultural Magazine* 29(2):54–57.

Skan, S. A. 1906. *Magnolia hypoleuca. Curtis's Botanical Magazine* 2 (4th ser.): 8077.

Slavin, Arthur D. 1937. Magnolias. *Bulletin of the Morris Arboretum* 1(7):90–93.

Small, John K. 1933. A magnolia as a new border plant. *Journal of the New York Botanical Garden* 34:150–152. [*M. macrophylla* var. *ashei.*]

Smith, John Donnell. 1909. Undescribed plants from Guatemala and other Central American republics. *Botanical Gazette* 47(4):253. [*M. guatemalensis.*]

Smith, Sir William Wright. 1920. Species novarum in herbario Horti Regii Botanici Edinburgensis. *Notes from the Royal Botanic Garden, Edinburgh* 12:212–213. [*M. nitida, M. rostrata.*]

Smithers, Sir Peter. 1986. An experiment with magnolias: updated. In *Rhododendrons 1985/86 with Magnolias and Camellias.* London: the Royal Horticultural Society. Pp. 5–13.

Spach, E. 1839. Magnoliaceae. In *Histoire Naturelle Végétaux, Phanérogames.* Paris. vol. 7. Pp. 427–490.

Spongberg, Stephen A. 1976. Magnoliaceae hardy in temperate North America. *Journal of the Arnold Arboretum* 57:250–312.

———. 1978. *Magnolia officinalis:* some questions. *Journal of the Magnolia Society* 14(1):3–7.

———. 1981. *Magnolia salicifolia:* an arboretum introduction. *Arnoldia* 41(2):50–59.

Stapf, Otto. 1923. *Magnolia wilsonii. Curtis's Botanical Magazine* 149:9004.

———. 1926. *Magnolia sprengeri* Diva. *Curtis's Botanical Magazine* 152:9116.

Steyermark, Julian. 1951. Botanical exploration in Venezuela I. *Fieldiana: Botany* 28(1):233. [*M. ptaritepuiana.*]

Steyermark, Julian, and Bassett Maguire. 1967. Botany of the Chimanta Massif II. *Memoirs of the New York Botanical Garden* 17:443–445. [*M. chimantensis.*]

On the stipules of *Magnolia fraseri. Proceedings of the National Academy of Science,* Philadelphia, 1887: 155–156

Stone, George E. 1913. *Magnolia tripetala* in Springfield, Massachusetts. *Rhodora* 15(171):63.

Templeton, Jewel W. 1965. *Magnolia grandiflora* 'Charles Dickens'. *Bulletin of the Morris Arboretum* 16(1):8–9.

Treseder, Neil G. 1972. Magnolias and their cultivation. *Journal of the Royal Horticultural Society* 97:336–46.

———. 1975. *Magnolia cylindrica.* In *Rhododendrons 1975 with Magnolias and Camellias.* London: The Royal Horticultural Society. Pp. 70–72.

Tucker, Shirley C. 1964. Terminal ideoblasts in Magnoliaceous leaves. *American Journal of Botany* 51(10):1051–1062.

———. 1974. Dedifferentiated guard cells in Magnoliaceous leaves. *Science* 185:445–447.

———. 1975. Wound regeneration in the lamina of Magnoliaceous leaves. *Canadian Journal of Botany* 53(14):1352–1364.

_____. 1977. Foliar sclereids in the Magnoliaceae. *Botanical Journal of the Linnean Society of London* 75:325–356.

Tucker, Shirley C., and Michael Postek. 1982. Foliar ontogeny and histogenesis in *Magnolia grandiflora* I. Apical organization and early development. *American Journal of Botany* 69(4):556–569.

Ueda, Kunihiko. 1980. Taxonomic study of *Magnolia sieboldii*. *Acta Phytotaxonomica et Geobotanica* 31:117–124.

_____. 1985. A nomenclatural revision of the Japanese *Magnolia* species together with two long-cultivated Chinese species 3. *M. heptapeta* and *M. quinquepeta*. *Acta Phytotaxonomica et Geobotanica* 36:149–158. [*M. denudata* and *M. liliiflora*.]

_____. 1986a. A nomenclatural revision of the Japanese *Magnolia* species together with two long-cultivated Chinese species 1. *M. hypoleuca*. *Taxon* 35(2):340–344.

_____. 1986b. A nomenclatural revision of the Japanese *Magnolia* species together with two long-cultivated Chinese species 2. *M. tomentosa* and *M. praecocissima*. *Taxon* 35(2):344–347. [*M. kobus* and var. *stellata*.]

_____. 1988. Star Magnolia—an indigenous Japanese plant. *Journal of the Arnold Arboretum* 69(3):281–288. [*M. kobus* var. *stellata*.]

Vasak, Vladimir. 1973. *Magnolia hypoleuca* in nature and cultivation. *Journal of the Magnolia Society* 9(1):3–6.

Vazquez-Garcia, José Antonio. 1990. Taxonomy of the genus *Magnolia* (Magnoliaceae) in Mexico and Central America. Unpublished Master's Thesis, University of Wisconsin, Madison.

Veitch, Peter C. M. 1921. Magnolias. *Journal of the Royal Horticultural Society* 46:315–322.

Vines, Robert A. 1960. Magnolia family. In *Trees, Shrubs and Woody Vines of the Southwest*. Austin, TX: University of Texas Press. Pp. 278–287.

Weatherby, C. A. 1926. A new magnolia from west Florida. *Rhodora* 28(326):35–36. [*M. macrophylla* var. *ashei*.]

Weaver, Richard E. Jr. 1981. *Magnolia fraseri*. *Arnoldia* 41(2):60–69.

Whitaker, Thomas W. 1933. Chromosome number and relationship in the Magnoliales. *Journal of the Arnold Arboretum* 14(4):376–385.

Williams, F. Julian. 1966. The garden at Caerhays. *Journal of the Royal Horticultural Society* 91:279–286.

Wilson, Ernest H. 1906. The Chinese Magnolias. *Gardeners Chronicle* 3, 39:234.

_____. 1926. Magnoliaceae collected by J. F. Rock in Yunnan and Indo-China. *Journal of the Arnold Arboretum* 7:235–239. [*M. campbellii* var. *mollicomata*, *M. delavayi*, *M. globosa*, *M. nitida*, *M. rostrata*, *M. wilsonii*.]

Witt, Joseph A. 1962. Magnolias of Japan. *Journal of the California Horticultural Society* 23(1):13–18.

Woodson, Robert E., and Russell J. Siebert. 1938. Flora of Panama II. *Annals of the Missouri Botanical Garden* 25:828–829. [*M. sororum*.]

Wyman, Donald. 1960. Magnolias hardy in the Arnold Arboretum. *Arnoldia* 20(3–4):17–28.

Yamamoto, Yoshimatsu. 1926. Magnoliaceae. In *Supplementa Iconum Plantarum Formosanarum 2*, Part 2: 13–16. [*M. kachirachirai*.]

Zuccarini, J. G. 1837. Plantarum novarum vel minus cognitarum, quae in horto botanico herbarioquew regio monacensi serantur, fasciculus secund. In *Abhandlungen der Mathematisch-Physikalischen Klasse der Königlich Bayerischen Akademie der Wissenschaften*, vol. 2. Pp. 309–380. [*M. macrophylla* var. *dealbata*.]

8

BREEDING MAGNOLIAS

There are over a hundred magnolia hybrids in cultivation today, and the number available to the gardener continues to increase. Many of these hybrids are the results of concerted efforts by magnolia breeders around the world, some of which are highlighted in chapter 9. As emphasized at the beginning of chapter 9, most magnolia hybrids have been produced by gardeners rather than by professional plant breeders. With practice, any amateur gardener can breed magnolias. All that is required is access to magnolia flowers, a few inexpensive tools, and the proper know-how. The following discussion is an attempt to provide the reader with the know-how.

MAKING THE CROSSES

The first step in breeding magnolias is to collect pollen from flowers of the plant to be used as the male parent of the cross. General guidelines for pollen collection and storage are given here. Several good articles describing various aspects of the breeding procedure are listed at the end of this chapter; the reader can refer to them for detailed information. The best of these is Savage (1978), from which much of this section is gratefully taken.

A difficult part of collecting pollen is knowing exactly when the flower is at the right stage of maturity. Ideally, flowers should be harvested about one day before they open. Lacking the ability to see into the future, one can only observe and make a reasonable estimate as to when that stage is reached. These estimations will become easier with a little practice. As discussed in chapter 2, an individual flower will have mature, receptive stigmas before the stamens are mature. Remembering this will help in the search for flowers at the right stage of maturity. Remember, also, that weather affects the maturation of the flower and that cool, cloudy, or rainy weather may slow the process. On the other hand, warm, sunny periods accelerate flower opening. The beginning magnolia breeder may want to collect flowers every day until becoming proficient at determining the right stage of maturity.

Flowers are best harvested in the middle of the day for pollen collection. At this time the stamens should be beginning to move away from the gynoecium rather than pressing tightly against it. Also, by this time of day the dew should have dried from the flowers, permitting the breeder to avoid the complications caused by wet stamens. Having determined the correct time to do so, the breeder may collect the pollen in two ways: by harvesting only the stamens, or by harvesting the entire flower. Most breeders prefer to harvest the entire flower, which makes handling and record-keeping easier. Therefore the collection of flowers rather than stamens is outlined here. If individual stamens are to be collected, the same procedure can be followed with only slight, but obvious, modification.

Begin by cutting the flowers from the plant and removing all tepals. Then cut the gynoecium straight across, above the tips of the anthers. Place the gynandrophore cut side down on white paper in a draft-free area until the pollen is shed. Glass or waxed paper can be substituted for white paper, but these make it more difficult to see the pollen. Record-keeping is important at all stages of plant breeding, and the paper or glass can serve as a place to record the name of the plant from which each flower came, as well as the date pollen was collected and any other pertinent information.

Pollen is usually shed the day following the cutting of the flower. Pollen will have fallen in a circle or halo around the gynoecium. Remove the gynandrophore from the paper once the pollen has been released. Carefully fold the pollen-covered paper to create a channel or funnel to aid in putting the pollen in a small, white, 2 × 3 in. (5 × 8 cm) envelope labeled with the plant name and date of pollen collection. Seal the envelope flap and seams with tape to keep pollen from leaking out. Envelopes should be placed in a tightly sealed jar about half full of a desiccant such as silica gel or calcium chloride (chloride of lime). These or other desiccants can usually be obtained at a florist shop, garden center, or craft supply store, sometimes under the name Drierite. If the pollen becomes damp it is susceptible to spoilage by molds and mildew, which will render the pollen inviable. Desiccants act to absorb water from the air and reduce the humidity within the jar. The jar containing the desiccant and envelopes of pollen can usually be stored in the refrigerator for short periods of time (a few months) or in the freezer at about 0°F (−18°C) for a year without much loss of viability. Again, this depends on how well the humidity is controlled.

Having successfully collected pollen, the magnolia hybridizer must next select receptive stigmas on which to dust the pollen. Looking at the magnolia plant chosen to be female parent, select several buds which are low enough to be reached and which are about two days away from opening. The bud should have shed all but a single perule. Clip the tip off the bud, leaving a hole about .25 in. (0.6 cm) in diameter. As much of the tepals should be left as possible since these will help protect the flower once it is pollinated.

Once the tip of the flower bud has been removed, the gynoecium should be visible. The stigmas should be reflexed (curled back) and should appear moist and slightly sticky. Open the envelope containing the pollen and dip a pipe cleaner into it. A small artist's watercolor paintbrush can be substituted for the pipe cleaner, but be sure to clean the bristles in alcohol between pollinations to avoid unwanted pollen contamination. Pipe cleaners are disposable, and new ones should be used for each pollination; alternatively, they should be cleaned between pollinations.

Quickly remove the pollen-coated pipe cleaner from the envelope and insert it into the opening in the flower bud. Cover the stigmas with pollen, being careful not to damage the gynoecium. After pollination, tie the flower's tepals shut at the tip of the bud with small twine or a rubber band. This will protect the gynoecium and keep unwanted pollen out. In addition, it seems to provide the perfect blend of humidity and temperature for pollen germination. Label the cross carefully with a tag and marker that will withstand weather. In a few days, the twine or rubber band must be removed. Failure to do so can result in the development of fungi which rot the flower under warm, moist conditions. At the same time, cover the flower with a mesh bag to keep the fruits from being eaten by animals as they develop. Pieces of nylon stocking work well for this. Keep the fruit covered for protection until the seeds are harvested.

Evaluation of hybrids is an important part of any breeding program. This usually means simply growing the hybrids from seed to maturity to see what their characteristics are and whether they are superior to what is currently in the trade. It is also a good time to test their true hybridity and determine whether or not they are fertile. If the hybrids are from species requiring many years to bloom, this maturation time may be shortened by grafting (chip-budding or side grafting) the hybrid onto limbs of a mature magnolia. Joe Hickman, Benton, Illinois, uses grafting and reports that the hybrids can usually be made to bloom in two to three years, decreasing the time required to see the results of hybridizing work. Selecting understock primarily involves selecting mature plants that are hardy, with strong root development.

Throughout the hybridization process, good record-keeping is essential. At the very least, parentage, direction of the cross, and other pertinent information about the parents should be recorded and made part of the label for the resulting progeny. The rule for recording a cross is to always list the seed (female) parent first, followed by the pollen (male) parent. Other useful information includes notes on date of cross, number of crosses, and the percentage which were successful, seed set of the cross, percentage of germination of seeds, percentage of seedlings which were hybrids, and descriptions of the hybrids and how they may differ from one another. While this information may not seem very important, especially if it is negative, it does provide a record of hybridization work on which other breeders can build.

BREEDING OBJECTIVES

The specific goals of different breeders and breeding programs vary, yet there are a few general objectives which might be considered by beginning magnolia breeders. Obvious objectives might include larger flowers, increased flower fragrance, flower durability, and desirable flower color or shape. The growth habit or form is also a characteristic which may be bred for—dwarf or slow-growing plants or trees with a particular form, such as pyramidal or rounded. Foliage characteristics can be exploited by breeding for leaf shape, size, glossiness, or persistence. Amount and quality of good seed set or the ornamental quality of the fruit may be objectives. Selections are often made for hybrids which are more hardy or easier to propagate, two characteristics which are not visible but are nonetheless important. In short, a hybridizer may breed for any desirable or marketable characteristics.

Some specific breeding objectives have been agreed on by most magnolia enthusiasts. It is generally recognized that plant hardiness is one of the most important limitations affecting the cultivation of magnolias. Therefore, increased plant hardiness, particularly in the precocious species, has been an objective of many breeders for years. One important aspect of hardiness in these spring-blooming species relates to flowering time. Delaying flowering by a week or two often allows these species to entirely escape frost damage. *Magnolia acuminata* has become important in breeding programs where delayed flowering is an objective. This species has the advantage of hardiness in addition to its late flowering.

Other recent breeding objectives have included the selection for a hardy, precocious, yellow-flowered hybrid. *Magnolia acuminata* has played an important role in this pursuit as well. This objective has been achieved with the release of several yellow cultivars, the most promising of which is 'Butterflies'. Several new forms are likely to emerge from breeding programs over the next few years.

Magnolia enthusiasts have for some time dreamed of the production of an evergreen hybrid with pink, purple, or yellow flowers. Several attempts have been made to achieve this, using *Magnolia grandiflora* as the evergreen parent and *M. liliiflora* or *M. acuminata* as the other parent. These attempts have so far met with frustration, and breeding along these lines continues.

BIOLOGICAL CONSIDERATIONS

There are several biological considerations to keep in mind when breeding magnolias. One of these is apomixis, when plants set seed from flowers which have not been fertilized. These apomictic seeds, when grown to maturity, produce plants that are exact clones of the parent plant, since all the genetic information the offspring received came from that one parent. If apomixis occurs in a flower used as part of a breeding program, the offspring will not be hybrids. Thus, it is best to grow offspring to maturity and observe them to be sure that they did, in fact, result from the desired cross and not from apomixis.

Another consideration is heterosis, or what is commonly described as hybrid vigor. Heterosis is the condition in which a hybrid exceeds the performance of its parents in one or more traits. Thus, offspring do not necessarily fall within the range of variation of the parents. For example, *Magnolia* × 'Orchid', a hybrid between *M. kobus* var. *stellata* and *M. liliiflora*, exhibits hardiness exceeding that found in either parent.

Frank Santamour, research geneticist at the U.S. National Arboretum, has done much research dealing with the genetic and cytological aspects of magnolia hybridization. His papers, referenced at the end of this chapter, are essential reading for anyone interested in breeding magnolias.

One of the most important biological considerations is chromosome number. The number of chromosomes in the ovule or pollen cells is referred to as the n number. Offspring receive an n number of chromosomes from each parent, thus always resulting in plants having $2n$ chromosomes. The $2n$ number of chromosomes in magnolias is most often 38 (diploid), 76 (tetraploid), or 114 (hexaploid). This information is given in the species descriptions in chapter 7. The chromosome number is important in breeding because it may affect the ease with which species cross, or the direction in which the cross may be made, the fertility of the resulting hybrid, or the inheritance of certain traits. In magnolias, species within a section are closely related and have the same number of chromosomes. Obtaining hybrids from parents with the same chromosome number is usually easier than hybridizing plants with different chromosome numbers. Therefore intrasectional (within a section) crosses are usually more successful than intersectional (between sections) crosses. Intersectional crosses often, but not always, result in hybrids which are partially sterile. Total hybrid sterility is uncommon.

Some hybridizers report that crosses between species of different chromosome numbers usually produce hybrids that most resemble the parent with the highest chromosome number. This has not been proven to be true. In some plant groups, hybrids resulting from two species of different ploidy levels invariably show intermediate characteristics. Carefully controlled hybridization studies in *Magnolia* are necessary to better understand this interaction.

It has been shown that most plants with a higher ploidy level (tetraploid as compared to diploid) have correspondingly larger flowers and leaves. Despite this it is noted that *Magnolia macrophylla*, a diploid, has the largest flowers and leaves of the genus. This fact suggests that the size of the flowers is not always a function of chromosome number only, but may be influenced by additional genetic factors. Kehr (1991) describes a polyploid form of *M. kobus* which has flowers of typical size but tepals that are thicker and wider, further supporting a genetic basis for flower size.

Increasing the chromosome number, or ploidy, of some magnolias may also allow magnolias to be crossed which otherwise would not be bred because of their different chromosome numbers. For example, doubling the chromosomes of a diploid (two sets of chromosomes), making it tetraploid (four sets of chromosomes), might allow it to be more easily crossed with another tetraploid, producing a tetraploid hybrid. This hybrid is also more likely to be fertile than a triploid (three sets of chromosomes) hybrid produced by a diploid × tetraploid cross.

Experienced magnolia breeders have noted that seemingly sterile hybrids, such as triploids, sometimes function with relative fertility when used as a pollen parent. Similarly, pentaploid (five sets of chromosomes) hybrids may be at least partially fertile. As discussed under *Magnolia × soulangiana*, some hybrids of that complex were inter-hybridized to develop subsequent *M. × soulangiana* derivatives. These latter hybrids have segregated into a huge range of chromosome numbers from triploid to octoploid. The selection pressure has always been in the direction of higher ploidy level due to the larger flowers and other desirable traits exhibited in those plants with higher chromosome numbers.

INDUCING POLYPLOIDY

Increasing a plant's chromosome number can be done by treating the plant with colchicine, an alkaloid extracted from *Colchicum autumnale*, the Autumn Crocus. Cells treated with colchicine do not divide in the usual manner. Normal cells undergo three main steps in dividing: (1) chromosomes double, (2) a new cell wall forms between the duplicate sets of chromosomes, and (3) the new cell wall splits to divide into two identical cells. In dividing cells which have undergone colchicine treatment, steps 2 and

3 are omitted. Only the doubling of the chromosomes occurs, leaving the cell with twice the normal number of chromosomes. This cell will then divide normally unless it too is treated with colchicine. If it does receive such treatment, or if a residue remains from the previous treatment, the cell may either die without further growth or division, or may repeat the process of chromosome multiplication for yet another cycle.

Since colchicine only affects a cell's ability to divide, non-dividing cells are not affected in any way by the treatment. A plant's terminal growth is dependent on the divisions of the apical meristem in the shoot tip. So although any dividing cell may become polyploid with a colchicine treatment, the plant's continued new growth will be polyploid only if the apical cells are affected. Therefore, two criteria must be met for colchicine treatment to be effective: (1) the apical cells must be dividing at the time of treatment, and (2) the amount of colchicine that is absorbed by the dividing cells must be sufficient to inhibit cell wall formation after the first division, yet not so much that it causes cell death.

Given these two criteria, it becomes obvious that colchicine treatment can be tedious. Dr. August Kehr of Hendersonville, North Carolina, has perfected a method of treating magnolias with colchicine. The discussion which follows is based on Dr. Kehr's work (Kehr 1985).

Colchicine is a yellow powder that may be obtained from chemical supply companies. When working with this chemical, it is important to be careful. Keep it away from eyes, avoid skin contact, and do not ingest. Kehr recommends making a 1% stock solution, from which the final solution will be prepared. To make the stock solution, dissolve .003 oz. (100 mg) of colchicine powder in 3 [U.S.] fl. oz. (100 cc) of distilled water (available at drug stores). Add a couple of paradichlorobenzene flakes (moth crystals) to prevent mold from growing in the solution before it is used. This stock should keep indefinitely if stored in the refrigerator.

The final colchicine solution can be prepared by mixing ¼ teaspoon (1 cc) of stock solution with 4¾ teaspoons (19 cc) water. This makes a .25% colchicine mixture. Add five drops of any commercial spreader/sticker or wetting agent to insure good coverage of the area treated, and five drops of dimethyl sulfide (DMSO) to increase absorption of the colchicine by the plant tissue. Read and follow the warning labels on all the above chemicals, and avoid skin contact with colchicine and DMSO.

Colchicine treatment is generally carried out using young magnolia seedlings. Large plants can be treated, but this method yields a decidedly lower success rate. For this reason, only colchicine application to seedlings will be addressed here. Kehr (1985) reports a method for treating large plants; the adventurous plant breeder is referred to his discussion.

Kehr recommends treating magnolia seedlings with the colchicine solution when the cotyledons are fully expanded and the first true leaves are barely visible. Place a drop of the solution on the growing tip of the seedling. Keep the plants in an area where humidity is high so that the solution can be more readily absorbed. If the treatment is successful, the seedling will stop growing for a month or two. Plants growing normally without a period of slowed growth have not had successful treatment.

When the seedling's growth resumes, curled leaves will develop. If these plants resume normal development, they may not have been successfully polyploidized. Seedlings with slowed growth rates probably reflect a successful treatment. Seedlings which appear to make no growth the first year, remaining leafless at the end of the year, are almost certainly polyploid.

As the plants become larger, polyploid forms may become more obvious. The leaves of these plants often have a rough surface and heavier texture than normal, and almost always have wider, more rounded leaves than normal. The flowers should also be larger than normal. Kehr (1991) found that tepals of flowers from polyploid *Magnolia kobus* are 35% wider than typical and have a thicker texture, which allows them to last longer.

To date, Dr. Kehr has created the following confirmed or probable polyploids: *Magnolia kobus, M. kobus* var. *stellata* 'Two Stones', *M. sieboldii* 'Genesis', *M. acuminata, M. virginiana, M. fraseri, M. sprengeri* 'Diva', 'Sundance', and several hybrids. As Dr. Kehr's selection and breeding program continues, it is likely that more polyploids will be produced and named.

If the creation of polyploid magnolias seems a bit too complicated, try starting a breeding program with a few simple crosses. The information given in this chapter, and that found in the additional reading, will start would-be plant breeders on their way to creating hybrid magnolias.

A comparison of the flowers of diploid (left) and tetraploid (right) forms of *Magnolia kobus*. The tetraploid flower, though not significantly larger in size, has wider tepals. Photo by August Kehr, Hendersonville, North Carolina.

REFERENCES AND ADDITIONAL READING

Demuth, Polly, and Frank S. Santamour. 1978. Carotenoid flower pigments in *Liriodendron* and *Magnolia*. *Bulletin of the Torrey Botanical Club* 105(1):65–66.

Greene, Thomas A. 1991. Family differences in growth and flowering in young southern magnolia. *HortScience* 26(3):302–304.

Hopkins, Harold C. 1979. Testing stored pollen. *Journal of the Magnolia Society* 15(1):24–26.

_____. 1990. A look backward at backcrossing. *Journal of the Magnolia Society* 25(2):1–5.

Jonsson, Lennarth. 1987. How easy is pollen collecting? *Journal of the Magnolia Society* 22(2):1–3.

Kehr, August E. 1985. Inducing polyploidy in magnolias. *Journal of the Magnolia Society* 20(2):6–9.

――――. 1991. Magnolia improvement by polyploidy. *Journal of the Magnolia Society* 26(2):26–28.

Ledvina, Dennis. 1984. Creating hybrids. *Journal of the Magnolia Society* 20(2):25–26.

McDaniel, Joseph C. 1974. Get in on the ground floor of magnolia hybridizing. *Journal of the Magnolia Society* 10(2):13–14.

――――. 1975. Hybridizing section *Rhytidospermum*. *Journal of the Magnolia Society* 11(1):9–11.

――――. 1979. Revisiting randomville with brush and bag. *Journal of the Magnolia Society* 15(2):29–30.

Rix, Martyn. 1978. The chromosomes of Camellia, Magnolia, and Rhododendron, and their significance in the breeding of hybrids in these genera. *Rhododendrons with Magnolias and Camellias* 1978: 66–68.

Santamour, Frank S. 1965a. Biochemical studies in *Magnolia* I. Floral anthocyanins. *Bulletin of the Morris Arboretum* 16(3):43–48.

――――. 1965b. Biochemical studies in *Magnolia* II. Leucoanthocyanins in leaves. *Bulletin of the Morris Arboretum* 16(4):63–64.

――――. 1966a. Biochemical studies in *Magnolia* IV. Flavonols and flavones. *Bulletin of the Morris Arboretum* 17(4):65–68.

――――. 1966b. Cytological notes III. *Magnolia pyramidata*, *M. cylindrica*, and *M. guatemalensis*. *Bulletin of the Morris Arboretum* 17(3):51.

――――. 1966c. Hybrid sterility in *Magnolia × thompsoniana*. *Bulletin of the Morris Arboretum* 17(2):29–30.

――――. 1967. A plea for quantification of breeding data in *Magnolia* hybridization. *Journal of the Magnolia Society* 4(2):6–7.

――――. 1969. Cytology of *Magnolia* hybrids 1. *Bulletin of the Morris Arboretum*. 20(4):63–65.

――――. 1970a. Cytology of *Magnolia* hybrids 2. *M. × soulangiana* hybrids. *Bulletin of the Morris Arboretum* 21(3):58–61.

――――. 1970b. Cytology of *Magnolia* hybrids 3. Intra-sectional hybrids. *Bulletin of the Morris Arboretum* 21(4):80–81.

――――. 1970c. Implications of cytology and biochemistry for magnolia hybridization. *Journal of the Magnolia Society* 7(2):8–10.

――――. 1976. Recent hybridizations with *Magnolia acuminata* at National Arboretum. *Journal of the Magnolia Society* 12(1):3–9.

――――. 1979a. Attempted intersubgeneric hybridization with *Magnolia virginiana*; an exercise in failure. *Journal of the Magnolia Society* 15(1):22–23.

――――. 1979b. Inter-subgeneric hybridization between *M. grandiflora* and *M. acuminata*. *Journal of the Magnolia Society* 15(2):11–13.

Santamour, Frank S., and John S. Treese. 1971. Cyanide production in *Magnolia*. *Bulletin of the Morris Arboretum* 22(3):58–59.

Savage, Philip J. 1978. Gathering gold dust. *Journal of the Magnolia Society* 14(1):11–18.

9

MAGNOLIA HYBRIDIZERS

Unlike many groups of ornamental plants, for which a few primary individuals or institutions produce the majority of the hybrids available, magnolia hybrids have been produced by many individuals. While a few breeding programs stand out, a large number of the cultivated hybrids listed in chapter 10 were produced by gardeners who probably do not consider themselves plant breeders. This is reflected in the great diversity of hybrids available and the lack of "breeding lines", such as exist in some other plant groups. Through the efforts of these people, numerous hybrid cultivars have been released, and much information has been gained as a result of their breeding work.

Because of the quality breeding efforts of so many magnolia enthusiasts, it was difficult to select which ones to highlight in this chapter. A handful of hybridizers stand out because of the quantity and/or quality of the hybrids they produced. Only these will be discussed here, but this is not meant to detract from the efforts of other magnolia breeders. This discussion of breeders includes a list of the major hybrids each has produced; descriptions and more information on these hybrids are given in chapter 10.

One of the earliest, and perhaps the most well-known of hybrids, *Magnolia × soulangiana*, was produced by Étienne Soulange-Bodin in his garden at Fromont, France, in 1820. Soulange-Bodin used *M. denudata* as the seed parent and *M. liliiflora* as the pollen parent, producing a hybrid intermediate in flower color and flowering time. Since its introduction into English gardens in 1827 or 1828, many *M. × soulangiana* cultivars have been bred and named. Forty one cultivars of *M. × soulangiana* are described in chapter 10. Flower color ranges from white to deep purple. Because of its ease of cultivation and its suitability for the home landscape, *M.* × *soulangiana* is likely to continue to be a popular magnolia choice. A more detailed plant description, together with cultivars of *M.* × *soulangiana*, is presented in chapter 10.

TODD GRESHAM

Perhaps the single largest magnolia hybridizing program was maintained by D. Todd Gresham (1909–1969) of Santa Cruz, California. Gresham was one of the most noted magnolia breeders and a founding member of the Magnolia Society. In 1955 Gresham used pollen from *Magnolia* × *veitchii* to fertilize flowers of *M. liliiflora* and *M.* × *soulangiana* 'Lennei Alba' in an attempt to combine the flower color of *M. campbellii* (through its hybrid *M.* × *veitchii*) with the hardiness and early flowering age of the seed parents. Two extremes in flower color were chosen for the female parents: *M. liliiflora*, with dark purple flowers, and *M.* × *soulangiana* 'Lennei Alba', with almost white flowers. In 1955 Gresham made only three crosses, all three using the above parents. In 1960–1964, however, he made 360 crosses, and in 1965–1968 an additional 220, using an incredible array of species and cultivars as parents.

In January 1966, Gresham arranged to have 1600 seedlings resulting from his crosses transported from his home at Hill of Doves in Santa Cruz, California, to Gloster Arboretum in Mississippi. Later in 1966, 1968, and 1969, a total of 13,900 hybrids were sent to Tom Dodd Nurseries, Semmes, Alabama. In all, some 15,500 hybrids were produced by Gresham.

Gresham himself introduced only about a dozen named selections from the 1955 crosses of *Magnolia liliiflora* × (*M.* × *veitchii*) and (*M.* × *soulangiana* 'Lennei Alba') × (*M.* × *veitchii*). Other seedlings resulting from these early crosses initially went to Gloster Arboretum, where people in the nursery trade and Magnolia Society members helped to evaluate and subsequently to introduce several cultivars. Many of these new selections were made by the late J. C. McDaniel of Urbana, Illinois; Ken Durio of Louisiana Nursery, Opelousas, Louisiana; Dr. John Giordano, Eight Mile, Alabama; and Dr. John Allen Smith of Magnolia Nursery, Chunchula, Alabama. The young plants resulting from Gresham's later crosses were trucked to Dodd Nurseries, where they were planted out for evaluation and release. John Allen Smith of Magnolia Nursery, Chunchula, Alabama, has made numerous selections from these plants. It is certain that the gardening community will benefit from Gresham's labor for some time to come as these new forms enter the market.

Gresham sometimes referred to plants of his (*M.* × *veitchii*) × *M. liliiflora* cross as the "svelte brunettes" because of the dark color and sleek form of the flowers. The range in flower color is from dark red to white, and flowers may reach 11 in. (28 cm) across. They have twelve tepals arranged in three whorls of four. These hybrids bloom in February–March in California and are hardy to Zone 6. Gresham (1962) gives a complete description of the plants resulting from this cross. The cultivars selected from this parentage include 'Dark Raiment', 'Heaven Scent', 'Peppermint Stick', 'Raspberry Ice', 'Royal Crown', 'Sayonara,' and 'Vin Rouge'.

Gresham referred to the results of his crosses between *Magnolia* × *soulangiana* and *M.* × *veitchii* as "buxom blondes" because of the light color and robust form of the flowers. The fragrant flowers are up to 12 in. (30.5 cm) across. The nine tepals are arranged in three whorls of three and are colored white with pink or rose tinges. These forms flower in February–March in California. Selections from this cross include 'Crimson Stipple', 'Cup Cake', 'Darrell Dean', 'Delicatissima', 'Elisa Odenwald', 'Joe McDaniel', 'Leather Leaf', 'Manchu Fan', 'Mary Nell', 'Peter Smithers', 'Rouged Alabaster', 'Royal Flush', 'Spring Rite', 'Sulphur Cockatoo', 'Sweet Sixteen', 'Tina Durio', and 'Todd Gresham'.

PHIL SAVAGE

Another important American magnolia breeder is Philip J. Savage of Bloomfield Hills, Michigan. Savage began breeding magnolias in 1960 and continues to produce superior hybrids while sharing knowledge about the plants he grows. He too was a founding member of the Magnolia Society and has held every office in the organization.

Savage has made almost every possible cross within section *Rhytidospermum* and has crossed nearly all species in that section with *Magnolia virginiana* of section *Magnolia*. He has focused his main efforts on creating late-blooming, cold-hardy precocious hybrids from the species in subgenus *Yulania*. Numerous Savage hybrids have been admired by magnolia enthusiasts. These include 'Big Dude', a cross between 'Wada's Picture' and *M. sprengeri* 'Diva'; 'Curly Head', a cross between *M. acuminata* and *M. × veitchii* 'Peter Veitch'; 'Fireglow', the result of a cross between *M. cylindrica* and *M. denudata* 'Sawada's Pink'; 'Goldfinch', a fine light yellow form which resulted from *M. acuminata* var. *subcordata* 'Miss Honeybee' × *M. denudata* 'Sawada's Cream'; 'Helen Fogg', a cross between *M. denudata* 'Sawada's Pink' and *M. × veitchii* 'Peter Veitch'; and 'Yellow Lantern', a cross between *M. acuminata* var. *subcordata* and *M. × soulangiana* 'Alexandrina'. The two named cultivars 'Karl Flinck' and 'Birgitta Flinck' are hybrids of *M. macrophylla* and *M. virginiana*. Savage's cross between *M. denudata* and *M. sargentiana* var. *robusta* produced 'Marj Gossler', which is doing well in nurseries on the West Coast of the United States. Much breeding has been done with cold hardiness and frost-free late flowering in mind using a northern Ohio form of *M. acuminata* (which Savage named 'Fertile Myrtle') as the seed parent and crossing with *M. sargentiana* var. *robusta*, *M. campbellii*, *M. sprengeri* 'Diva', a darker flowered seedling of 'Diva', *M. × 'Picture'*, and many others. All these have flowered in recent years. *Magnolia × wieseneri* pollen, which, unlike its seed, is fertile, has been put to good use in crosses with *M. virginiana*, *M. fraseri*, and *M. hypoleuca*.

Perhaps Savage's most notable hybrid to date is 'Butterflies', a cross between *Magnolia acuminata* and *M. denudata* 'Sawada's Cream'. It is an exceptional, precocious yellow form superior to any of the yellow forms available today. The flowers have a deeper color than those of many creamy yellow cultivars such as 'Elizabeth'. 'Butterflies' is also quite cold-hardy. (See chapter 10 for more information on this excellent plant and on all other hybrids by Savage.) Through his breeding work, Savage has made enormous contributions to magnolia enthusiasts.

THE U.S. NATIONAL ARBORETUM

Among institutions with magnolia breeding programs, the United States National Arboretum (USNA) in Washington, D.C., has made especially significant contributions. Over the years, the staff of the arboretum have produced the following important hybrids.

The Freeman Hybrids, controlled crosses between *Magnolia virginiana* ($2n = 38$) and *M. grandiflora* ($2n = 114$), were made in 1930 by Oliver M. Freeman. The seed parent was a large specimen of *M. virginiana*; a superior specimen of *M. grandiflora* was chosen as the pollen parent. Both trees were growing in Washington, D.C. The hybrids ($2n = 76$) most closely resemble *M. grandiflora*, the evergreen character being dominant in this cross. The hybrids are quite variable, as are both parents. Freeman (1937) offers a complete description of the hybrids.

Dr. William Kosar of the USNA reported that second-generation seedlings from this cross showed morphological abnormalities, possibly due to incomplete pairing of chromosomes. Dr. J.C. McDaniel, Urbana, Illinois, repeated the cross (*Magnolia virginiana* × *M. grandiflora*) but carried the program even further. In order to "dilute" the dominant effect of *M. grandiflora* ($2n = 114$), he backcrossed a hybrid such as 'Freeman' ($2n = 76$) onto *M. virginiana* ($2n = 38$) and got triploid offspring ($2n = 57$) which were more intermediate in character between *M. virginiana* and *M. grandiflora* than the previous generation. The cultivars 'Freeman' and 'Maryland' were selected and registered from Freeman's original cross.

The Kosar/deVos hybrids, or "Eight Little Girls," resulted from crosses at the USNA between *Magnolia liliiflora* and *M. kobus* var. *stellata*. In 1955 Dr. Francis deVos, research geneticist at the U.S. National Arboretum, began the breeding work using *M. liliifora* 'Nigra' as the seed parent and *M. kobus* var. *stellata* 'Rosea' as the pollen parent. 'Nigra' was selected for its hardiness and late-blooming characteristics. *Magnolia kobus* var. *stellata* 'Rosea' was selected on the basis of its habit, prolific flowering, fragrance, and mildew resistance. Dr. deVos' crosses began flowering in 1962 at seven years of age. Selections from this cross include 'Ann', 'Judy', 'Randy', and 'Ricki'.

In 1956 Mr. William Kosar, horticulturist at the USNA, continued deVos's work using pollen of *Magnolia kobus* var. *stellata* 'Rosea' and 'Waterlily' on *M. liliiflora* 'Nigra' and 'Reflorescens'. The four selections made from Kosar's work are 'Betty', 'Jane', 'Pinkie', and 'Susan'.

These eight Kosar/deVos hybrids have larger flowers than those of *Magnolia kobus* var. *stellata* and are later-blooming, thus avoiding some early frost damage. The flowers resemble those of *M. liliiflora*, but the tepals are usually narrower. Basic flower color ranges from pink to reddish purple, but the color and tepal number may vary from year to year in response to environmental conditions. These hybrids grow to about 10–12 ft. (3–3.7 m) tall. Their growth habit is shrubby, intermediate between the spreading habit of *M. liliiflora* and the densely branched habit of *M. kobus* var. *stellata*. They inherited mildew resistance from the pollen parent. Hybrids are triploid ($2n = 57$ chromosomes) and are all sterile yet are easily propagated from cuttings. (These cultivars are discussed further in chapter 10; also see Dudley and Kosar [1968] for complete descriptions.)

'Galaxy' and 'Spectrum' were created by William Kosar of the USNA in 1963. *Magnolia liliiflora* 'Nigra' was crossed with *M. sprengeri* 'Diva' to produce these hardy plants with large purple flowers. These plants are spectacular and are some of my personal favorites. 'Nimbus' is the result of a cross also made by William Kosar in 1956 between *M. hypoleuca* and *M. virginiana*.

As mentioned in chapter 8, Frank Santamour, research geneticist at the USNA, has done extensive research on genetics and inheritance in magnolias. Some of his published work is included in reference list at the end of that chapter.

THE BROOKLYN BOTANIC GARDEN

The Brooklyn Botanic Garden (BBG) in New York has also been a leader in quality research and hybridization of magnolias. Although the BBG Research Center in Ossining, New York, is no longer in operation, the hybrids resulting from the breeding program there will be regarded as valuable landscape plants for years to come. Three active magnolia researchers at this center were Evamaria Sperber, Doris Stone, and Lola Koerting, also a former president of the Magnolia Society.

Probably today the best-known hybrid to emerge from this program is 'Elizabeth', a cross between *Magnolia acuminata* and *M. denudata*. This cross was made by Evamaria Sperber in 1956. 'Elizabeth' was among the first hardy, yellow-flowered precocious magnolias available, and it received Plant Patent #4145 in 1978.

Another of Sperber's crosses, *Magnolia* × *brooklynensis*, has recently been receiving attention as well. *Magnolia* × *brooklynensis* results from a cross between *M. liliiflora* and *M. acuminata* made in 1954, and two selections from the original cross are the cultivars 'Evamaria' and 'Hattie Carthan'. *M.* × *brooklynensis* was later crossed with *M. acuminata* var. *subcordata* by Doris Stone in 1967 to produce 'Yellow Bird', a yellow-flowered selection blooming as the leaves emerge.

Perhaps the last introduction by the Brooklyn Botanic Garden is 'Marillyn', a cross between *Magnolia kobus* and *M. liliiflora* 'Nigra', also made by Sperber in 1954. This hardy precocious hybrid has large, reddish purple flowers. In 1991 the BBG discontinued its breeding program.

OTHER MAGNOLIA BREEDERS

Various other magnolia breeders have contributed a great deal to our knowledge and appreciation of magnolia hybrids. One of these is Dr. Frank Galyon of Knoxville, Tennessee. Dr. Galyon has produced several superior hybrids including 'Emma Cook', a cross between *Magnolia denudata* and *M. kobus* var. *stellata* 'Waterlily'; 'Paul Cook', a cross between *M.* × *soulangiana* 'Lennei' and *M. sprengeri* 'Diva'; 'Purple Prince' and 'Purple Princess', results of a cross between *M. liliiflora* 'Darkest Purple' and *M.* × *soulangiana* 'Lennei'; and 'Raspberry Swirl', a cross between *M.* × *liliiflora* 'Darkest Purple' and *M. sprengeri* 'Diva'.

Dr. August Kehr, Hendersonville, North Carolina, has also contributed a great deal to our knowledge of magnolia breeding. Dr. Kehr has introduced 'Sundance' (*M. acuminata* × *M. denudata*); *M.* × *brooklynensis* 'Golden Girl'; and 'Daybreak', a cross between *M.* × *brooklynensis* 'Woodsman' and *M.* 'Tina Durio. In addition, Kehr has worked extensively with the creation of polyploids in his breeding work (see chapter 8). His most recent polyploid cultivars are *M. acuminata* 'Patriot' and 'Laser'.

Two magnolia breeders in New Zealand have produced several superior hybrids. Oswald Blumhardt, Whangarei, has released 'First Flush', a cross between *Magnolia campbellii* and *M.* × *soulangiana* 'Amabilis'; 'Early Rose', a cross between *M. campbellii* and *M. liliiflora*; and 'Star Wars', a highly praised *M. campbellii* × *M. liliiflora* hybrid. Felix Jury, North Taranaki, Waitara, is a well-known camellia and rhododendron hybridizer making a name for himself worldwide with his magnolia hybrids. Jury began hybridizing work in the early 1960s. His goal was to produce large-flowered magnolias that bloom at a young age. Many of his hybrids include *M.* × *soulangiana* as a parent because of its floriferous habit. Jury named 'Mark Jury', a presumed hybrid between *M. campbellii* var. *mollicomata* 'Lanarth' and *M. sargentiana* var. *robusta*. 'Mark Jury' is parent to numerous Jury hybrids including 'Atlas' and Iolanthe', from crosses between 'Mark Jury' and *M.* × *soulangiana* 'Lennei'. 'Athene', 'Lotus', and 'Milky Way' result from crosses between 'Mark Jury' and *M.* × *soulangiana* 'Lennei Alba'. 'Serene' is selected from a cross between 'Mark Jury' and *M. liliiflora*. 'Apollo' and 'Vulcan' are selected from presumed crosses of *M. liliiflora* and *M. campbellii* var. *mollicomata* 'Lanarth'.

The late J. C. McDaniel also provided many magnolia enthusiasts with knowledge through his hybridizing experience. He worked mostly with native North American species, not only to produce new plants for the garden, but also to understand the natural variability and relationships between the plants.

Many Magnolia Society members, too numerous to mention, have added to our understanding of magnolias through their breeding work. Magnolia gardeners and enthusiasts for generations to come can benefit from their efforts.

REFERENCES AND ADDITIONAL READING

Also see references listed in chapter 10.

Dudley, Theodore R., and William F. Kosar. 1968. Eight new hybrid *Magnolia* cultivars. *Bulletin of the Morris Arboretum* 19(2):26–29.

Freeman, Oliver M. 1937. A new hybrid magnolia. *National Horticultural Magazine* 16(3):161–162.

Gresham, D. Todd. 1962. Deciduous magnolias of Californian origin. *Bulletin of the Morris Arboretum* 13(3):47–50.

_____. 1966. *Magnolia wilsonii* × *M. globosa*: a new hybrid. *Bulletin of the Morris Arboretum* 17(4):70–73.

_____. 1967. Trial by the royal family of Magnoliaceae. *Journal of the Magnolia Society* 4(2):7–8.

Kalmbacher, George. 1972. *M.* × *brooklynensis* 'Evamaria'. *Journal of the Magnolia Society* 8(2):7–8.

Koerting, Lola. 1977. A tree born in Brooklyn. *Journal of the Magnolia Society* 13(2):21–22. [*Magnolia* 'Elizabeth'.]

Santamour, Frank S. 1980. 'Galaxy' Magnolia. *HortScience* 15(6):832.

Savage, Philip J. 1989. Magnolias in Michigan: Part IV. *Journal of the Magnolia Society* 23(1):5–10.

10

HYBRIDS

This chapter contains information on named hybrids currently found in the literature and in nursery catalogs. Crosses from which no named selections have been made are given in chapter 7 under the parent species. Information given for each named hybrid is based entirely on what is available. Nursery catalogs usually give quite brief descriptions, and written accounts elsewhere do not always have much to add. As mentioned in chapter 5, cultivar registration is encouraged so that pertinent information regarding new hybrids may be available. Plant breeders are encouraged to submit as complete a description as possible to the registration authority.

Magnolia hybrids may be named in one of two ways. They may receive a Latin name, such as *Magnolia* × *soulangiana*, or a cultivar name, such as *M.* × 'Elizabeth'. Latin hybrid names, like species names, must be published with a Latin description and type specimen. For this reason, few hybrid combinations are given scientific names; most are known by a cultivar name. When a Latin name is given, it refers to a group of hybrids, including the results of all crosses made between those species, regardless of who made the cross or when. Thus all hybrids resulting from crosses between *M. denudata* and *M. liliiflora* belong to *M.* × *soulangiana*. When a hybrid is given a cultivar name rather than a scientific name, only that plant is included. As is the case with cultivars selected from a species, only vegetatively propagated offspring may carry that cultivar name. Thus, only plants vegetatively propagated from 'Elizabeth' can carry the 'Elizabeth' name.

Some hybrid combinations have been given scientific names in the literature, yet their names were not validly published since the proposal was not accompanied by a Latin description and citation of a type specimen. Whenever this is the case, such hybrids are given under the cultivar names, and the proposed Latin name is included in the description of the hybrid.

The first section of this chapter includes hybrids which have received Latin names. These hybrid descriptions are given in much the same format as the species descriptions in chapter 7, including a list of the cultivars selected from that cross. The second section describes hybrids which have been given cultivar names rather than Latin names. In both cases, hybrids are arranged alphabetically.

For each hybrid, parents are listed *alphabetically* rather than in the preferred form of female parent first. Because the direction of the cross is not always known, it seems preferable to establish consistency through use of the alphabetical method. Whenever possible the direction of the cross is given in the description of the hybrid.

HYBRIDS WITH LATIN NAMES

Magnolia × *brooklynensis* Kalmbacher.
  Newsletter of the American Magnolia Society 8(2):7–8. 1972. Brooklyn Magnolia.

All offspring of crosses between *M. acuminata* and *M. liliiflora* are included in *M.* × *brooklynensis*. These plants are generally hardier than *M. liliiflora* and flower over a longer period of time, from about mid May to mid June.

This cross was first made in 1954 by Mrs. Evamaria Sperber at the Brooklyn Botanic Garden, Brooklyn, New York, using *Magnolia acuminata* as the seed parent and *M. liliiflora* as the pollen parent. Both parents being tetraploid ($2n = 78$), a tetraploid hybrid was produced. The cross was also repeated by J. C. McDaniel of Urbana, Illinois, in the 1960s. McDaniel reported that the reciprocal cross failed.

The following selections are available:

'Evamaria' A Brooklyn Botanic Garden selection from the original cross. Flowers are about 4 in. (10 cm) across with 6 tepals, purple suffused with yellow-green. The leaves are oval with an acuminate apex, the lower surface having silvery pubescence. 'Evamaria' flowers at nineteen years of age and is fertile. Named in honor of Evamaria Sperber, who made the cross. Registered in 1970 by the Brooklyn Botanic Garden. U.S. Plant Patent #2820. Flowers May–June. **Plates 84–85.**
'Golden Girl' This selection is the result of a cross made by August Kehr, Hendersonville, North Carolina, using *Magnolia acuminata* as the seed parent. It is similar to 'Evamaria', but the flower color is almost entirely yellow with slight purple stains. Registered by Kehr in 1991.
'Hattie Carthan' A Brooklyn Botanic Garden selection from a cross between 'Evamaria' and *Magnolia* × *brooklynensis* BBG #209. Flowers are bright yellow with magenta midveins. The outer whorl of tepals has a green tinge. The cross was made in 1968 by Doris Stone of the Brooklyn Botanic Garden. This form was named in honor of Mrs. Hattie Carthan, president of the Bedford-Stuyvesant Restoration Corporation and founder of the Magnolia Tree Earth Center in Brooklyn, New York. The cultivar was registered in 1984.
'Woodsman' A selection from a cross made by J. C. McDaniel, Urbana, Illinois, of *M. acuminata* 'Klassen' × *M. liliiflora* 'O'Neill', and registered in 1974. Flowers are larger and darker than 'Evamaria' with a combination of yellow, green, and purple coloring in tepals. A fertile hybrid. **Plates 86–87.**

HYBRIDS

M. acuminata × (*M.* × *brooklynensis* 'Evamaria') = 'Yellow Bird'
(*M.* × *brooklynensis* 'Woodsman') × *M.* 'Tina Durio' = 'Daybreak'

Magnolia × *kewensis* Pearce.

> [*M. salicifolia* 'Kewensis' (Pearce) Spongberg.] *Gardeners Chronicle* 3, 132:154. 1952. Kew Hybrids.

Magnolia × *kewensis* is the grex name for all crosses between *M. kobus* and *M. salicifolia*. The tree appears to be the result of natural fertilization of *M. kobus* seed with *M. salicifolia* pollen. The original seedling was discovered by C. F. Coates at the Royal Botanical Garden at Kew in 1938. Dandy (unpublished) cites a 1958 report of natural hybridization between these two species in Aomori Prefecture, Japan. The flower has 9 tepals, the inner 6 broader than those of *M. salicifolia*. The outer 3 tepals are calyx-like. Flower color is pure white. *Magnolia* × *kewensis* is a diploid ($2n = 38$), as are both parents. Flowers in March–April. Hardy to Zone 5.

As with *Magnolia* × *proctoriana*, Spongberg (1976a) included *M.* × *kewensis* within *M. salicifolia*, contending that the variation of the latter species accommodated those plants known as *M.* × *kewensis*. However, leaves of *M.* × *kewensis* appear to most closely resemble those of *M. kobus*, while the flowers are very similar to flowers of *M. salicifolia* (Pearce 1952). Until further study is completed, these plants are considered as the hybrid, *M.* × *kewensis*.

'Kew Clone' ('Kewensis')   Plants from the original tree at Kew, described above.
'Parson's Clone'   Flowers larger than typical, tepals broader. Listed by Treseder's Nurseries, Cornwall, England, ca. 1973.
'Wada's Memory   Flowers white, drooping, reaching 7 in. (18 cm) across. Tends to be precocious. Young leaves are red, becoming green as they mature. Received at the University of Washington Arboretum, Seattle, Washington, in a batch of *Magnolia kobus* seedlings from K. Wada, Japan. Registered by Brian Mulligan of the Washington Arboretum in 1959. Hardy to Zones (4)5. **Plate 88.**

HYBRIDS

Magnolia × *kewensis* has been successfully crossed with *M. acuminata* and with *M. kobus* var. *stellata*, but no selections have yet been made.

Magnolia × *proctoriana* Rehder.

> [*M.* × *slavinii* Harkness, *M. stellata* × ? E. Keiper.] *Journal of the Arnold Arboretum* 20:412. 1939. Proctor Hybrids. **Plate 89.**

Magnolia × *proctoriana* includes the offspring of all crosses between *M. salicifolia* and *M. kobus* var. *stellata*. The original selection was made at the Arnold Arboretum of Harvard University in 1928 from a group believed to be *M. salicifolia* seedlings. The seeds had been collected from a tree growing in a small private arboretum in Massachusetts established by T. E. Proctor. The seedling selected had a pyramidal habit with smaller and narrower leaves than is typical for *M. salicifolia*. It also flowered at an earlier age than that species, and bore the showy, white, multi-tepaled flowers of *M. kobus* var. *stellata*. It is assumed that the parent tree received pollen from a nearby *M. kobus* var. *stellata*. Flowers March–April. Hardy to Zone 5.

Spongberg (1976a) included *Magnolia* × *proctoriana* in *M. salicifolia*, asserting that the degree of natural variation in *M. salicifolia* easily accommodates specimens of *M.* × *proctoriana*. I have found specimens of *M.* × *proctoriana* to produce pubescent pedicels, unlike *M. salicifolia* but like *M. kobus* var. *stellata*. In addition, in some cases the leaves and flowers are more like those of *M. kobus* var. *stellata* than those of *M. salicifolia*. The hybrid is also reported from the wild in Japan, where the native ranges of the two parents overlap. Until additional specimens are examined, it is best simply to recognize these hybrids as *M.* × *proctoriana*.

'Slavin's Snowy'   A small, fast-growing tree. Flowers are white with a pink tinge at the base, reaching 6 in. (15 cm) across, with 6–9 tepals. This large-flowered form was first described as *M.* × *slavinii* and thought to be a cross between *M. salicifolia* and *M.* × *soulangiana*. Its flower characteristics have led to its placement within *M.* × *proctoriana*

HYBRIDS

No hybrids involving *Magnolia* × *proctoriana* are in cultivation.

Magnolia × *soulangiana* Soulange-Bodin.

[*M. conspicua* var. *soulangiana* (Hamelin) Loud., *M. obovata* var. *soulangiana* (Hamelin) Ser., *M. yulan* var. *soulangiana* (Hamelin) Lindl.] *Transactions of the Linnean Society of Paris* 1826: 269. 1826. Saucer Magnolia. **Plate 90.**

Magnolia × *soulangiana* resulted from a cross between *Magnolia denudata* (seed) and *Magnolia liliiflora* (pollen) made by Étienne Soulange-Bodin in 1820. Soulange-Bodin, a former diplomat in the French army, established and was the first director of the Royal Institute of Horticulture near Paris, where he produced this hybrid. It was probably introduced into English gardens in 1827 or 1828.

It is likely that Soulange-Bodin's cross was also made independently in Japanese nurseries, possibly occurring spontaneously when the two parent plants were grown in close proximity. After the introduction of the hybrid by Soulange-Bodin, a plethora of named forms appeared on the scene, including seedlings from backcrosses and open pollinations. These forms cover the complete range between the two parents in flower color, shape, and size. It is now practically impossible to keep track of all the forms and account for their origins.

Treseder (1978) provides an excellent discussion of the variation shown within *Magnolia* × *soulangiana*. Perhaps the most interesting of these is the range of chromosome numbers found in the hybrids. *Magnolia denudata* is hexaploid ($2n = 114$), and *M. liliiflora* is tetraploid ($2n = 76$). A hybrid between the two might be intermediate, that is, pentaploid ($2n = 95$), but incomplete pairing of chromosomes or fusion of unreduced gametes can produce chromosome numbers lower or higher, respectively. With ploidy levels as high as they are in the two parents (six sets of chromosomes in *M. denudata* and five sets in *M. liliiflora*), a vast number of variations are possible. In view of the emormous amount of genetic material available and all the possibilities that conceivably exist from outcrossing and second- and third-generation backcrossing, the great and wide-ranging differences in forms of *M.* × *soulangiana* are hardly surprising. Polyploid forms of the hybrid are likely to be more vigorous and have larger flowers, leading almost inevitably to preferential selection for cultivation. The repeated introduction of these polyploids into the gene pool in turn further increases the possibility of variation. Although *M.* × *soulangiana* hybrids are usually sterile, a few seeds can develop, opening the door to the production of the hybrids of the next generation.

Magnolia × *soulangiana* is one of the most commonly cultivated magnolias in the world. Indeed, many North American gardeners know no other magnolia. These shrubs or small trees are very hardy, flower at an early age, tolerate a wide array of soil and climatic conditions, and bloom profusely before the leaves appear in the spring, sometimes continuing to bloom after the leaves are produced. *Magnolia* × *soulangiana* tends to bloom earlier in the season than either of its parents. Flower color is usually intermediate between that of the parents, although there is great variation, ranging from pale pink or white, approaching *M. denudata*, to deep red-purple, similar to *M. liliiflora*. Leaves are also variable but are usually obovate and 6–8 in. (15–20.3 cm) long.

The following selections have been made.

'Adral'   A late-blooming form, flowers red-purple outside, almost white inside. Hardy to −10°F (−23°C). Purchased by Ralph Smith as 'Lennei' and named by him in 1964.

'Alba'   Flowers almost white; outer base of tepals faint rose. Grown by Louis van Houtte of Belgium and named by him in 1867. Still in cultivation.

'Alba Superba'   Flowers white, flowering 10–14 days after *Magnolia denudata*. It is doubtful whether this form is significantly different from 'Alba'.

'Alexandrina'   There appear to be at least two forms grown under this name. The one most commonly seen in North America has deep red-purple, cup-shaped flowers appearing approximately two weeks later than typical. Treseder (1978) describes the cultivar as having white flowers tinged with purple, which is presumably the form commonly grown in England. Galyon (1990) presents an excellent historical account of the various forms known by this name. (See also 'Big Pink'.) Introduced by Cels of Paris in 1831. Zones 6–9. **Plate 91.**

'Alexandrina Alba'   Flowers larger and whiter than those of 'Alexandrina', and plant blooms about a week later. Origin unclear.

'Alexandrina Variegata'   Leaves variegated creamy yellow; otherwise like 'Alexandrina Alba'. Originated in the Netherlands ca. 1893.

'Amabilis'   Flowers almost white, with a yellowish cast; similar to those of 'Alba'. Listed by Baumann Nursery in France, 1865. Still in cultivation.

'André Leroy'   Flowers dark pink to purple. Flower color not quite as dark as 'Lennei'. Named after a nurseryman in Angiers, France, in 1892.

'Big Pink'   A superior cultivar originating in Japan and sent to K. Sawada of Overlook Nurseries in Mobile, Alabama. It was sold in the United States for some time as 'Alexandrina' and has recently been marketed as 'Big Pink'. It blooms later than most cultivars of *Magnolia* × *soulangiana*, making it better for use in northern climes. The tree is very floriferous, with flowers less likely to be damaged by wind or rain.

'Brozzoni'   Flowers mostly white, but rose color extends up the veins of the 6 tepals. Flowers reach about 10 in. (25.4 cm) across. Named in honor of Camillo Brozzoni, Brescia, Italy, in 1873.

'Burgundy'   Flowers deep purple, flowering earlier than most cultivars. Received by Clarke Nursery, San José, California, from an unknown source. Named in 1943.

'Candolleana'   Flowers white, large, blooming about a week later than others. A seedling of *Magnolia denudata*, probably crossed with *M. liliiflora* or *M.* × *soulangiana*. Named ca. 1893. Probable origination France. No longer in cultivation.

'Coates'   Flowers dark purple, lighter colored than those of 'Rustica Rubra'. A fast-growing seedling of 'Rustica Rubra' from Leonard Coates Nursery. Named by Gossler Farms Nursery, Springfield, Oregon, ca. 1973.

'Cyathiformis'   Flowers cup-shaped, white, with light purple tinge at the base. From Germany ca. 1850. Probably no longer in cultivation.

'Deep Purple Dream' ('Purple Dream')   Beautiful flowers very dark purple with the bowl shape of 'Lennei' flowers. Blooms later than 'Lennei'. Selected at Gloster Arboretum, Gloster, Mississippi, from seedlings of open-pollinated 'Lennei' purchased from Tom Dodd Nursery, Symmes, Alabama. **Plate 92.**

'Dorsopurpurea'   Flowers like those of *Magnolia denudata* but with pink tinge at the base of the tepals. From Hakoneya Nurseries, Yokohama, Japan, ca. 1925.

'Dottie Grosse'   Low-growing, dense shrub, flowers typical, leaves smaller than typical. Registered ca. 1988 by J. Dilworth of Oxford, Pennsylvania.

'Early Lennei'   An early-flowering form selected from 'Lennei'. Blooms approximately two weeks earlier. Named by Clarke and Company, San José, California, in 1958.

'Globuliflora'   Flowers small, globose. Cultivated in Italy in 1889. Probably no longer in cultivation.

'Grace McDade'   Pink flowers larger than typical; hardier than most *Magnolia* × *soulangiana* cultivars. Named by Clint McDade of Semmes Nursery, Semmes, Alabama, 1945.

'Grandis'   Flowers white with red-purple along the midrib on the outer surface. Cultivated in Germany ca. 1855. Probably no longer in cultivation.

'Highland Park'   Flowers purple, earlier than typical, fragrant. Extremely profuse bloomer. Original tree at Highland Park, Rochester, New York. Registered in 1961 by Bernard Harkness of the Division of Parks and Recreation, Rochester, New York.

'Late Soulangiana'   See 'Speciosa'

'Lennei'   Hardy, vigorous form. Tulip-shaped flowers, magenta purple outside, white inside. Probably has the darkest flower color available. Very common in cultivation. 'Lennei' probably originated in Italy ca. 1850 and was named in honor of German botanist Peter Joseph Lenne. Chromosome number: $2n = 133$. Zones 6–9. **Plates 93–94.**

'Lennei Alba'   Flowers large, white, goblet-shaped. Named by Froebel, Switzerland, ca. 1905. A backcross between 'Lennei' and *Magnolia denudata*. Zones 6–9.

'Lilliputian'   Small tree. Flowers white with pink tinge, smallest of all *Magnolia* × *soulangiana* cultivars. Named by Semmes Nursery, Semmes, Alabama, in 1946.

'Lombardy Rose'   Flowers large, rose colored, blooming into the summer. Possibly a seedling of 'Lennei'. Named by Semmes Nursery, Semmes, Alabama, in 1946. Chromosome number: $2n = 123$. **Plate 95.**

'Niemetzii'   Fastigiate form with purple flowers. Named by F. W. Niemetz, a Romanian nurseryman, in 1907. Rare in cultivation. **Plate 96.**

'Norbertiana'   See 'Norbertii'

'Norbertii' ('Norbertiana')   Two forms appear to be grown under this name. The form originally described by Loudon in 1838 is late-blooming with red-purple flowers. Treseder (1978) describes a form with white flowers tinged with purple.

'Picture' ('Wada's Picture')   Flowers very dark purple outside, white inside. A quite vigorous form with leaves and flowers larger than typical. Flowers sometimes reach 14 in. (35.6 cm) across. A commonly cultivated form and very desirable. Named by K. Wada, Hakoneya Nurseries, Yokohama, Japan, ca. 1925. Chromosome number: $2n = 143$. Zones 5–9.

'Pink Alba Superba'   Very similar to 'Alba Superba' but with deep pink flowers. Named in 1956 by Wister.

'Purple Dream'   See 'Deep Purple Dream'

'Purpliana'   Flowers reddish purple, early blooming. Named by Sawada of Overlook Nurseries, Mobile, Alabama, in 1950.

'Purpurea'   Flowers purple outside, white inside. 'Purpurea' may represent a collective name for various purple-flowered seedlings. Named by George Madlinger, Memphis, Tennessee, in 1960.

'Richeneri'   Flowers small and soft purple. Plants bloom later than typical. Listed by Ellwanger and Barry, Rochester, New York, 1886. Probably no longer in cultivation.

'Rosea'   Flowers large, white with red center. Named by Pampanini in 1915. Still found in cultivation.

'Rubra'   See 'Rustica Rubra'

'Rustica'   See 'Rustica Rubra'

'Rustica Rubra' ('Rubra', 'Rustica')   Flowers deeper red than those of 'Lennei', small and globose. Vigorous grower. Often used as rootstock. Seedling of 'Lennei' with a chromosome number of $2n = 156$. Named by Boskoop Nurseries, the Netherlands ca. 1893. **Plates 97–98.**

'San José'   Flowers large, fragrant, rosy purple, almost as dark as 'Lennei'; blooms very early. Named by Clarke and Company, San José, California, in 1940. A form of unknown origin is also in cultivation under this name; flowers are white with dark pink at the base of the tepals.

'Speciosa' ('Late Soulangiana')   Flowers nearly white with slight pink tones. Upright plant, smaller than typical *Magnolia* × *soulangiana*, late blooming. Originated in France and Belgium ca. 1830. Still in cultivation today.

'Spectabilis'   Flowers very large and nearly pure white. Plants bloom into midsummer. Named by Mouillefert, France, in 1891. Seldom seen in cultivation.

'Stricta'   Foliage narrower than the type, flowers purple. Named by Mouillefert, France, in 1891. Seldom cultivated.

'Triumphans'   White buds opening to white flowers with red stripes down the center of the tepals and at the base. A French clone named ca. 1891.

'Veitchii Rubra'   Flowers bright pink. Thought to have originated at the former W. B. Clarke Nursery, San José, California. **Plate 99.**

'Verbanica'   Flowers deep pink, tepals narrower than typical; late blooming, hardy. Named by André Leroy, Angiers, France, 1873. Common in cultivation. **Plate 100.**

'Wada's Picture'   See 'Picture'

'White'   Flowers pure white inside, tinted lavender outside. Named by Armstrong Nurseries, Ontario, California, in 1948.

'White Giant'   A seedling of 'Picture' with very large white flowers opening almost flat. Selected by K. Wada of Hakoneya Nurseries, Yokohama, Japan, and registered by him ca. 1980.

HYBRIDS

Hybrids between *Magnolia* × *soulangiana* and *M. campbellii* were made by Charles Raffill at Kew in 1943. A plant thought to be from this cross is cultivated at Lanarth. Flowers are white inside and crimson outside with 8 tepals. Apparently no hybrids from this cross have been named.

M. acuminata var. subcordata × (M. × soulangiana 'Alexandrina') = 'Yellow Lantern'
M. campbellii(?) × (M. × soulangiana 'Amabilis') = 'First Flush'

M. liliiflora 'Nigra' × (*M.* × *soulangiana* 'Lennei') = 'Fenicchia Hybrid'

M. liliiflora 'Darkest Purple' × (*M.* × *soulangiana* 'Lennei') = 'Purple Prince' and 'Purple Princess'

M. liliiflora 'Nigra' × (*M.* × *soulangiana* 'Rustica Rubra') = 'Dark Splendor', 'Orchid Beauty', and 'Red Beauty'

'Mark Jury' × (*M.* × *soulangiana* 'Lennei') = 'Atlas', 'Iolanthe'

'Mark Jury' × (*M.* × *soulangiana* 'Lennei Alba') = 'Athene', 'Lotus', and 'Milky Way'

(*M.* × *soulangiana*) × *M. sprengeri* 'Diva' = 'Paul Cook'

(*M.* × *soulangiana* 'Wada's Picture') × *M. sprengeri* 'Diva' = 'Big Dude'

(*M.* × *soulangiana*) × (*M.* × *veitchii*) = Gresham Hybrids; see chapter 9.

(*M.* × *soulangiana* 'Picture') × ? = Pickard Selections

Magnolia × *thompsoniana* Cels ex Jaume St.-Hilaire.
> [*M. glauca* var. *major* Sims, *M. glauca* var. *thompsoniana* Don.] *La flore et la pomone françaises* 5:451. 1832. Thompson's Magnolia.

Hybrids between *Magnolia virginiana* and *M. tripetala* were named *M.* × *thompsoniana* since the first such hybrid occurred naturally in 1808 in Archibald Thompson's nursery near London. Although *M. virginiana* rarely sets seed in the English climate, it must have done so that year, for in discussing this plant Sabine remarked in 1818, "It was raised from seed gathered from an old tree [of *M. virginiana*] in the garden of Mr. Thompson, of Mile End, in 1808, in which year the magnolias flowered and ripened their seeds in perfection." A tree of *M. tripetala* which was the pollen parent grew nearby. The seeds from the seed parent, *M. virginiana*, were collected by Thompson, who planted them out in his nursery. One particular seedling produced larger leaves and flowers than the others, which led Thompson to believe it to be a larger, more vigorous form of *M. virginiana*, a view supported by Sabine, who published it in 1820 as *M. virginiana* var. *major*. In 1832 it was described by Jaume Saint-Hilaire as the hybrid *M.* × *thompsoniana* based on Thompson's original plant.

J. C. McDaniel (1966), Urbana, Illinois, developed a group of *Magnolia* × *thompsoniana* hybrids using hardier forms of both parents, in a successful attempt to produce a hardy hybrid. His success in crossing these two species proved the parentage of Thompson's seedling, as McDaniel's clones fit the original description quite well. One of his clones, 'Urbana', is currently available in nurseries. All forms of this grex flower in May–June.

'Thompsoniana'   The form of the original cross found in Thompson's nursery in 1808. Flowers have 12 tepals, the outer 3 calyx-like, and are similar to flowers of *Magnolia virginiana* in color and shape, but three times as large. The flowers produce the pleasant fragrance of *M. virginiana*, and the semievergreen leaves are much like those of that species, but reach 10 in. (25.4 cm) in length.

'Urbana'   The result of *Magnolia virginiana* × *M. tripetala*, a cross made by J. C. McDaniel, Urbana, Illinois, in 1960 and registered in 1969. 'Urbana' is a multi-stemmed shrub to 20 ft. (6 m) in height. Leaves and flowers are similar to those of *M. tripetala*, yet the creamy white flowers are fragrant, produced in June. This form is sterile. The original tree is in Urbana, Illinois. This form is probably hardier (to Zone 4) than 'Thompsoniana' since McDaniel used the hardier northern form of *M. virginiana* in his cross. The reciprocal cross has not been successful. **Plate 101.**

Magnolia × *veitchii* Bean.

> In Veitch *Journal of the Royal Horticultural Society* 46:321, fig. 190. 1921. Veitch's Magnolia.
> **Plates 102–103.**

Peter Veitch of the Royal Nurseries at Exeter, England, successfully crossed *Magnolia campbellii* and *M. denudata* in 1907. This cross emerged from a hybridizing program using several magnolia species as pollen parents and *M. campbellii* as the seed parent. The breeding objectives were to produce a hybrid as beautiful as *M. campbellii*, but hardier and flowering at an earlier age. The flowers of the resulting hybrids appear in about ten years from seed or as early as seven years from layers, and are similar to those of *M.* × *soulangiana*. The general growth habit of the plant is similar to that of *M. campbellii*, but the branchlets are much more brittle. Six seedlings developed from this cross, of which the first to flower, a pink form, was named 'Veitchii' and the second to flower 'Isca'. The remaining four plants were not thought to be better than existing plants and were destroyed. *Magnolia* × *veitchii* is probably hardy to Zone 6. Flowers March–April. The following cultivars have been released.

'Alba' Flowers white. Named by S. A. Pearce, Royal Botanic Gardens, Kew, in 1959. Origin unknown.

'Isca' Flowers white, light pink in bud. Blooms approximately one week earlier than 'Veitchii'. A shorter, spreading plant. One of the six original seedlings of Veitch's cross. Named after the town of Isca in Exeter, England. **Plate 104.**

'Peter Veitch' See 'Veitchii'

'Veitchii' ('Peter Veitch') Flowers salmon-pink in color, shaped like an inverted pear. Growth habit is much like that of its seed parent, *Magnolia campbellii*, but with larger leaves. 'Veitchii' is a vigorous form and flowers profusely. This was the first seedling to flower from the Veitch cross.

HYBRIDS

Crosses between *Magnolia denudata* and *M.* × *veitchii* were made in the 1960s by Frank Santamour of the U.S. National Arboretum. These hybrids were more vigorous than either parent and produced large white to pink flowers. Plants were subsequently distributed by the arboretum to various arboreta for testing, but none have yet been named. Oswald Blumhardt of New Zealand has also made this cross. Blumhardt reported in 1982 that he had a few promising hybrids, yet none of these have been named either. Named hybrids involving *M.* × *veitchii* include the following.

> *M. acuminata* × (*M.* × *veitchii* 'Peter Veitch') = 'Curly Head'
> *M. cylindrica* × (*M.* × *veitchii*) = 'Albatross'
> *M. denudata* 'Sawada's Pink' × (*M.* × *veitchii* 'Peter Veitch') = 'Helen Fogg'
> *M. liliiflora* × (*M.* × *veitchii*) = Gresham Hybrids; see chapter 9.
> (*M.* × *soulangiana*) × (*M.* × *veitchii*) = Gresham Hybrids; see chapter 9.

Magnolia × *wieseneri* Carrière.

> (*Magnolia* × *watsonii* Hook f.) *Revue Horticole* 62:406. 1890. **Plate 105.**

A natural cross between *Magnolia sieboldii* and *M. hypoleuca* in Japan produced the hybrids known as *M.* × *wieseneri*. This original plant of this hybrid was found in the garden of Mr. Wiesener, who had purchased it as *M. sieboldii* from Mr. Tokada of Japan. The plant was described in 1890 by Carrière and named by him *M.* × *wieseneri*.

A similar plant was acquired by the Royal Botanic Gardens at Kew from the Japanese Court at the time of the International Exposition in Paris in 1889. J. D. Hooker, unaware of Carrière's description, named the Kew plant *Magnolia* × *watsonii* in 1891 in honor of William Watson, the former assistant curator at Kew. This name has often been used in the trade, but Carrière's name predates it and is therefore correct.

Carrière described the color of the anther filaments of this hybrid as white, not maroon as in the form most commonly cultivated. It is to be expected, however, that different crosses between the two parent species as well as different offspring from a single cross will produce a variety of forms. K. Wada of Hakoneya Nurseries in Yokohama, Japan, suggested that hybrids between these two species continue to occur naturally from time to time. This appears to be the case, for Treseder (1978) identified three forms in cultivation in England differing only in minor characters such as stamen color and shape of the flower bud. If these forms are all hybrids of *Magnolia sieboldii* and *M. hypoleuca*, then they would be included under the grex name *M.* × *wieseneri*, regardless of divergences from the original description. Such hybrids should be given cultivar names to distinguish them. As yet, none of these forms having minor differences have been named.

Magnolia × *wieseneri* hybrids are intermediate between the parents. Plants are small, shrubby trees with thick, bright green, obovate leaves and yellowish veins. Leaves grow to 7 in. (17.8 cm) long and are sometimes crowded into false whorls at the branch ends as in *M. hypoleuca*. The creamy white flowers are 6 in. (15 cm) across and face upward, as in *M. hypoleuca*, thus differing from the drooping flowers of *M. sieboldii* and other members of section *Oyama*. The shapes of the buds and flowers, however, resemble those of *M. sieboldii*, as do the crimson stamens and green gynoecium. Plants flower in May–June and are hardy to Zones (5)6. The Japanese vernacular name is Ukezaki-oyama-renge, which translates as "upward-facing oyama magnolia."

As this hybrid seldom sets seed, it is of special interest that Sir Peter Smithers' plant at Vico Morcote in Switzerland set twenty-nine seeds in one aggregate in 1982. At least nine of these seeds germinated, eight seedlings closely resembling those of the vigorous *Magnolia hypoleuca*, while one looked like *M.* × *wieseneri*. Smithers reports that a specimen of *M. hypoleuca* is growing near the *M.* × *wieseneri* that set seeds, suggesting that the seeds might be from a backcross onto that species.

The plants which resembled *Magnolia hypoleuca* (which can reach 100 ft. [30 m] in the wild) are quite vigorous, with larger leaves than typical *M.* × *wieseneri*. One of these specimens flowered in 1988, only six years from seed, and was 16 ft. (5 m) tall. The flowers were larger than those of *M.* × *wieseneri* and had the pleasant fragrance of that hybrid. Although *M.* × *wieseneri* matures at about 10 ft. (3 m), this seedling became a large tree similar in proportion to *M. hypoleuca*. Smithers suggests its use be restricted to large gardens.

HYBRIDS WITH CULTIVAR NAMES

'Albatross'. *Magnolia cylindrica* × (*M.* × *veitchii* 'Peter Veitch').

In 1970, Michael Taylor, head gardener at Trewithen Gardens, Cornwall, England, grew several *Magnolia cylindrica* seedlings from seed collected from Trewithen trees. One of these seedlings was given to Peter Borlase, head gardener at Lanhydroc Gardens, Cornwall, in 1974. This seedling was ultimately the only one of those grown by Taylor to survive. The Lanhydroc seedling first bloomed in 1981 at eleven years of age. At fourteen years of age it was 20 ft. (6 m) tall and the number of flowers had increased as well.

When the tree flowered, it became obvious that it was a hybrid. *Magnolia* × *veitchii* 'Peter Veitch', growing near the *M. cylindrica* from which the seed was collected, is assumed to be the pollen parent. This plant appears to have inherited flower size and tepal number from the pollen parent. 'Albatross' was named in 1985.

Leaves are obovate, 5–7 in. (12.7–17.8 cm) long, and 3–4 in. (7.6–10.2 cm) wide, glabrous above, and pubescent beneath. Flowers are 8–12 in. (20–30 cm) across with 11 rather floppy tepals. The tepals are white inside, white sometimes tinged with pink outside, or tinted green at the base. The outer 3–4 tepals are usually shorter than the inner tepals. Stamens are pink. Flowers in April in Cornwall, England. **Plate 106.**

'Ann'. *M. kobus* var. *stellata* × *Magnolia liliiflora* 'Nigra'.

This is a U.S. National Arboretum (USNA) introduction resulting from a cross made by Dr. Francis deVos, research geneticist at the arboretum, in 1955. Flowers are red-purple and 2–4 in. (5–10 cm) in diameter, with 6–8 tepals. Tepals are red purple at the base, gradually paling toward the apex. Flower buds are erect and tapered. Blooms the earliest of the "Little Girls," mid April at the USNA, Washington, D.C. (Zone 7). The leaves are about 4 in. (10 cm) long and leathery, with rippled margins. For more information regarding this cross see chapter 9. **Plates 107–108.**

'Ann Rosse'. *Magnolia denudata* × *M. sargentiana* var. *robusta*.

This plant results from the natural hybridization between the two parent plants growing at Nymans Gardens, Sussex, England. The flowers are 6–8 in. (15–20.3 cm) across at maturity with 9 pointed tepals. The tepals are white with a purple tinge at the base and flushed with pink at the apex. Stamen filaments are dark reddish purple.

'Apollo'. *Magnolia campbellii* var. *mollicomata* 'Lanarth' (?) × *M. liliiflora* 'Nigra'.

This hybrid was created by Felix Jury, North Taranaki, New Zealand. It produces an exceptionally heavy crop of very large, star-like flowers. The early flowers are deep violet outside, paler inside; later flowers are deep rose-pink. Flowered at two years from seed. This hybrid blooms after 'Athene' and before 'Serene'. Released by Mark Jury Nursery, North Taranaki, New Zealand, in 1990.

'Athene'. 'Mark Jury' × (*Magnolia* × *soulangiana* 'Lennei Alba').

This Felix Jury hybrid is a floriferous, upright tree bearing flowers similar in shape to those of *M. campbellii*. Flowers have heavy textured tepals that are white with violet pink undertones. Plant protection (patent) applied for by Mark Jury Nursery, North Taranaki, New Zealand. Sister seedlings from this cross are 'Lotus' and 'Milky Way'. **Plate 109.**

'Atlas.' 'Mark Jury' × (*Magnolia* × *soulangiana* 'Lennei').

Another hybrid by Felix Jury, North Taranaki, New Zealand. Flowers are up to 14 in. (35.6 cm) across with 6 in. (15.2 cm) wide tepals. It produces lilac-pink, cup-and-saucer flowers at an early age. **Plate 110.**

'Betty'. *M. kobus* var. *stellata* 'Rosea' × *M. liliiflora* 'Nigra'.

This hybrid was introduced by the U.S. National Arboretum (USNA). It was selected from a cross made in 1956 by William Kosar, horticulturist at the arboretum. Tepals are red-purple at the base, gray-purple at apex. Flowers are about 8 in. (20.3 cm) in diameter with 12–19 tepals. Flowers mid to late April at the USNA, Washington, D.C. (Zone 7). Flower buds are pointed and sometimes curve. Leaves are broadly ovate, about 5 in. (12.7 cm) long, and glabrous. For more information regarding this cross see chapter 9.

'Big Dude'. (*M.* × *soulangiana* 'Wada's Picture') × *M. sprengeri* 'Diva'.

Phil Savage of Bloomfield Hills, Michigan, created this hybrid using 'Diva' as the pollen parent. 'Big Dude' lives up to its name: the fragrant flowers are quite large, often reaching 14 in. (35.6 cm) in diameter with 9–12 tepals. The tepals are reddish purple outside and white inside, fading to rose-pink and white. Flowers are an elongated cup-shape, fragrant, and nodding due to their large size. Savage reports that 'Big Dude' has bloomed after a winter low of −29°F (−34°C). Registered by Savage in 1989. **Plate 111.**

'Birgitta Flinck'. *M. macrophylla* × *M. virginiana*.

This hybrid was also produced by Phil Savage of Bloomfield Hills, Michigan, and named in 1989. Savage used *Magnolia virginiana* as the seed parent. These plants are quite vigorous. Leaves are glossy, 14 in. (35.6 cm) long and 4 in. (10.2 cm) wide, and resemble those of *M. virginiana*. This is a hardy clone with flowers intermediate between the parents. The inner tepals are pure white, lacking the purple blotches of sister seedling 'Karl Flinck'. This cross was given the name *Magnolia* × *flinckii* by Savage, but the name has not yet been published with the required Latin description.

'Butterflies'. *M. acuminata* × *M. denudata* 'Sawada's Cream'.

This hybrid (U.S. Plant Patent #7456) was produced by Phil Savage of Bloomfield Hills, Michigan, using a particularly fertile form of *Magnolia acuminata* as the seed parent. 'Butterflies' is a neat-growing, upright tree with precocious, deep yellow flowers. Flowers have 10–16 tepals intermediate in size between those of the parents. The stamens are red as in the pollen parent. This was a favorite yellow form among Magnolia Society members at the 1988 meeting hosted by Savage. The yellow color is the darkest of presently available precocious yellow cultivars, and, as it becomes more widely available, this plant will probably become one of the most popular magnolia hybrids. It is being propagated and distributed by Klehm Nursery, Barrington, Illinois. **Plates 112–113.**

'Buzzard'. *M. campbellii* × *M. sargentiana* var. *robusta*.

Similar to, but generally considered inferior to, 'Hawk', yet not much information is currently available. The original cross was made by Charles Michael at Caerhays Castle, Cornwall, England. 'Buzzard' is a seedling of this cross grown at Chyverton and named by Nigel Holman.

'Caerhays Belle'. *M. sargentiana* var. *robusta* × *M. sprengeri* 'Diva'.

    This hybrid was made by Charles Michael, head gardener at Caerhays Castle in Cornwall, England, in 1951. The first seedling of the cross took fourteen years to flower; grafted plants take about five years to bloom. Flowers are large, about 12 in. (30.5 cm) across, held erect, and salmon-pink in color. The 12 broad tepals have wavy margins. The gynoecium is green with yellow stigmas, and the stamens are cream with red anthers. The plant produces flowers even on the lowermost branches and can be a fast grower. Flowers February–March. Considered marginally hardy to Zone 6.

'Caerhays Surprise'. *M. campbellii* var. *mollicomata* × *M. liliiflora* 'Nigra'.

    Philip Tregunna, head gardener at Caerhays Castle, Cornwall, England, made this cross in 1959. For his efforts he was able to realize two seeds, the seedling from one of which germinated but died several years later. This form, the one remaining seedling, bloomed at eight years of age, producing bright violet-red flower buds. Flowers develop into a pinkish lavender color and reach about 8 in. (20.3 cm) in diameter with 9 tepals. The outer two whorls of tepals are reflexed. The pedicel is pubescent as in *Magnolia campbelli* var. *mollicomata*.

'Candy Cane'. Gresham Hybrid of unknown parentage.

    Flowers with 9 erect tepals in three whorls, each whorl having a distinct coloration. The outer whorl is also somewhat reduced. Outer tepals are rose at base fading to white at tip, with a rose stripe running from base to tip. Middle whorl of tepals has similar coloration, but slightly darker. The inner whorl is even deeper rose-purple at the base. Selected and named by John Allen Smith, Magnolia Nursery, Chunchula, Alabama, from Todd Gresham's hybrids at Gloster Arboretum, Gloster, Mississippi.

'Charles Coates'. *M. sieboldii* × *M. tripetala*.

    In 1946 Charles Coates, former plant propagator at the Royal Botanic Gardens at Kew, discovered three seedlings growing in the garden. He suspected the seedlings were hybrids, collected them, and took them to the nursery. They flowered twelve years later and were identified as hybrids between *Magnolia sieboldii* and *M. tripetala*. One of the seedlings was selected and named by S. A. Pearce, of the Royal Botanic Gardens at Kew, London, in honor of Coates.

    This form is intermediate between the parents in several characters. It is semi-bushy and about as wide as it is tall. Leaves are 10 in. (25.4 cm) long and 5 in. (12.7 cm) wide and broadly elliptic. The leaf apex is acute, the base cuneate. Flowers are 4 in. (10.2 cm) across and creamy white and upward-facing, resembling those of *Magnolia tripetala*. From *M. sieboldii* the flowers inherited maroon stamens and a pleasant fragrance. Tepals are usually crumpled and have wavy margins. This hybrid has been repeated and confirmed by August Kehr, Hendersonville, North Carolina. Flowers (April) May–June. Hardy to Zone 6.

'Charles Raffill'. See *Magnolia campbellii*, chapter 7.

'Crimson Stipple'. (*Magnolia* × *soulangiana* 'Lennei Alba') × (*M.* × *veitchii*).

    This Gresham Hybrid is similar to 'Rouged Alabaster' (which see), but with crimson stippling on the backs of the tepals. Flowers are fragrant and reach 12 in. (30.5 cm) across, with 9 white tepals with pink or rose tinges. Tepals are arranged in three whorls of 3 tepals. Flowers February–March in Santa Cruz, California, Zone 9. Upright growth habit. Named by Gresham in 1962. For more information regarding this cross see chapter 9.

'Cup Cake'. (*Magnolia* × *soulangiana* 'Lennei Alba') × (*M.* × *veitchii*).

This Gresham Hybrid has cream-colored flowers. The flowers are fragrant and reach 12 in. (30.5 cm) across. Named by Ken Durio of Louisiana Nursery, Opelousas, Louisiana. For more information regarding this cross see chapter 9.

'Curly Head'. *Magnolia acuminata* × (*M.* × *veitchii* 'Peter Veitch').

This plant is the result of the breeding work of Phil Savage, Bloomfield Hills, Michigan. Savage used pollen of 'Peter Veitch' which had been collected and frozen by Harold Hopkins, Gaithersburg, Maryland. This is a tall, upright tree with dense habit. Leaves have revolute edges, and flowers are described as pastel pink and yellow on white. Named by Savage in 1990, formerly published under the name 'Editor Hopkins'.

'Dark Raiment'. *Magnolia liliiflora* × (*M.* × *veitchii*).

This is one of Gresham's hybrids, named by him in 1962. It received its name because of its dark, coriaceous foliage. The lower leaves sometimes show fall coloration. Flowers are colored a deep red-violet, with 12 tepals, the outer 8 forming a cup around the inner 4, as in *Magnolia campbellii*. A vigorous tree, blooming at about four years of age. For more information regarding this cross see chapter 9.

'Dark Shadow'. Gresham Hybrid of unknown parentage.

A compact tree with deep reddish purple flowers. The flowers have 9 tepals, the outer whorl reduced and sepaloid. The inner two whorls contain tepals 4 in. (10.2 cm) long. Selected from Todd Gresham's hybrids planted out at Gloster Arboretum, Gloster, Mississippi. Named by John Allen Smith, Magnolia Nursery, Chunchula, Alabama.

'Dark Splendor'. *Magnolia liliiflora* 'Nigra' × (*M.* × *soulangiana* 'Rustica Rubra').

Otto Spring of Otto Spring Nursery, Okmulgee, Oklahoma, backcrossed *Magnolia* × *soulangiana* 'Rustica Rubra' with *M. liliiflora* 'Nigra', producing a dark-flowered form. He describes it as a tall, shapely, everblooming tree. Flowers are dark velvety red with little or no purple shading. Introduced by Spring in 1966, registered in 1970. Sometimes listed as a cultivar of *M.* × *soulangiana*. See also 'Orchid Beauty' and 'Red Beauty'.

'Darrell Dean'. (*M.* × *soulangiana* 'Rustica Rubra') × (*M.* × *veitchii*).

A Gresham Hybrid with large, wine-red flowers to 12 in. (30.5 cm) across with 9–12 tepals. Flowers are poised horizontally on the branches. Later to bloom than most Gresham Hybrids. Selected from Gresham's material at Gloster Arboretum and registered by Ken Durio of Louisiana Nursery, Opelousas, Louisiana, in 1984. Named in honor of the Durios' late son. For more information regarding this cross see chapter 9. **Plate 114.**

'Daybreak'. (*M.* × *brooklynensis* 'Woodsman') × 'Tina Durio'.

This cross was made by August Kehr, Hendersonville, North Carolina, using

'Woodsman' as the seed parent. Flower size is similar to that of 'Tina Durio', but flowers are light rose pink in color. Flowers are extremely fragrant. This upright form blooms later than the Gresham Hybrids, mid May in Hendersonville, North Carolina, Zone 6. Registered by Kehr in 1991.

'Delicatissima'. (*Magnolia* × *soulangiana* 'Lennei Alba') × (*M.* × *veitchii* 'Rubra').

This Gresham Hybrid is a compact grower. It has fragrant, campanulate flowers which are smaller than most from this cross. Flowers have 9 white tepals tinged with rose-pink. Tepals are arranged in three whorls of 3 tepals. Flowers February–March in Santa Cruz, California, Zone 9. Named by Gresham in 1962. For more information regarding this cross see chapter 9.

'Early Rose'. *Magnolia campbellii* × *M. liliiflora.*

This cross was made by Oswald Blumhardt of Whangarei, New Zealand in the 1970s. Flowers are pink, much like those of *Magnolia campbellii,* but plants flower at about three years from seed. No other information is available at this time.

'Elisa Odenwald'. (*M.* × *soulangiana* 'Lennei Alba') × (*Magnolia* × *veitchii*).

Flowers are creamy white with light pink tones at the base of the inner tepals. The flowers are fragrant and develop to 12 in. (30.5 cm) across. Named and registered by Ken Durio of Louisiana Nursery, Opelousas, Louisiana, in 1984. Parent material is from Gresham's original material at Gloster Arboretum, Gloster, Mississippi. For more information regarding this cross see chapter 9.

'Elizabeth'. *Magnolia acuminata* × *M. denudata.*

This cross was made by Evamaria Sperber of the Brookyn Botanic Garden in 1956. It was selected as BBG#391 in 1977 and named 'Elizabeth' in 1978. Growth habit is pyramidal to 20 ft. (6.1 m). Young leaves are copper-colored when unfolding, becoming dark green and obovate. The fragrant, precocious flowers are uniformly yellow, with 6–9 spatulate tepals. Stamens are rose colored as in *Magnolia denudata.* Flower size is similar to that of *M. denudata,* but flower shape is intermediate between that of the parents. The flowers of 'Elizabeth' are not as deep in color as some of the new forms, but they are larger. This is a pentaploid form ($2n = 95$). This hybrid is usually sterile, but the plant is easily propagated from cuttings. 'Elizabeth' blooms in April and May in New York City (Zone 6), late enough to avoid frost damage. When mature, tree form and leaves most resemble those of *M. acuminata.* Hardy to Zone 5.

'Elizabeth' is named in honor of Elizabeth Scholz, director of the Brooklyn Botanic Garden, and was granted U.S. Plant Patent #4145 in 1978. This was one of the first precocious yellow-flowered magnolias to become available. Although many new forms are now coming into the market, it is still one of the finer ones, with its large flowers, rose-red stamens, and hardiness quality. **Plates 115–116.**

'Emma Cook'. *Magnolia denudata* × *M. kobus* var. *stellata* 'Waterlily'.

Dr. Frank Galyon of Knoxville, Tennessee, bred this plant which is a small tree with growth habit similar to that of *Magnolia denudata,* the seed parent. Flowers are about 6 in. (15.2 cm) across and lavender-pink in color. The inner 9–11 tepals are broader than those of *M. kobus* var. *stellata.* The outer 3 tepals are sepaloid. This hybrid was registered by Galyon in 1975. J. C. McDaniel, Urbana, Illinois, made a reciprocal cross. McDaniel's plants closely resemble *M. denudata*.

'Fenicchia Hybrid'. *M. liliiflora* 'Nigra' × (*Magnolia* × *soulangiana* 'Lennei').

This cross was made in 1953 by Richard A. Fenicchia of Rochester, New York. 'Fennicchia Hybrid' flowers later than *Magnolia* × *soulangiana*, thus escaping some frost damage. Flowers are reddish purple and larger than those of *M. liliiflora*. The tree is more vigorous than either parent, displaying hybrid vigor.

'Fireglow'. *Magnolia cylindrica* × *M. denudata* 'Sawada's Pink'.

This hybrid was bred by Phil Savage of Bloomfield Hills, Michigan, using *Magnolia cylindrica* as the pollen parent. 'Fireglow' is an upright tree with thick, leathery leaves and white flowers. The 6 tepals are tinged with magenta on the lower third and have a magenta stripe down the midrib. It was registered by Savage ca. 1985.

'First Flush'. *Magnolia campbellii* × (*M.* × *soulangiana* 'Amabilis').

This form is the result of hybridizing work by Oswald Blumhardt of Whangarei, New Zealand. 'First Flush' blooms quite early in the spring. Flowers are white and flushed with pink on the lower half of the outside of the tepals. Named by Blumhardt in 1982.

'Frank Gladney'. *Magnolia campbellii* × pink Gresham Hybrid (?).

This form was named and registered by Ken Durio of Louisiana Nursery, Opelousas, Louisiana, in 1984. It has cup-and-saucer shaped flowers 10–12 in. (25.4–30.5 cm) across. The 12 tepals are pink outside and creamy white inside. This form blooms later than *Magnolia campbellii*.

'Freeman'. *Magnolia grandiflora* × *M. virginiana*.

Oliver M. Freeman of the U.S. National Arboretum (USNA), Washington, D.C., made a series of hybrids using these two parents in 1930. The hybrids most closely resemble *Magnolia grandiflora*. 'Freeman' flowers are fragrant and about 4–6 in. (10–15 cm) across, which is intermediate between the two parents. Flowers appear in May–June and sporadically throughout the remainder of the summer. 'Freeman' is evergreen, developing as a tree or shrub with a columnar growth habit. It is mostly sterile. This form was named by William Kosar, USNA, Washington, D.C., in 1962 in honor of Freeman. See "U.S. National Arboretum," chapter 9, for more on this hybrid.

'Full Eclipse'. Gresham Hybrid of unknown parentage.

An early-blooming Gresham Hybrid with 9 erect tepals, pointed and reflexed slightly. The outer surfaces of the tepals are red-purple, with a dark purple stripe in the center of each tepal. Inner surfaces are white. Growth habit is upright and columnar. Selected and named by John Allen Smith, Magnolia Nursery, Chunchula, Alabama, from Todd Gresham's hybrids at Gloster Arboretum, Gloster, Mississippi.

'Galaxy'. *Magnolia liliiflora* 'Nigra' × *M. sprengeri* 'Diva'.

This hybrid was created by William Kosar of the U.S. National Arboretum in 1963, using 'Nigra' as the seed parent. 'Galaxy' is a single-stemmed, upright tree with a pyramidal or almost columnar shape. At twenty-five years of age, the parent tree is 30 ft. (9.1 m) tall by 20–25 ft. (6.1–7.6 m) wide. Flowers are 8–10 in. (20.3–25.4 cm) across and dark red-purple in bud, opening to a lighter shade of reddish purple outside, light rose-purple inside. Flowers have 12 tepals and are slightly fragrant. The flowers open after most frosts have passed. The original tree first flowered at nine years from seed. Hardy to Zones (4)5–8. 'Galaxy' is a partially sterile pentaploid ($2n = 95$), easily reproduced from cuttings. Named by Frank Santamour of the U.S. National Arboretum in 1963. **Plate 117.**

'George Henry Kern'. *M. kobus* var. *stellata* × *Magnolia liliiflora.*

This *Magnolia kobus* var. *stellata* seedling was discovered in Cincinnati, Ohio, by Carl E. Kern of Wyoming Nursery, Cincinnati, Ohio. Its name, given in 1948, commemorates Carl Kern's son. *Magnolia* × *soulangiana* has been listed as the pollen parent in some sources, but it is most likely a spontaneous hybrid cross between *M. kobus* var. *stellata* and *M. liliiflora.* It differs from *M. kobus* var. *stellata* in having 8–10 broad tepals of a rich red-rose outside, lighter in color inside. It blooms later than *M. kobus* var. *stellata* yet does not appear to be any hardier. Reportedly it begins flowering around April and continues through July. Awarded U.S. Plant Patent #820 in 1949.

'Goldfinch'. *Magnolia acuminata* var. *subcordata* 'Miss Honeybee' × *M. denudata* 'Sawada's Cream'.

This hybrid was created by Phil Savage, Bloomfield Hills, Michigan, using 'Miss Honeybee' as the seed parent. It is a tall, upright form which produces precocious flowers very early in the season. Flowers are light yellow. Savage reports that 'Goldfinch' bloomed at four years from seed. Registered in 1989.

'Griffin'. *M. grandiflora* × *M. virginiana.*

This form is often attributed to *Magnolia grandiflora* but is probably the result of introgressive hybridization between *M. grandiflora* and *M. virginiana* var. *australis.* 'Griffin' is a compact tree with a spreading growth habit. Leaves are small, thick, and a lustrous dark green. Flowers are large with 12 tepals. This form is fertile. The original tree is in the city park in Griffin, Georgia. It was named by J. C. McDaniel, Urbana, Illinois, in 1965.

'Hawk'. *Magnolia campbellii* × *M. sargentiana* var. *robusta.*

This form is the result of a cross made in 1951 by Charles Michael, gardener at Caerhays Castle, Cornwall, England. 'Hawk' was a seedling from Michael's cross which was grown at Chyverton and named about 1972 or 1973 by Nigel Holman, Cornwall. 'Hawk' produces abundant flowers at about ten years of age. Flower color is rose-red and similar to that of *Magnolia campbellii* var. *mollicomata* 'Lanarth'.

'Heaven Scent'. *M. liliiflora* × (*M.* × *veitchii*).

This is a Gresham Hybrid with fragrant, pinkish lavender flowers similar to those of 'Royal Crown' (which see) but lighter in color. Smaller in stature than others of this cross. For more information regarding this cross see chapter 9.

'Helen Fogg'. *Magnolia denudata* 'Sawada's Pink' × (*M.* × *veitchii* 'Peter Veitch').

This tree is another bred by Phil Savage, Bloomfield Hills, Michigan. 'Helen Fogg' is a vigorous, symmetrical tree bearing white flowers. The lower half of the tepals are pink. Savage reports that this is the hardiest 'Peter Veitch' hybrid he has grown. It is fertile. Named by Savage in 1989.

'Iolanthe'. 'Mark Jury' × (*M.* × *soulangiana* 'Lennei').

Felix Jury of North Taranaki, New Zealand, used pollen of 'Mark Jury' to pollinate flowers on the fertile *Magnolia* × *soulangiana* 'Lennei'. Jury introduced 'Iolanthe' in 1974. This is a vigorous, fast-growing form showing the strong apical dominance of *M. campbellii*, a pollen parent of 'Mark Jury'. 'Iolanthe' produces none of the bushy, straggly growth of *M.* × *soulangiana* 'Lennei'. It flowers at an amazingly early four years from seed. Flowers are 10–11 in. (25.4–27.9 cm) in diameter with 9 broad tepals. They are a clear pink color, lighter than 'Lennei' outside and white inside. The flowers are poised sideways on the branches much like those of the form's grandparent *M. sargentiana* var. *robusta*, with a tendency toward the cup-and-saucer shape of *M. campbellii*. The large buds and hairy perules are also reminiscent of *M. campbellii*. Flowering occurs January–February in the southern U.S., March in the northwest U.S., May in the northeast U.S. 'Iolanthe' is known for flowering over a long period of time. Hardy to Zone 6. This plant has survived brief periods of −12°F (−25°C). **Plate 118.**

'Ivory Chalice'. *Magnolia acuminata* × *M. denudata*.

This form was registered in 1985 by David Leach of Madison, Ohio. It is precocious, with yellow to yellow-green tepals. Anthers are tipped with red. Flowers are about 6 in. (15 cm) across. Hardy to −22°F (−30°C), Zone 4.

'Jane'. *M. kobus* var. *stellata* 'Waterlily' × *M. liliiflora* 'Reflorescens'.

This is the result of a cross made in 1956 by William Kosar, horticulturist at the U.S. National Arboretum (USNA). The fragrant flowers are cup-shaped and 3–4 in. (7.6–10.2 cm) in diameter, with 8–10 tepals. The tepals are colored red-purple outside, white inside. It is late-blooming, early to mid May at the USNA (Zone 7). The margins of the tepals roll inwards as in *Magnolia liliiflora*. Flower buds are erect and slender, a uniform red-purple in color. Leaves are much like those of *M. liliiflora* in size (about 6 in. [15.2 cm] long), shape, and texture. For more information regarding this cross see chapter 9.

'Jersey Belle'. *Magnolia sinensis* × *M. wilsonii*.

This form is the result of a natural cross. Violet Lort-Phillips of Jersey grew the original plant from seed of *Magnolia wilsonii* in 1970. 'Jersey Belle' has large white flowers 6–8 in. (15.2–20.3 cm) across with 9 tepals and violet-purple stamens. Leaves are similar to those of *M. wilsonii*, being elliptic, about 9 in. (23 cm) long by 5 in. (12.7 cm) wide. Growth habit is upright, similar to that of *M. wilsonii*, and 'Jersey Belle' will mature at probably 10–15 ft. (3–4.6 m). The fruits are pink and ovoid.

'Jersey Belle' shares the same parentage as the putative hybrid *Magnolia* × *highdownensis*. The Highdown Magnolia is now considered a form of *M. wilsonii* for reasons discussed under that species. The invalidation of the Highdown Magnolia as a hybrid leaves 'Jersey Belle' as the first definite hybrid between these two parents. Named and registered by Violet Lort-Phillips in 1981. **Plate 119.**

'Joe McDaniel'. (*M. × soulangiana* 'Rustica Rubra') × (*Magnolia × veitchii*).

A Gresham Hybrid with tulip-shaped buds, opening to bowl-shaped flowers, deepest purple of this cross. Registered by Ken Durio of Louisiana Nursery, Opelousas, Louisiana, in 1984 and selected from Gresham's material at Gloster Arboretum, Gloster, Mississippi. For more information regarding this cross see chapter 9.

'Jon Jon'. Gresham Hybrid of unknown parentage.

A late-flowering Gresham Hybrid which flowers profusely. Tepals deep reddish purple at base, becoming lighter toward the tip. Tepals 5.5 to 6 in. (14.0 to 15.2 cm) long. Selected and named by John Allen Smith, Magnolia Nursery, Chunchula, Alabama, from hybrids of Todd Gresham planted at Gloster Arboretum, Gloster, Mississippi.

'Judy'. *M. kobus* var. *stellata* × *Magnolia liliiflora* 'Nigra'.

This hybrid was part of the Kosar/deVos breeding program at the U.S. National Arboretum (USNA). 'Judy' resulted from a cross made in 1955 by Dr. Francis deVos, research geneticist at the arboretum. It is a slow-growing form with a fastigiate growth habit. Flowers are small with 10 tepals, red-purple, and white inside. Flower buds are candle-like, pointed, and red-purple at base, gradually paling toward the apex. Blooms are 2–3 in. (5–7.6 cm) in diameter appearing in mid to late April at the USNA, Washington, D.C. (Zone 7). Leaves are broadly ovate, about 6 in. (15.2 cm) long, and glabrous. For more information regarding this cross see chapter 9. **Plate 120.**

'Karl Flinck'. *Magnolia macrophylla* × *M. virginiana*.

This hybrid was produced by Phil Savage of Bloomfield Hills, Michigan, and named in 1989. Savage used *Magnolia virginiana* as the seed parent. These plants are quite vigorous. Leaves are glossy, 14 in. (35.6 cm) long and 4 in. (10.2 cm) wide, and resemble those of *M. virginiana*. This is a hardy clone with flowers intermediate between those of the parents. The inner tepals have purple blotches at the base as in *M. macrophylla*. Sister seedling 'Birgitta Flinck' has pure white flowers. Savage refers to this cross as *Magnolia × flinckii*, but that name has not yet been validly published with an accompanying Latin description. Hardy to at least Zone 5.

'Kew's Surprise'. See *Magnolia campbellii*, chapter 7.

'Lacey'. *Magnolia denudata* × ?

This plant apparently originated before 1955 in France and was brought to the Lacey Estate in Louisiana. It was introduced into the trade by Overlook Nurseries of Semmes, Alabama. Flowers are much larger than those of *Magnolia denudata*, opening to 8 in. (20.3 cm) across. Tepals are white with a pink spot at the base and a pink midrib. Leaves are typical of *M. denudata*. 'Lacey' has a bushier habit than *M. denudata* and is often considered unmanageable. It is not particularly outstanding and is not widely grown.

'Leather Leaf'. (*Magnolia × soulangiana* 'Lennei Alba') × (*M. × veitchii*).

This Gresham Hybrid has large, crinkled, coriaceous leaves. The white flowers are fragrant and 12 in. (30.5 cm) across. Tepals are arranged in three whorls of 3 tepals. Selected by Todd Gresham, date unknown. For more information regarding this cross see chapter 9.

'Legacy'. *Magnolia denudata* × *M. sprengeri* 'Diva'.

This hybrid was produced by David Leach, North Madison, Ohio. The flowers are 9 in. (22.9 cm) in diameter with 8–11 tepals. Flower color is red-purple outside and white inside, giving the garden effect of a soft, clear pink. The stamens are pink and ivory, the gynoecium green. The original tree is 26 ft. (8.1 m) tall and uninjured at −24°F (−31°C). 'Legacy' blooms mid to late April in North Madison, Ohio. Registered by Leach in 1991. **Plate 121.**

'Lotus'. 'Mark Jury' × (*M.* × *soulangiana* 'Lennei Alba').

'Lotus' was created by Felix Jury, North Taranaki, New Zealand. It is a small, upright pyramidal tree with large cream-colored flowers resembling those of the lotus flower (*Nelumbo* spp.). The individual tepals are spatulate. This hybrid is not as floriferous or as precocious as other Felix Jury hybrids. Sister seedlings are 'Athene' and 'Milky Way'. **Plate 122.**

'Manchu Fan'. (*M.* × *soulangiana* 'Lennei Alba') × (*Magnolia* × *veitchii*).

A Gresham Hybrid with flowers similar to those of *M. denudata*, but cream white with thicker tepals. Flowers smaller than those of other Gresham Hybrids. For more information regarding this cross see chapter 9.

'Marillyn'. *Magnolia kobus* × *M. liliiflora* 'Nigra'.

Magnolia liliiflora 'Nigra' was used as the seed parent in this 1954 cross by Evamaria Sperber of the Brooklyn Botanic Garden (BBG), Brooklyn, New York. 'Marillyn' has a multistemmed, shrubby habit and has grown to 15 ft. (4.6 m) in height in about thirty-five years. The fragrant flowers are held upright and semi-open throughout most of the flowering period. The 6 tepals are reddish purple outside and lighter inside with dark veins. Stamens are purple. Leaves are elliptic, about 6 in. (15 cm) long, and copper-colored when expanding. 'Marillyn' resembles the Kosar/deVos "Eight Little Girls" and, carrying a female cultivar name, is likely to be confused with them. It differs from the "Girls" in its looser growth habit and greater hardiness (Zones 4–5). This form is a sterile triploid (2n = 57). 'Marillyn' is named in honor of a Brooklyn Botanic Garden friend and benefactor. This plant was registered in 1989 by Lola Koerting of the BBG. **Plate 123.**

'Marj Gossler'. *Magnolia denudata* × *M. sargentiana* var. *robusta*.

This cross was produced by Phil Savage of Bloomfield Hills, Michigan, using pollen of *M. sargentiana* var. *robusta* collected by Marjory and Roger Gossler of Gossler Farms Nursery, Springfield, Oregon. This form was selected by Savage and named by him in 1989. Flowers of 'Marj Gossler' closely resemble those of *Magnolia* × *veitchii* in tepal color, size, and shape. Flowers are large and fragrant with 7–8 tepals. Tepals are white with pink at the base, paling at the tips on the outside. Flowers are 10–12 in. (24.5–30.5 cm) across. Flower buds are colored a reddish pink. This plant grows about 2 ft. (0.6 m) per year. Hardy to Zone 5. Savage named this cross *Magnolia* × *gossleri*, but this name has not yet been validly published with an accompanying Latin description.

'Mark Jury'. *Magnolia campbellii* var. *mollicomata* 'Lanarth' × *M. sargentiana* var. *robusta* (?).

This plant was named by Felix Jury of North Taranaki, New Zealand, from a naturally occurring cross between these presumed parents. It was imported from Hillier and Sons Nursery in England by Jury in 1950 as a seedling of *Magnolia campbellii* var. *mollicomata* 'Lanarth'. The parent from which this seedling came grows adjacent to *M. sargentiana* var. *robusta* and apparently was pollinated by that plant. It flowered in nine years from seed, producing heavily textured flowers much like *M. campbellii* var. *mollicomata* in size and shape but of a more lavender color. Flowers are 10–11 in. (2.45–27.9 cm) across, cream with lavender undertones. This hybrid has not been widely grown because of its large size and long time to flower. Breeders at the Mark Jury Nursery regard it as "a better breeder parent than a garden plant." It is female sterile but has been used as a pollen parent for the following hybrids created by Felix Jury: 'Athene', 'Atlas', 'Iolanthe', 'Lotus', 'Milky Way', and 'Serene'.

'Maryland'. *Magnolia grandiflora* × *M. virginiana.*

This is the result of a cross made by Oliver M. Freeman of the U.S. National Arboretum (USNA), Washington, D.C. 'Maryland' is a tree with a spreading growth habit. Its flowers are slightly larger than those of sister seedling 'Freeman'; leaves are lighter green. Selected mainly for the ease of propagation by cuttings. Named by Frederick Meyer, plant taxonomist at the USNA, Washington, D.C. in 1971. See chapter 9 for more information on this cross. **Plates 124–126.**

'Mary Nell'. (*Magnolia* × *soulangiana* 'Lennei Alba') × (*M.* × *veitchii* 'Peter Veitch').

This Gresham Hybrid blooms prolifically with cup-shaped, 10 in. (25.4 cm) diameter flowers. The flower color is white with purple coloration at the base of the thick tepals. 'Mary Nell' is a vigorous shrubby form with olive-green foliage. Selected by Ken Durio of Louisiana Nursery, Opelousas, Louisiana, from one of Gresham's crosses. Named in 1986 in honor of Mary Nell McDaniel, wife of the late Dr. J. C. McDaniel. For more information regarding this cross see chapter 9.

'Melanie'. See 'Purple Princess'.

'Michael Rosse'. *Magnolia campbellii* var. *alba* × *M. sargentiana* var. *robusta.*

This hybrid is the result of a chance cross between adjacent trees in the garden at Caerhays Castle, Cornwall, England. The seeds were grown out at Hillier and Sons Nursery, and this seedling was purchased by Nymans Gardens in Sussex. A sister seedling is thought to have been sent to Windsor Great Park (see 'Princess Margaret'). 'Michael Rosse' is an erect-growing tree with elliptic to obovate leaves similar to those of *Magnolia campbellii*. Flowers have the typical shape of *M. campbellii* and are of a deep pink color. Treseder (1978) reports a similar hybrid from his nursery (see 'Moresk').

'Milky Way'. 'Mark Jury' × (*Magnolia* × *soulangiana* 'Lennei Alba').

This Felix Jury hybrid has heavy-textured white flowers with a soft pink base. The form is very floriferous and blooms at an early age. It is a sister seedling to 'Athene' and 'Lotus'.

'Moresk'. *Magnolia campbellii* var. *alba* × *M. sargentiana* var. *robusta.*

This hybrid was purchased by Treseder from Caerhays Castle, Cornwall, England. Named by Treseder. 'Moresk' may have resulted from the same cross at Caerhays which produced 'Michael Rosse' and 'Princess Margaret'. Flowers have a cup-and-saucer shape resembling that of *Magnolia campbellii* flowers. Flowers are deep rose-pink, with

11–12 tepals arranged in three whorls, 8–10 in. (20.3–25.4 cm) across. Elliptic leaves are 6–8 in. (15.2–20.3 cm) long, also similar to those of *M. campbellii*.

'Mossman's Giant'. *Magnolia campbelli* var. *mollicomata* (?) × *M. sargentiana* var. *robusta* (?).

This plant originated as a chance seedling at Iufer Nursery in Salem, Oregon. The seedling was purchased from the nursery by Frank Mossman of Vancouver, Washington, about 1963. The parentage of this hybrid is unknown, but flower and leaf characteristics suggest a cross between *Magnolia sargentiana* var. *robusta* and *M. campbellii* var. *mollicomata*. 'Mossman's Giant' refers to the flower size, which can be 14–15 in. (35.6–38 cm) in diameter. Flowers resemble those of *M. campbellii* var. *mollicomata* in their somewhat cup-and-saucer shape, but the heavy tepals flop after a few days. Flower texture is like that of *M. sargentiana* var. *robusta*. Tepals are pale rose-violet outside and creamy white inside. Stigmas and pistils are reddish purple. The leaves resemble those of *M. campbellii* var. *mollicomata*. Mossman's plant bloomed at twelve or thirteen years of age and appears to be hardy to −12°F (−24°C), about Zone 5. Registered by Mossman in 1986.

'Nimbus'. *Magnolia hypoleuca* × *M. virginiana*.

This cross was made by William Kosar in 1956 at the U.S. National Arboretum (USNA), Washington, D.C. 'Nimbus' was registered by Frank Santamour of the USNA in 1980. This hybrid resembles *Magnolia* × *thompsoniana* in overall appearance. It is an upright, partially evergreen, multistemmed tree which flowers at eight years from seed. Leaves are elliptic, up to 12 in. (30.5 cm) long, and arranged in false whorls at the ends of the branches, arrangement and size being much like those of the seed parent, *M. hypoleuca*. They are dark green and glossy above and glaucescent beneath, as in the pollen parent, *M. virginiana*. The fragrant flowers are made up of 11 tepals in four whorls and are about 6 in. (15.2 cm) across, which makes them intermediate in size between the parents. The outer whorl of tepals is green and calyx-like. The inner three whorls are creamy white. Stamen filaments are reddish purple. This form is sterile, vigorous, and hardy to Zone 6. Flowers in May in Washington, D.C. (Zone 7). **Plate 128.**

'Northwest'. *Magnolia liliiflora* 'Nigra' × *M. sprengeri* 'Diva'.

This form is of the same parentage as 'Galaxy' and 'Spectrum'. It was named by Irene Burden of Hazel Dell Gardens, Canby, Oregon. Flowers are much like those of 'Galaxy' in color, but the tepals are less floppy. It is not quite as bushy as 'Galaxy' in growth habit but develops at about the same rate. Burden reported that 'Northwest' has proven quite hardy in Oregon, undamaged by a drop to 5°F (−15°C) after about a month of warm temperatures. This weather pattern damages many of the old standby *Magnolia* × *soulangiana* cultivars. Hardiness may be similar to that of 'Galaxy', which is Zones (4)5.

'Orchid'. *Magnolia liliiflora* × *M. kobus* var. *stellata*.

This hybrid arose as a chance seedling of *Magnolia kobus* var. *stellata* at Hillenmeyer's Nursery in Lexington, Kentucky. It was named by Louis Hillenmeyer, Jr., in 1961. Based on the characteristics of the hybrid, parentage is presumed to be *M. liliiflora* with *M. kobus* var. *stellata* as the pollen parent. In his study of floral anthocyanins of magnolias, Santamour (1965) reported that 'Orchid' could not be of that parentage. In his study of flavonoids in magnolias the following year, however, he suggested that 'Orchid', as well as some cultivated forms of *M. liliiflora* 'Nigra', may be a cross between *M. liliiflora* and *M. kobus* var. *stellata*. Comparison of 'Orchid' with the Kosar/deVos "Eight Little Girls" leaves little doubt that this is the correct parentage.

'Orchid' is a round, shrubby plant with obovate leaves. Flower buds are curved as in *Magnolia liliiflora* and colored a dark reddish purple. The flowers have 9 tepals that remain extended and do not open very widely. The 6 inner tepals are uniformly reddish purple and sometimes appear twisted or have inrolled margins as in *M. liliiflora*. The outer 3 tepals are small and sepaloid. Stigmas and anthers are reddish purple. The flowers commonly produce a very faint fragrance. The principle virtue of this hybrid is that it is hardier than either parent, being hardy to at least Zone 5 and maybe Zone 4. **Plate 129.**

'Orchid Beauty'. *Magnolia liliiflora* 'Nigra' × (*M.* × *soulangiana* 'Rustica Rubra').

Otto Spring of Okmulgee, Oklahoma, backcrossed 'Rustica Rubra' with *Magnolia liliiflora* 'Nigra' to produce this hybrid. It has short branches and large, light purple flowers. Spring introduced 'Orchid Beauty' in 1966 and registered it in 1970. See sister cultivars 'Red Beauty' and 'Dark Splendor'.

'Osaka'. *Magnolia liliiflora?* × ?.

This form originated in Japan and has been cultivated in that country for some years. Its parental origin is unrecorded, and it is doubtful if 'Osaka' is grown anywhere outside Japan. The leaves and flowers are similar to those of *Magnolia liliiflora,* but the flowers are smaller with 8–9 white tepals with violet flushes inside and at the tips on the outer surface. Stamens have rose-colored filaments. Elliptic leaves are 4–8 in. (10.2–20.3 cm) long by 2–4 in. (5.1–10.2 cm) wide. The plant forms a spreading bush. It is possible that this form is a hybrid of the *M.* × *soulangiana* grex.

'Paul Cook'. (*Magnolia* × *soulangiana* 'Lennei' seedling) × *M. sprengeri* 'Diva'.

This hybrid results from the breeding program of Dr. Frank Galyon of Knoxville, Tennessee. It is a vigorous form, hardier than 'Diva' (Zones [6]7–9), and flowers at an earlier age. Flowers are about 10 in. (25.4 cm) across. The 6–9 tepals are lavender-pink outside and white inside; flower color seems a lighter version of the color of 'Diva'. 'Paul Cook' is a fertile form and produces vigorous, hardy seedlings. Registered by Galyon in 1975. **Plate 130.**

'Peppermint Stick'. *Magnolia liliiflora* × (*M.* × *veitchii*).

This is a Gresham Hybrid from one of his original crosses. Flowers are white, with violet shading on the midrib and base of the tepals. Inner tepals enclose the gynoecium as in the hybrid's grandparent *Magnolia campbellii*. Named by Gresham in 1962. For more information regarding this cross see chapter 9.

'Peter Smithers'. (*M.* × *soulangiana* 'Rustica Rubra') × (*Magnolia* × *veitchii*).

The flowers of this Gresham Hybrid are dark rose-red and 10 in. (25.4 cm) across with 9 tepals. Stamens and gynoecium are reddish. Named by Ken Durio of Louisiana Nursery, Opelousas, Louisiana, in 1985 in honor of Sir Peter Smithers of Switzerland. For more information regarding this cross see chapter 9.

'Phelan Bright'. Gresham Hybrid of unknown parentage.

Fragrant flowers 12–14 in. (30.5–35.6 cm) across and clear white with 12 tepals. The flower shape is similar to the cup-and-saucer shape of *Magnolia campbellii* flowers. The tree is upright and spreading with a single trunk and blooms later than most Gresham Hybrids. Zones (7)8–9. Selected by Tina Durio, Louisiana Nursery, Opelousas, Louisiana, and introduced in 1992. Named in honor of the secretary of the Magnolia Society.

Pickard Selections

The following hybrids were registered about 1980 by A. A. Pickard of Magnolia Gardens, Canterbury, Kent, England. The seed parent is *Magnolia* × *soulangiana* 'Picture'. The pollen parent is unknown. All are vigorous trees, hardy to Zone 5. Growth habit is upright unless noted otherwise.

'Pickard's Coral'   Fragrant, tulip-shaped flowers, white with pink undertones.

'Pickard's Cornelian'   Goblet-shaped flowers, dark wine-red in color.

'Pickard's Crystal'   Goblet-shaped flowers, white with pinkish flush at the base of the tepals.

'Pickard's Firefly'   Fragrant, goblet-shaped flowers, wine-red in color. Leaves are similar to those of 'Picture' but are more elongated and pointed.

'Pickard's Garnet'   Fragrant bowl-shaped flowers, sometimes double, with broad tepals. Flower color is dark wine-red, darker than 'Pickard's Ruby' or *Magnolia* × *soulangiana* 'Lennei'.

'Pickard's Glow'   Fragrant flowers wine-red, fading to white.

'Pickard's Maime'   Large, fragrant flowers, goblet-shaped, similar to flowers of *Magnolia* × *soulangiana* 'Picture' but with broader tepals and darker less brilliant color. **Plate 131.**

'Pickard's Opal'   Goblet-shaped flowers, white with slight tint of purple at the base of the tepals.

'Pickard's Pearl'   Fragrant goblet-shaped flowers, white tinged with pink.

'Pickard's Pink Diamond'   Fragrant tulip-shaped flowers with broad tepals, white with pink undertones.

'Pickard's Ruby'   Fragrant goblet-shaped flowers, wine red in color. Flower color similar to that of *Mangolia* × *soulangiana* 'Lennei', but flowers of 'Pickard's Ruby' are larger, about 11 in. (28 cm) across. Growth habit nearly fastigiate.

'Pickard's Schmetterling'   Fragrant, elongated flowers with narrow tepals. Flower color wine-red.

'Pickard's Snow Queen'   Large, pure white flowers; 6 broadly spatulate tepals, slightly reflexed margins.

'Pickard's Sundew'   Fragrant, goblet-shaped flowers white with pink tinges, in April–May. Fast grower. **Plate 132.**

'Pink Goblet'. Gresham Hybrid of unknown parentage.

This Gresham Hybrid produces goblet-shaped flowers. The 9 tepals are rose pink outside and white inside, giving an overall effect of pink. A vigorous tree selected from Gresham's hybrids at Gloster Arboretum, Gloster, Mississippi, by John Allen Smith, Magnolia Nursery, Chunchula, Alabama.

'Pinkie'. *M. kobus* var. *stellata* 'Rosea' × *Magnolia liliiflora* 'Reflorescens'.

This cultivar is a result of the Kosar/deVos breeding work at the U.S. National Arboretum. William Kosar made the cross in 1956 which resulted in 'Pinkie', a form with the lightest colored flowers of this group of hybrids. Flowers are cup-shaped and pinkish purple, 5–7 in. (12.7–17.8 cm) across, with 9–12 tepals. The inner surface of the tepals is white. Flower buds are short and stout with a blunt tip. Leaves are about 7 in. (17.8 cm) long and glabrous. 'Pinkie' blooms one to two weeks later than the other "Little Girls." For more information regarding this cross see chapter 9.

'Porcelain Dove'. *Magnolia globosa* × *M. virginiana*.

This hybrid was created by the late D. Todd Gresham in 1965 at his home, Hill of Doves, in Santa Cruz, California. Gresham used *Magnolia globosa* as the seed parent. 'Porcelain Dove' was selected and named by Tom and Bill Dodd from hybrids Gresham shipped to their nursery in Semmes, Alabama. The leaves are much like those of *M. virginiana* and are semievergreen (Gresham must have used *M. virginiana* var. *australis* in his cross, although he did not list it as such). The flowers also resemble those of *M. virginiana* but are larger and quite fragrant. This hybrid received its name for its porcelain-white flower color and in honor of Todd Gresham's home, Hill of Doves. Named in 1986.

'Princess Margaret'. *Magnolia campbellii* var. alba × *M. sargentiana* var. *robusta*.

This hybrid is the result of a chance crossing of adjacent plants at Caerhays Castle, Cornwall, England. The seeds from *M. campbellii* var. *alba* were grown out at Hillier and Sons Nursery. This seedling was purchased from Hilliers by the Crown Estate at Windsor. Flowers reach 11 in. (27.9 cm) across and have 11 tepals. Tepals are rose-colored outside and cream-colored inside. See also 'Michael Rosse' and 'Moresk'.

'Pristine'. *Magnolia denudata* × *M. kobus* var. *stellata* 'Waterlily'.

This hybrid was made in 1968 by J. C. McDaniel, Urbana, Illinois, using *Magnolia kobus* var. *stellata* 'Waterlily' as the seed parent. The pollen parent was an established specimen of *M. denudata* which consistently bloomed later than was typical. Only one seedling was produced, and it displayed signs of a greater influence from the pollen parent, particularly in foliage characteristics and growth habit. Grafted plants bloom in about five years. The hybrids resemble *M. denudata* in growth habit and are easily propagated by cuttings. The flowers are pure white and are held upright on the branchlets as in *M. denudata*. Tepal number is intermediate between that of the parents. A vigorous form hardy to −15°F (−26°C). 'Pristine' was registered by McDaniel in 1979.

'Purple Eye'. *Magnolia denudata* × ?.

This form is sometimes listed as a cultivar of *Magnolia denudata*, but it is probably a backcross of *M.* × *soulangiana* with *M. denudata*. It probably originated as a seedling from

Caerhays Castle, Cornwall, England. This seedling was later given to Peter Veitch and distributed by Veitch and Sons Nurseries. Currently it is seldom seen in trade. Flowers are white with dark purple blotches at the base of the inner 3 tepals and lighter purple blotches on the outer 6. Stamen filaments are also purple. Tepals are broader and fleshier than those of *M. × soulangiana,* as are the leaves. 'Purple Eye' flowers about a week later than *M. denudata*— in February in the southern U.S., March–April in the Northwest, April–May in the Northeast.

'Purple Prince' *Magnolia liliiflora* 'Darkest Purple' × (*Magnolia* × soulangiana 'Lennei').

This cross was made by Dr. Frank Galyon of Knoxville, Tennessee, using 'Darkest Purple' as the seed parent. The flowers are about 10 in. (25 cm) across and are similar in shape to those of 'Lennei'. Tepals are dark purple on both sides, but slightly lighter inside. Tepal number is 6, with 3 sepaloid tepals similar to those of 'Lennei'. Each tepal is about 3 in. (8 cm) wide and 5.5 in. (14 cm) long. Original tree grew at the University of Tennessee Arboretum, Oak Ridge, Tennessee. Registered by Frank Galyon in 1976.

'Purple Princess'. *Magnolia liliiflora* 'Darkest Purple' × (*M. × soulangiana* 'Lennei').

This cross was made by Dr. Frank Galyon of Knoxville, Tennessee, using 'Darkest Purple' as the seed parent. It is a sister seedling to 'Purple Prince'. The flowers are 10 in. (25.4 cm) across, the tepals dark purple inside and outside. The flower shape is that of *Magnolia liliiflora.* Registered in 1975 as 'Melanie'; the name was changed to 'Purple Princess' by Galyon in 1976.

'Randy'. *M. kobus* var. *stellata* × *Magnolia liliiflora* 'Nigra'.

'Randy' is one of the "Eight Little Girls" or Kosar/deVos hybrids which were released from the U.S. National Arboretum (USNA). This hybrid results from a cross in 1955 by Dr. Francis deVos, research geneticist at the USNA. Flowers are 3–5 in. (7.6–12.7 cm) in diameter, with 9–11 red-purple tepals. The inner surface of the tepals is white. Flower buds are erect, pointed, and dark red-purple at the base, paling toward the apex. Growth habit is erect, almost columnar, and very floriferous. Flowers late April at the USNA, Washington, D.C. (Zone 7). For more information regarding this cross see chapter 9. **Plate 133.**

'Raspberry Ice'. *Magnolia liliiflora* × (*M. × veitchii*).

This is a Gresham Hybrid with flowers about 9 in. (23 cm) across, pinkish white with violet shading at the base of the 12 tepals. Shrubby habit; flowers about ten days after 'Royal Crown' (February–March in Santa Cruz, California). Named by Gresham in 1962. For more information regarding this cross see chapter 9. **Plate 134.**

'Raspberry Swirl'. *Magnolia liliiflora* 'Darkest Purple' × *M. sprengeri* 'Diva'.

Frank Galyon, Knoxville, Tennessee, made this cross using 'Diva' pollen. This hybrid has flowers which are dark purple inside and outside. Galyon suggests grafting or budding this form since it sends up multiple trunks on its own roots. Registered in 1990.

'Red Beauty'. *Magnolia liliiflora* 'Nigra' × (*M. × soulangiana* 'Rustica Rubra').

Otto Spring of Okmulgee, Oklahoma, backcrossed *Magnolia × soulangiana* 'Rustica Rubra' with *M. liliiflora* 'Nigra' to produce this rather tall, shapely tree. Flowers are large and almost scarlet in color. 'Red Beauty' was introduced in 1968 and registered in 1970. This form is sometimes listed as a cultivar of *M. × soulangiana*. See also 'Orchid Beauty' and 'Dark Splendor'.

'Ricki'. *M. kobus* var. *stellata* × *Magnolia liliiflora* 'Nigra'.

This U.S. National Arboretum (USNA) introduction resulted from a cross made in 1955 by Dr. Francis deVos, research geneticist at the USNA. Flowers appear reddish purple to purple and develop to 4–6 in. (10–15 cm) across, with 10–15 twisted tepals. The inner surface of the tepals is white to pale purple. The slender, pointed flower buds are red-purple at the base, fading to pale purple at the apex. 'Ricki' flowers in late April at the USNA, Washington, D.C. (Zone 7). For more information regarding this cross see chapter 9. **Plate 135.**

'Rouged Alabaster'. (*Magnolia × soulangiana* 'Lennei Alba') × (*M. × veitchii*).

This is a Gresham Hybrid with fragrant flowers to 12 in. (30.5 cm) in diameter, cup-shaped with 9 white tepals tinged rose-pink to purple at the base. The exceptionally thick tepals are arranged in three whorls of 3 tepals. Tepals are constricted just above the base and flared at the top and are 6.5 in. (16.5 cm) long by 4.5 in. (11.4 cm) wide. This is a vigorous form which sometimes sets seed. Flowers in February–March in Santa Cruz, California. Named by Gresham in 1962. The seed parent of 'Todd's Forty-Niner'. For more information regarding this cross see chapter 9.

'Royal Crown'. *Magnolia liliiflora* × (*M. × veitchii*).

This is a popular Gresham Hybrid clone with dark red to violet flowers, white on the inside. Flowers reach 12 in. (30.5 cm) across, with 12 tepals. The outer tepals have inrolled margins like those of *Magnolia liliiflora*. Blooms at an early age. Hardy to Zones (5)6. Named by Gresham in 1962. The pollen parent of 'Todd's Forty-Niner'. For more information regarding this cross see chapter 9. **Plate 136.**

'Royal Flush'. (*Magnolia × soulangiana* 'Lennei Alba') × (*M. × veitchii*).

This white Gresham Hybrid has a dark reddish rose color at the base of the tepals. Flowers are fragrant and reach 12 in. (30.5 cm) across. Tepals are arranged in three whorls of 3 tepals. Flowers in February–March in Santa Cruz, California. Named by Gresham in 1962. For more information regarding this cross see chapter 9.

'Sangreal'. Gresham Hybrid of unknown parentage.

Large red-purple flowers are produced on a vigorous, floriferous tree. The 9 tepals are 3.5 in. (8.9 cm) to 4 in (10.2 cm) long. Selected by John Allen Smith, Magnolia Nursery, Chunchula, Alabama, from Gresham's hybrids planted out at Gloster Arboretum, Gloster, Mississippi. **Plate 137.**

'Sayonara'. *Magnolia liliiflora* × (*M. × veitchii*).

This is a Gresham Hybrid with globular flowers, a pure white with rose-pink at the base of the tepals. Flowers are beautifully shaped, fragrant, and reach 12 in. (30.5 cm) across with 9 tepals arranged in three whorls of 3 tepals. Blooms later than most Gresham Hybrids—in March in Santa Cruz, California. Registered in 1966 by Gresham. 'Sayonara' received an Award of Merit from the Royal Horticultural Society in 1990. For more information regarding this cross see chapter 9.

'Serene'. *Magnolia liliiflora* × *M.* 'Mark Jury'.

This hybrid is the result of crosses made by Felix Jury of North Taranaki, New Zealand, in the 1970s. It is a small, upright, pyramidal tree flowering at about the same time as *Magnolia liliiflora*. The cup- or bowl-shaped flowers are held erect. The short, broad tepals are bright rose, similar to the flower color of *M. liliiflora* yet brigher. Flowers at an early age.

'Shirley Curry'. *Magnolia coco* × *M. grandiflora*.

S. Christopher Early of Atlanta, Georgia, selected this form in 1985 from seedlings of a cross between *Magnolia coco* and *M. grandiflora*. 'Shirley Curry' is a vigorous tree with flowers intermediate in size between the flowers of the parents. The leaves are also intermediate in size, and are shiny on top and rufous on the lower surface. Seed coats are bright red and ornamental. This plant most closely resembles *M. grandiflora* in its general appearance, as would be expected since *M. grandiflora* is hexaploid ($2n = 114$) compared to the diploid ($2n = 38$) *M. coco*. The leaves and flowers are smaller than those of *M. grandiflora*, however, suggesting a dwarfing influence deriving from *M. coco*. 'Shirley Curry' is fertile.

Hybridity has not been proven, but Early's crosses were controlled. With the exception of possible apomictic seed set, it is probable that the cross is as he reported. A similar putative hybrid was introduced as *Magnolia grandiflora* 'Hartwicus' by Baumann Nursery in France in 1842. It is doubtful that 'Hartwicus' is still in cultivation.

'Silver Parasol'. *Magnolia hypoleuca* × *M. tripetala*.

This form arose as the result of a natural cross between the parent species at the Arnold Arboretum, Harvard University, Cambridge, Massachusetts. It was grown from the seed of a specimen of *Magnolia hypoleuca* (its female parent) that was introduced by Sargent. 'Silver Parasol' has a pyramidal growth habit and smooth, silver-gray bark. Leaves are elliptic to obovate, 12–16 in. (30.5–40.7 cm) long by 5–8 in. (12.7–20.3 cm) wide. The fragrant flowers are 8–10 in. (20.3–25.4 cm) across with 9 (sometimes to 12) tepals. The outer 3 tepals are calyx-like, greenish, and reflexed, while the inner tepals are creamy white. Stamen filaments are reddish purple, anthers are white, and the gynoecium is rose-colored. The follicetum is 2–4 in. (5–10 cm) long and reddish pink. This form flowers in May–June at the Arnold Arboretum (Zone 6). Registered by Stephen Spongberg in 1981.

'Snow White'. See 'Wada's Snow White'.

'Spectrum'. *Magnolia liliiflora* 'Nigra' × *M. sprengeri* 'Diva'.

This hybrid, a sister seedling of 'Galaxy', results from a cross made in 1963 by William Kosar of the U.S. National Arboretum (USNA), Washington, D.C. Flowers are deep reddish purple, to 10–12 in (25.4–30.5 cm) across, similar to those of 'Galaxy' but larger, and more sparse, produced at ten years of age. The growth habit of 'Spectrum' is more broadly oval than that of 'Galaxy'. 'Spectrum' is not as cold hardy as 'Galaxy' yet performs better than 'Galaxy' in warmer climates. Zones 5–8. Registered in 1984 by Frank Santamour of the USNA. **Plate 138.**

'Spring Rite'. (*Magnolia* × *soulangiana* 'Lennei Alba') × (*M.* × *veitchii*).

This Gresham Hybrid has snowy white flowers tinged with a light rose at the base of the tepals. Flowers are fragrant and reach 12 in. (30.5 cm) across. The 9 tepals are arranged in three whorls of 3 tepals. Flowers February–March in Santa Cruz, California. A fast-growing form named by Gresham in 1962. For more information regarding this cross see chapter 9.

'Star Wars'. *M. campbellii* × *M. liliiflora.*

Oswald Blumhardt of New Zealand developed and named this hybrid in the 1970s. The large flowers are bright pink though darker in color than 'Early Rose', another Blumhardt hybrid of the same parentage. The flower display of 'Star Wars' is effective for almost a month. Blumhardt reports that the outer tepals of 'Star Wars' are rolled inward as in *Magnolia liliiflora*; other growers report "normal" tepals. 'Star Wars' is fertile and flowers at about four years of age. It is easily propagated from cuttings. Growth habit is similar to that of its seed parent, *M. campbellii*, with a strong central leader. This hybrid currently seems to be increasing in popularity, although its hardiness has not been widely tested.

'Sulphur Cockatoo'. (*Magnolia* × *soulangiana* 'Lennei Alba') × (*M.* × *veitchii*).

This Gresham Hybrid has large white flowers, the inner tepals tinted violet at the base. A slight yellowish tinge to the white is evident, especially in bud. Flowers reach 12 in. (30.5 cm) across with 9 tepals arranged in three whorls of 3 tepals. Flowers February–March in Santa Cruz, California. Named by Gresham in 1962. For more information regarding this cross see chapter 9.

'Sundance'. *M. acuminata* × *M. denudata.*

This form was selected by August Kehr of Hendersonville, North Carolina, as a seedling resulting from a cross made by J. C. McDaniel, Urbana, Illinois, using *M. acuminata* as the seed parent. The precocious, barium yellow flowers are about 8 in. (20.3 cm) across, appearing in late April to early May in Hendersonville (Zone 6). 'Sundance' is partly fertile, occasionally producing seed from open- or hand-pollination. Its pollen is quite viable. This is a vigorous form easily propagated from cuttings. It flowers at about four years of age and is hardy to at least −6°F (−21°C). Registered by Kehr in 1985.

'Susan'. *M. kobus* var. *stellata* 'Rosea' × *Magnolia liliiflora* 'Nigra'.

This U.S. National Arboretum (USNA) introduction results from a 1956 cross made by Mr. William Kosar, horticulturist at the USNA. Flowers are 4–6 in. (10–15 cm) across and red-purple. The 6 tepals are red-purple on both surfaces and sometimes twisted. Flower buds are erect and slender. Flowers mid to late April at the USNA, Washington, D.C. (Zone 7). Leaves are ovate, about 5 in. (12.7 cm) long, and glabrous. For more information regarding this cross see chapter 9.

'Sweet Sixteen'. (*Magnolia* × *soulangiana* 'Lennei Alba') × (*M.* × *veitchii*).

A Gresham Hybrid with fragrant flowers in pure white, opening in a wide cup-shape. Blooms later than *Magnolia denudata*—February to May depending on location. This is a more suitable white-flowered form if late frosts brown the tepals of *M. denduata*. Named by Ken Durio of Louisiana Nursery, Opelousas, Louisiana, in 1984. For more information regarding this cross see chapter 9.

'Sweet Summer'. *Magnolia grandiflora* 'Samuel Sommer' × *M. virginiana* var. *australis*.

This hybrid was created by Dr. Frank Galyon, Knoxville, Tennessee, in 1962, using *Magnolia virginiana* var. *australis* as the female parent. This form has large *grandiflora*-like flowers with 9–12 tepals. It is an allotetraploid that is fertile in both directions. Overall, the plant most closely resembles its male parent. Temperatures of −24°F (−31°C) resulted in bark splitting. Registered in 1990 by Frank Galyon.

'Tina Durio'. (*M.* × *soulangiana* 'Lennei Alba') × (*Magnolia* × *veitchii*).

This Gresham Hybrid's flowers are similar to those of *Magnolia campbellii*, 10–12 in. (25.4–30.5 cm) across with 9–12 tepals. The white flowers have a slight pink tinge at the base of the tepals. Named in 1984 by Ken Durio of Louisiana Nursery, Opelousas, Louisiana, in honor of his daughter. For more information regarding this cross see chapter 9. **Plates 139–140.**

'Todd Gresham'. (*M.* × *soulangiana* 'Rustica Rubra') × (*Magnolia* × *veitchii*).

This is a Gresham Hybrid with flowers colored red-purple outside and white inside, with reddish stamens and gynoecium. Flowers open wide to 10 in. (25.4 cm), with 9 tepals in three whorls of 3 tepals. This is a fast-growing form with decorative, bright red fruits. Named by Ken Durio of Louisiana Nursery, Opelousas, Louisiana, in 1984. For more information regarding this cross see chapter 9. **Plate 141.**

'Todd's Forty-Niner'. (*Magnolia* × 'Rouged Alabaster') × (*M.* × 'Royal Crown').

The late D. Todd Gresham, Santa Cruz, California, made this cross in 1964 between two of his hybrids using 'Rouged Alabaster' as the seed parent. Flowers are dark purple in bud and lighter in color when open, with 12 tepals. The outer 4 tepals are reflexed at anthesis, red-purple outside and lighter toward the tips. The inside of the tepals is a translucent white. The inner 8 tepals are held erect. Selected after Gresham's death from hybrids he sent to Tom Dodd Nurseries, Semmes, Alabama. Named by Bill Dodd in 1986 and registered by John Allen Smith, Magnolia Nursery, Chunchula, Alabama, in 1991. This was the forty-ninth cross made by Gresham in 1964.

'Treve Holman'. *Magnolia campbellii* var. *mollicomata* × *M. sargentiana* var. *robusta*.

This cross was made at Caerhays Castle, Cornwall, England, then purchased by Treseder's Nurseries and sold—incorrectly labeled as *Magnolia campbellii* × (*M.* × *soulangiana*)—to Nigel Holman in 1964. Holman grew the tree at Chyverton, Cornwall, England, where it became a vigorous tree of 30 ft. (9 m) in nine years. At twenty-one years of age it was a narrow tree 45 ft. (13.7 m) tall. In 1973, at eleven years of age, it produced a single flower in mid April. The flower was large, pink with purple undertones, and had the cup-and-saucer shape characteristic of *M. campbellii* var. *mollicomata*. This hybrid is considered hardier than its parents. It was named by Nigel Holman in 1973 and commemorates his father.

'Vin Rouge'. *Magnolia liliiflora* × (*M.* × *veitchii*).

This is a Gresham Hybrid named by Gresham in 1962. Young foliage growth is bronze with red veining and red stipules. Flowers are wine-red. For more information regarding this cross see chapter 9.

'Vulcan'. *Magnolia campbellii* var. *mollicomata* 'Lanarth' × *M. liliiflora* hybrid.

This recent Felix Jury introduction has brilliant ruby-red flowers of the *Magnolia campbellii* shape. It is a smaller tree than other Jury hybrids ('Athene', 'Atlas', 'Lotus', 'Milky Way', and 'Serene') and flowers at an early age. Introduced in 1990 by Mark Jury Nursery, North Taranaki, New Zealand. **Plate 142.**

'Wada's Snow White'. *M. denudata* × *M. salicifolia*.

This hybrid was developed by K. Wada of Hakoneya Nurseries, Yokohama, Japan. The plant displays hybrid vigor and flowers an an early age, with pure white, fragrant flowers. Registered by K. Wada, Hakoneya Nurseries, Japan, 1979.

'Winelight'. Gresham Hybrid of unknown parentage.

Flowers are large, opening later than those of most Gresham hybrids. Flowers have 9 thick tepals which are reddish purple at the base, becoming lighter at the tip, and 4–5 in. (10.2–12.7 cm) long. Selected from Todd Gresham's hybrids at Gloster Arboretum, Gloster, Mississippi, by John Allen Smith of Magnolia Nursery, Chunchula, Alabama.

'Yellow Bird'. *Magnolia acuminata* × (*M.* × *brooklynensis* 'Evamaria').

This is a hybrid developed by Doris Stone, plant breeder at the Brooklyn Botanic Garden, New York, in 1967. This pyramidal form was selected from the resulting seedlings for its flower color and consistency of bloom. Flowers are yellow with a slight greenish tinge at the base of the outer tepals, and are held upright on the branches. Tepals are 3–3.5 in. (7.6–8.9 cm) long and about 2 in. (5.1 cm) wide. 'Yellow Bird' flowers earlier in the year than *Magnolia acuminata* but later than the yellow-flowered 'Elizabeth', another Brooklyn Botanic Garden introduction. The flowers appear at the same time as the leaves and continue for two to three weeks. Leaves are elliptic, dark green, and glabrous. Bark is furrowed like that of *M. acuminata*. Introduced by the Brooklyn Botanic Garden in 1981.

'Yellow Fever'. *M. acuminata* × *M. denudata*.

This yellow form was registered by Ken Durio of Louisiana Nursery, Opelousas, Louisiana, about 1980. It is a vigorous, upright tree with precocious flowers blooming late enough in the season to escape most frost damage. Flower buds are yellow, fading to cream-color as they open. The fragrant flowers are 6–8 in. (15–20.3 cm) across. Tepals are light yellow with a light pink tinge on the outside, fading after opening.

'Yellow Garland'. *Magnolia acuminata* × *M. denudata*.

This hybrid was registered by David Leach of North Madison, Ohio, in 1985. The precocious flowers are 8 in. (20.3 cm) across with 6 tepals. Tepals are yellow with yellow-green midribs. Anthers are yellow with maroon bases.

'Yellow Lantern'. *Magnolia acuminata* var. *subcordata* × (*M.* × *soulangiana* 'Alexandrina').

This hybrid is another precocious yellow form bred by Phil Savage, using *M. acuminata* var. *subcordata* as the seed parent. 'Yellow Lantern' is an upright, single-stemmed tree which is fertile. The large flower buds are tomentose, opening to a uniform lemon-yellow, cup-shaped flower similar in form and size to the flower of the hybrid's parent 'Alexandrina'. Color is without greenish tinges. Registered by Savage in 1985.

REFERENCES AND ADDITIONAL READING

Callaway, Dorothy J. 1991. *Magnolia* 'Galaxy'. *American Nurseryman* 174(2):122.

Carrière, E. A. 1890. *Magnolia wieseneri*. *Revue Horticole* 62:406–407.

Dirr, Michael A. 1980. A new magnolia on the landscape. *Horticulture* 58(5):66. ['Ann'.]

———. 1991. Bright, bold and saucy, these magnolias can dress up streets and shade. *Nursery Manager* 7(6):38–40. ['Galaxy' and 'Spectrum'.]

Domoto, Toichi. 1962. *Magnolia* × *soulangiana* hybrids. *Journal of the California Horticultural Society* 23(1):45–57.

Dudley, Theodore R., and William F. Kosar. 1968. Eight new hybrid *Magnolia* cultivars. *Bulletin of the Morris Arboretum* 19(2):26–29.

Early, S. Christopher. 1986. Magnolia 'Shirley Curry'. *Journal of the Magnolia Society* 21(2):29.

Flinck, Karl E., and Stephen A. J. Spongberg. 1984. Is *Magnolia sinensis* × *wilsonii* 'Jersey Belle' the first hybrid of its kind? *International Dendrology Society Yearbook* 1984: 118–119. (Reprinted in *Journal of the Magnolia Society* 24[2]:11–12.)

Fogg, John. 1965. *Magnolia* × *soulangiana* 'Adral'. *Journal of the Magnolia Society* 2(1):7–8.

Freeman, Oliver M. 1937. A new hybrid magnolia. *The National Horticultural Magazine* 16(3):161–162.

Galyon, Frank B. 1990. Magnolia 'Alexandrina'. *Journal of the Magnolia Society* 25(2):16–18.

Gresham, D Todd. 1962. Deciduous magnolias of Californian origin. *Bulletin of the Morris Arboretum* 13(3):47–50. (Reprinted in the *Journal of the Royal Horticultural Society* 89:327–332.)

———. 1966. *Magnolia wilsonii* × *M. globosa*: a new hybrid. *Bulletin of the Morris Arboretum* 17(4):70–73.

Grey-Wilson, C., and M. Lear. 1985. *Magnolia* 'Albatross'. *Kew Magazine* 2(1):201–204.

Harkness, Bernard. 1954. A new hybrid magnolia. *National Horticultural Magazine* 33:118–120. [*M.* × *slavinii* = *M.* × *proctoriana*.]

———. 1961. Magnolia notes from Rochester. *Bulletin of the Morris Arboretum* 12(2):19. [*M.* × *proctoriana*.]

Hooker, Joseph D. 1891. *Magnolia watsonii*. *Curtis's Botanical Magazine* 117:t 7157.

Kalmbacher, George. 1972. *Magnolia* × *brooklynensis* 'Evamaria'. *Journal of the Magnolia Society* 8(2):7–8.

———. 1973. More on *Magnolia* × *brooklynensis* 'Evamaria'. *Journal of the Magnolia Society* 9(1):13–14.

Koerting, Lola E. 1977. A tree born in Brooklyn. *Journal of the Magnolia Society* 13(2):21–22. ['Elizabeth'.]

———. 1981. Introducing *Magnolia* × 'Yellow Bird'. *Journal of the Magnolia Society* 17(2):30.

———. 1984. *Magnolia* 'Hattie Carthan'. *Journal of the Magnolia Society* 20(2):17–18.

———. 1985. Introducing *Magnolia* 'Yellow Bird'. In *Rhododendrons 1984/85 with Magnolias and Camellias*. London: The Royal Horticultural Society. Pp. 51–52.

———. 1989. A new magnolia introduction by the Brooklyn Botanic Garden. *Journal of the Magnolia Society* 25(1):12–13. ['Marillyn'.]

Lindley, J. 1828. *Magnolia yulan* var. *soulangiana*. *Edwards Botanical Register* 14:t 1164.

Lort-Phillips, Violet. 1981. A Magnolia problem. *International Dendrology Society Yearbook* 1981. ['Jersey Belle'.]

Magnolia 'George Henry Kern'. In American Association of Nurseryman Registration of New Woody Plants. *American Nurseryman* 89(5)(1949):33–34.

Magnolia glauca var. *major*. Thomson's new Swamp Magnolia. *Curtis's Botanical Magazine* 47(1820):t 2164. [*M.* × *thompsoniana*.]

McDaniel, Joseph C. 1963. Recent hybridization with American magnolias. *Proceedings of the International Plant Propagators Society* 13:124–132.

_____. 1966. A new-old magnolia hybrid. *Illinois Research* 8(4):8–9.

_____. 1973. Illinois clones of *Magnolia* × *brooklynensis*. *Journal of the Magnolia Society* 9(1):13–14.

_____. 1974. Breeding the Woodsman magnolia. *American Nurseryman* CXL(3):88–90.

_____. 1975. Some Asiatic-American magnolia hybrids. *American Horticulturist* 54(1):10–13. [*M.* × *brooklynensis*.]

McDaniel, Joseph C., and S. D. Groves. 1969. 'Griffin' evergreen magnolia. *Proceedings of the International Plant Propagators Society* 19:376–377.

Pearce, S. A. 1952. *Magnolia kewensis*. *Gardeners Chronicle* 3, 132:154.

Rehder, Alfred. 1939. *Magnolia proctoriana*. New species, varieties and combinations from the collections of the Arnold Arboretum. *Journal of the Arnold Arboretum* 20:412–413.

Santamour, Frank S. 1965. Biochemical studies in *Magnolia* I. Floral anthocyanins. *Bulletin of the Morris Arboretum* 16(3):43–48.

_____. 1966a. Biochemical studies in *Magnolia* II. Leucoanthocyanins in leaves. *Bulletin of the Morris Arboretum* 16(4):63–64.

_____. 1966b. Hybrid sterility in *Magnolia* × *thompsoniana*. *Bulletin of the Morris Arboretum* 17(2):29–30.

_____. 1980a. 'Galaxy' Magnolia. *HortScience* 15(6):832.

_____. 1980b. 'Nimbus'—a new hybrid magnolia cultivar. *Journal of the Magnolia Society* 16(1):35–36.

_____. 1984. 'Spectrum': a new hybrid magnolia cultivar. *Journal of the Magnolia Society* 20(1):1–2.

Savage, Philip J. 1989. Magnolias in Michigan: Part 4. *Journal of the Magnolia Society* 23(1):5–10

Smith, Ralph H. 1964. A new late-flowering cultivar of *Magnolia* × *soulangiana*. *Bulletin of the Morris Arboretum* 15:64. [*M.* × *soulangiana* 'Adral'.]

Spongberg, Stephen A. 1976a. Magnoliaceae hardy in temperate North America. *Journal of the Arnold Arboretum* 57(3):250–312.

_____. 1976b. Some old and new interspecific magnolia hybrids. *Arnoldia* 36(4):129–145.

Spongberg, Stephen A., and Richard W. Weaver, Jr. 1981. 'Silver Parasol': a new magnolia cultivar. *Arnoldia* 41(2):70–77.

Stern, F. C. *Magnolia sinensis* × *wilsonii*. *New Flora and Silva* 10:105–107.

Treseder, Neil G. 1969. *Magnolia* × 'Lennei'. *Journal of the Magnolia Society* 6(2):5–6.

_____. 1978. *Magnolias*. London: Faber and Faber.

Williams, F. Julian. 1974. *Magnolia* 'Caerhays Surprise'. *Journal of the Royal Horticultural Society* 99:364–365.

APPENDIX A

MAGNOLIA INVALID NAME CROSS-REFERENCE

Invalid Name	Valid Name
M. acuminata var. alabamensis	M. acuminata var. subcordata
M. angustifolia	M. grandiflora
M. ashei	M. macrophylla var. ashei
M. aulacosperma	M. biondii
M. auricularis	M. fraseri
M. auriculata	M. fraseri
M. auriculata var. pyramidata	M. fraseri var. pyramidata
M. biloba	M. officinalis 'Biloba'
M. borealis	M. kobus 'Borealis'
M. candolii	M. acuminata
M. conspicua	M. denudata
M. conspicua var. elongata	M. sprengeri var. elongata
M. conspicua var. emarginata	M. sargentiana
M. conspicua var. fargesii	M. biondii
M. conspicua var. purpurascens	M. sprengeri
M. conspicua var. rosea	M. denudata
M. cordata	M. acuminata var. subcordata
M. cordifolia	M. acuminata var. subcordata
M. cubensis subsp. acunae	M. cubensis
M. cubensis var. baracoensis	M. cubensis
M. dealbata	M. macrophylla var. dealbata
M. decandollii	M. acuminata

Invalid Name	Valid Name
M. denudata var. *elongata*	*M. sprengeri* var. *elongata*
M. denudata var. *emarginata*	*M. sargentiana*
M. denudata var. *fargesii*	*M. biondii*
M. denudata var. *purpurascens*	*M. denudata*
M. denudata var. *purpurascens*	*M. sprengeri*
M. denudata var. *purpurea*	*M. liliiflora*
M. denudata var. *typica*	*M. liliiflora*
M. discolor	*M. liliiflora*
M. diva	*M. sprengeri* 'Diva'
M. elliptica	*M. grandiflora*
M. elongata	*M. sprengeri* var. *elongata*
M. emarginata	*M. sargentiana*
M. exoniensis	*M. grandiflora*
M. famasiha	*M. salicifolia*
M. fargesii	*M. biondii*
M. ferruginea	*M. grandiflora*
M. foetida	*M. grandiflora*
M. fragrans	*M. virginiana*
M. frondosa	*M. tripetala*
M. glauca	*M. virginiana*
M. glauca var. *kobus*	*M. kobus*
M. globosa var. *sinensis*	*M. sinensis*
M. halleana	*M. kobus* var. *stellata*
M. hartwegii	*M. grandiflora*
M. heptapeta	*M. denudata*
M. javanica	*M. macklottii*
M. kobus var. *borealis*	*M. kobus* 'Borealis'
M. lacunosa	*M. grandiflora*
M. lanceolata	*M. grandiflora*
M. liliifera	*M. coco*
M. liliifera var. *championi*	*M. championii*
M. liliifera var. *taliensis*	*M. wilsonii*
M. longifolia	*M. grandiflora*
M. maxima	*M. grandiflora*
M. membranacea var. *pealiana*	*M. pealiana*
M. microphylla	*M. grandiflora*
M. mollicomata	*M. campbellii* var. *mollicomata*
M. nicholsoniana	*M. sinensis*
M. nicholsoniana	*M. wilsonii*
M. obovata	*M. biondii*
M. obovata	*M. denudata*
M. obovata	*M. grandiflora*
M. obovata	*M. hypoleuca*
M. obovata	*M. liliiflora*
M. obovata var. *denudata*	*M. denudata*
M. obovata var. *discolor*	*M. liliiflora*
M. obovata var. *liliiflora*	*M. liliiflora*
M. obovata var. *purpurea*	*M. liliiflora*
M. obtusifolia	*M. grandiflora*
M. officinalis var. *biloba*	*M. officinalis* 'Biloba'
M. oyama	*M. sieboldii*

Invalid Name	Valid Name
M. palustris	*M. virginiana*
M. parviflora	*M. sieboldii*
M. parviflora **var.** *wilsonii*	*M. wilsonii*
M. praecocissima	*M. kobus*
M. praecox	*M. grandiflora*
M. pravertiana	*M. grandiflora*
M. precia	*M. denudata*
M. pumila	*M. coco*
M. pumila **var.** *championi*	*M. championii*
M. purpurea	*M. liliiflora*
M. purpurea **var.** *denudata*	*M. denudata*
M. pyramidata	*M. fraseri* **var.** *pyramidata*
M. quinquepeta	*M. liliiflora*
M. roraimae	*M. ptaritepuiana*
M. sieboldii **subsp.** *sinensis*	*M. sinensis*
M. soulangiana **var.** *nigra*	*M. liliiflora*
M. sprengeri **subsp.** *diva*	*M. sprengeri* 'Diva'
M. sprengeri **var.** diva	*M. sprengeri* 'Diva'
M. stellata	*M. kobus* **var.** *stellata*
M. stricta	*M. grandiflora*
M. taliensis	*M. wilsonii*
M. tardiflora	*M. grandiflora*
M. thurberi	*M. kobus*
M. tomentosa	*M. grandiflora*
M. tomentosa	*M. kobus*
M. tomentosa	*M. kobus* **var.** *stellata*
M. tsarongensis	*M. globosa*
M. umbellata	*M. tripetala*
M. umbrella	*M. tripetala*
M. verecunda	*M. sieboldii*
M. virginiana **var.** *acuminata*	*M. acuminata*
M. virginiana **var.** *foetida*	*M. grandiflora*
M. virginiana **var.** *tripetala*	*M. tripetala*
M. yulan	*M. denudata*
M. × *highdownensis*	*M. wilsonii*
M. × *loebneri*	*M. kobus* **var.** *loebneri*
M. × *slavinii*	*M.* × *proctoriana*

GLOSSARY OF
BOTANICAL TERMS

The following glossary includes definitions of botanical terms used in this book. Many of these terms are explained more fully in chapter 2. Following the glossary are drawings of the various leaf shapes, apices (tips), and bases which are used in species descriptions in chapter 7.

Abaxial. The lower surface (of leaves, tepals, and other plant parts)

Adaxial. The upper surface (of leaves, tepals, and other plant parts)

Adnate. United or fused with another plant part (Figure 2.6).

Alternate. The foliar arrangement when leaves occur singly on the stem at different heights.

Allotetraploid. A tetraploid plant having more than two sets of chromosomes which are genetically different; results from hybridization between species.

Androecium. The male portion of the flower; the stamens.

Annular scar. A scar in the shape of a ring circling the stem.

Anther sacs. The lobes of an anther containing pollen.

Anthesis. The time at which a flower comes into full bloom.

Anthocyanins. Natural pigments especially common in petals of flowers. Usually blue, purple, or red.

Apical. At the apex, point, or tip. Usually refers to stem tip. Leaf apices are illustrated in Figure B2. See also **Meristem**.

Apomixis. The occasional setting of seed from flowers which have not been fertilized. Apomictic seed produce exact clones of the parent plant.

Asexual propagation. See **Vegetative propagation**.

Axillary buds. Buds which occur in the axil of a leaf, that is, where the leaf attaches to the stem.

Bisexual. Having both male and female reproductive structures in a flower.

Blade. The flat, expanded portion of a leaf.

Bract. A modified, often reduced, leaf at the base of a flower. (Figure 2.3)

Callus. Undifferentiated tissue which often forms as a step in propagation by cuttings, tissue culture, or grafting.

Calyx. A collective term for all the sepals of a flower; the lowermost whorl of floral organs.

Cambium. A thin layer of expanding tissue between the bark and the wood.

Capsule. A dry, dehiscent fruit resulting from the maturation of a compound ovary.

Carpel. The structure that bears and encloses the ovule.

Character. A feature of the plant used in comparisons among taxa. For example, flower color is a character by which *Magnolia denudata* and *M. liliiflora* may be differentiated.

Chartaceous. Papery and thin in texture.

Chip-budding. A method of grafting propagation in which only a bud and a small portion of the surrounding tissue are grafted onto a stock plant. (Figure 4.1)

Chlorosis. Yellowing of plant tissue caused by loss of chlorophyll. Usually the result of drought, extreme temperatures, or nutrient deficiency.

Clone. A plant which is a genetic duplicate of its parent; a result of vegetative propagation.

Colchicine. An alkaloid extracted from *Colchicum autumnale* (the meadow saffron). When applied to a growing meristem, induces polyploid growth of new tissue in the treated plant.

Concrescent. Characterized by the growing together of parts which were originally separate.

Cone. An elongated fruiting structure bearing overlapping scales.

Connective. An extension of the filament, connecting the two cells of an anther.

Convariety. A term sometimes used in botanical nomenclature to indicate a level beneath variety.

Coriaceous. Thick, tough, and leathery.

Cotyledon. The seed leaf. The primary leaves of the embryo. These leaves serve as food storage organs for the seed and may become photosynthetic when the seed germinates.

Cultivar. A cultivated variety; a variety which has arisen in cultivation.

Cytology. The study of morphological and physiological aspects of cells.

Damping-off. A fungal disease which causes seedlings to rot and shrivel at soil level or to die before emerging.

Deciduous. Characterized by the falling of parts at the end of a growing period, as with leaves on nonevergreen plants that are shed in autumn.

Dehisce. To burst open when fully ripe to release seeds or pollen (Figure 2.4).

Diagnosis. A brief discussion of a plant and how it differs from the Latin description of a separate species previously published.

Diploid. An organism which has two sets of chromosomes.

DNA (deoxyriboneucleic acid). The carrier of genetic information in cells.

Endemic. A plant native to a particular, restricted region.

Endosperm. The nutritive tissue surrounding the embryo in a seed.

Epithet. See **Specific epithet.**

Ethnobotany. The study of people's interaction with and use of the surrounding vegetation.

Explant. A tissue or organ fragment used to initiate cell tissue culture (Figure 4.4).

Fastigiate. Erect and upright.

Filament. In a stamen, the stalk supporting the anthers.

Flavonoids. A group of plant compounds which may be pigments, waste products, or regulators. Flavonoids are often used in taxonomic studies of plants since some flavonoids are specific to certain plant groups.

Follicetum. An aggregate of follicles. In magnolias, refers to the entire fruit "cone" (Figure 2.5).

Follicle. A dry, dehiscent fruit that splits along one suture (Figure 2.5). In magnolias, splitting sometimes occurs along two sutures.

Free. Separate, distinct, not united.

Funicular thread. The threadlike stalk attaching the seed to the placenta or ovary wall (Figure 2.5); also known as the raphe.

Glabrous. Not hairy or pubescent.

Glaucous. Covered with a waxy coating which gives the surface a whitish or bluish cast.

Glaucescent. Slightly glaucous.

Grex. A group of hybrids having the same two species as parents. See chapter 5 for discussion.

Gynandrophore. In magnolias, the central axis of the flower, including the male and female floral parts.

Gynoecium. A collective term for the female parts of the flower.

Heterosis. The condition in which a hybrid exceeds the performance of its parents in one or more characteristics. Sometimes called "hybrid vigor."

Hexaploid. Having six sets of chromosomes.

Hybrid vigor. See **Heterosis.**

Indehiscent. Not dehiscing; remaining closed at maturity.

Indumentum. Hairs; pubescence.

Introgressive hybridization. Gradual genetic modification of one species by another through intermediate hybrids.

Introrse. Stamen dehiscence along two sutures on the inside of the anthers. (Figure 2.4)

Lateral. Stamen dehiscence along two sutures on the sides (edges) of the anther. (Figure 2.4)

Leader. The main growing stem of a tree or shrub.

Lectotype. A specimen selected from the original collection of plant material when the type is missing or was not designated.

Membranaceous. Thin, soft, flexible; sometimes translucent.

Meristem. Undifferentiated tissue capable of developing into various organs or tissues. The apical meristem is at the growing tip of a shoot.

Nomenclature. The naming of organisms; the correct usage of scientific names in taxonomy.

Octaploid. Having eight sets of chromosomes.

Ovule. The egg-containing unit of the ovary; after fertilization it becomes the seed.

Paratype. One of a group of specimens cited in a description of a new species.

Pedicel. The stalk of an individual flower within an inflorescence (Figure 2.3). In magnolias sometimes used to mean peduncle.

Peduncle. The stalk of an inflorescence or a solitary flower.

Pentaploid. Having five sets of chromosomes.

Perule. Modified stipules enclosing the flower bud (Figure 2.3).

Pericarp. The wall of a ripened ovary; the wall of a fruit.

Petaloid. Resembling a petal.

Petiole. The stem of a leaf; the stalk attaching the leaf blade to the stem of the plant.

pH. A measure of acidity or alkalinity of a substance based on the relative concentration of hydrogen ions. pH values range from 0 to 14; ph 7 is neutral, below 7 is acid, above 7 is alkaline.

Phenols. A group of chemical compounds found in plants. Some phenols have no known function, but many are pigments, flavonoids, tannins, and so on.

Pistil. The female reproductive organ of a flower, composed of an ovary, style, and stigma.

Ploidy. Referring to the number of chromosomes in a cell or organism. See also **Diploid, Hexaploid, Pentaploid, Tetraploid, Triploid.**

Precocious. Flowering early in the spring before the appearance of leaves on the plant.

Pubescence. Short, soft hairs.

Pubescent. Having short, soft hairs on plant parts.

Raphe. See **Funicular thread.**

Reciprocal cross. In plant breeding, a cross in which the pollen and seed parents are reversed. For example, *Magnolia denudata* × *M. liliiflora* is the reciprocal cross of *M. liliiflora* × *M. denudata.*

Recurved. Bent or curved backward.

Reflexed. Curved backward.

Rootstock. In grafting, the rootstock serves as the roots onto which a desirable shoot is grafted (Figure 4.1).

Rufous. Reddish brown in color.

Samara. A one- or two-seeded, simple, indehiscent fruit. The pericarp bears winglike outgrowths that aid in wind dissemination. Maple fruits are an example.

Scion. A portion of a stem that is transferred to a new root stock in grafting (Figure 4.1). The scion becomes the new plant, growing on the roots of another.

Sepal. One of the separate parts of the calyx. Usually green and leaflike.

Sepaloid. Resembling sepals.

Sessile. Without a stalk; sitting directly on its base.

Side-grafting. A method of propagation by grafting in which the scion is grafted onto the side of the stem of the stock plant (Figure 4.2).

Spathaceous. Resembling a spathe, or sheathing bract.

Spathe. A bract or leaf surrounding a flower cluster.

Specific epithet. The Latin name given to a species; the second part of a scientific name, sometimes referred to simply as the epithet.

Stamen. The pollen-bearing (male) organ in the flower; includes the filament and anther (Figure 2.3).

Stigma. The part of the pistil that receives pollen and on which the pollen germinates (Figure 2.3).

Stipitate. Borne on a stalk.

Stipule. A small appendage found at the base of leaf petioles; usually leaflike and present in pairs (Figure 2.6).

Stock. See **Rootstock.**

Sub-. A prefix meaning "almost."

Sucker. A shoot which develops from underground stems.

Suture. The seam at which dehiscence or splitting occurs in a fruit; the seam resulting from fusion of parts.

Taxa. Plural of *taxon.*

Taxon. Any taxonomic unit into which living organisms are classified, such as variety, species, or genus.

Taxonomy. The science of classifying organisms.

Tepal. A term used for floral parts which are not differentiated into sepals and petals (Figure 2.3).

Tetraploid. Having four sets of chromosomes.

Tomentose. Very hairy.

Triploid. Having three sets of chromosomes.

True leaves. The leaves of a newly germinated seedling, as differentiated from the cotyledons, which resemble leaves.

Type specimen. The specimen or entity to which the name of a taxon corresponds. A permanent record of the specific plant the author intended to name.

Unisexual. Applied to flowers that have either stamens (male) or pistils (female) but not both in the same flower.

Understock. See **Rootstock.**

Variety. A subdivision of a species; plants differ as a group from the rest of the species in some minor definable character. Usually a naturally occurring variation.

Vegetative Propagation. Propagation by cuttings, grafting, layering, tissue culture, or some other method not involving propagation by seed. Vegetative propagation produces a plant which is a clone of the parent plant.

Vitreous. Growth appearing waterlogged and·crisp.

Whorl. A group of three or more parts (leaves, branches, tepals) arising from a single node.

Witches'-broom. An abnormality manifested by excessive proliferation of shoots within a small area. Usually occurs on branches and may be genetic or caused by fungi.

Fig. B1. Leaf Shapes

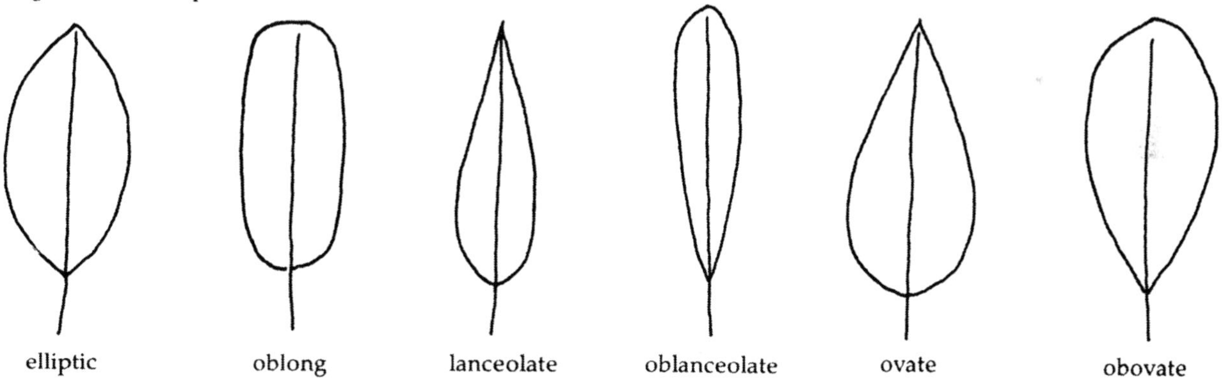

elliptic   oblong   lanceolate   oblanceolate   ovate   obovate

Fig. B2. Leaf Apices

acuminate   acute   cuspidate   emarginate   rounded

Fig. B3. Leaf Bases

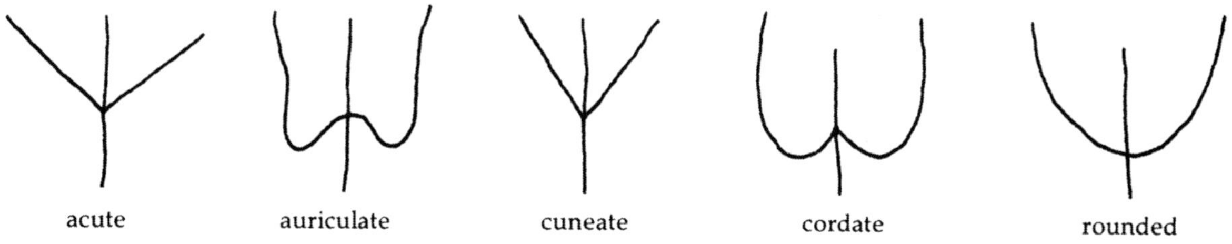

acute   auriculate   cuneate   cordate   rounded

RELATIVE BLOOM PERIODS

The following tables present relative bloom periods for some commonly cultivated magnolias. Since bloom dates vary greatly depending on location, both the relative dates and the length of bloom are approximate and variable. The tables give sequence and duration of bloom and illustrate that with careful selection and hospitable climate, one can have magnolias in bloom for as long as thirteen weeks in the spring and summer.

In the tables, week 1, the first week of bloom, corresponds to late January to early February in USDA Zones 7–8, late March to early April in Zones 5–6, and mid April to early May in Zone 4.

Relative Bloom Periods Listed Earliest to Latest

	WEEK												
	1	2	3	4	5	6	7	8	9	10	11	12	13
M. kobus var. stellata	▬												
M. kobus var. loebneri	▬												
M. denudata		▬											
Gresham Hybrids		▬▬▬▬											
M. × veitchii		▬											
M. × soulangiana		▬▬▬▬											
M. sargentiana		▬											
M. kobus		▬▬											
M. × proctoriana		▬▬											
M. campbellii		▬▬											
M. sprengeri		▬▬											
M. salicifolia		▬▬											
"Little Girls"		▬▬▬											
M. × kewensis			▬										
M. liliiflora			▬▬▬▬▬										
M. × brooklynensis					▬▬▬▬								
M. acuminata							▬▬						
M. virginiana									▬▬▬▬				
M. grandiflora										▬▬▬▬			
Freeman Hybrids										▬▬▬			
M. macrophylla										▬▬			
M. macrophylla var. ashei										▬▬			
M. fraseri										▬▬			
M. sieboldii										▬▬▬▬			
M. hypoleuca										▬▬			
M. tripetala										▬▬			

Relative Bloom Periods Listed Alphabetically by Plant Name

	WEEK												
	1	2	3	4	5	6	7	8	9	10	11	12	13
M. acuminata								▬					
M. × brooklynensis					▬	▬	▬						
M. campbellii		▬	▬										
M. denudata		▬											
M. fraseri											▬	▬	
Freeman Hybrids											▬	▬	
M. grandiflora										▬	▬	▬	
Gresham Hybrids		▬	▬	▬									
M. hypoleuca											▬		
M. × kewensis			▬										
M. kobus		▬	▬										
M. kobus var. loebneri	▬												
M. kobus var. stellata	▬												
M. liliiflora			▬	▬	▬	▬	▬						
M. macrophylla											▬	▬	
M. macrophylla var. ashei											▬	▬	
M. × proctoriana		▬											
M. salicifolia		▬	▬										
M. sargentiana		▬											
M. sieboldii											▬	▬	▬
M. × soulangiana		▬	▬	▬									
M. tripetala											▬	▬	
M. × veitchii		▬											
M. virginiana									▬	▬	▬		

APPENDIX D

NURSERY LIST

This is a list of some nurseries worldwide which carry a large selection of magnolias. It is not meant to be a comprehensive list of all magnolia sources. Appearance on this list does not imply a recommendation of the nursery by the author, and any omission of nurseries specializing in magnolias is unintentional.

Blumhardt, Oswald
No. 9 Road
Whangarei
New Zealand

Bowood Garden Centre
Bowood Estate
Calne
Wiltshire
United Kingdom

Burncoose and Southdown Nursery
Gennap Redruth
Cornwall RE 166BJ
United Kingdom

Peter G. G. Chappell
Spinners
School Lane
Boldge, Lymington
Hampshire SO41 5QE
United Kingdom

Duncan and Davies Nursery, Ltd.
P.O. Box 340
New Plymouth
New Zealand
North American office:
P.O. Box 648
Lynden, WA 98264-0648
U.S.A.

Eisenhut, Otto
6575 San Nazzaro/TI
Switzerland

Esveld, Firma C.
Rijneveld 72
2771 XS Boskoop
The Netherlands

Fairweather Gardens
P.O. Box 330
Greenwich, NJ 08323
U.S.A.

Gossler Farms Nursery
1200 Weaver Road
Springfield, OR 97478-9663
U.S.A.

Greer Gardens Nursery
1280 Goodpasture Island Road
Eugene, OR 97401-1794
U.S.A.

Grootendorst, F. J. and Sons
Laag Boskoop 16
P.O. Box 130
2770 AC Boskoop
The Netherlands

Hillier Nurseries (Winchester) Ltd.
Ampfield House
Ampfield, Romsey
Hants SO51 9PA
United Kingdom

Klehm and Son Nursery
Rt. 5, Box 197
Penny Road
South Barrington, IL 60010
U.S.A.

Louisiana Nursery
Rt. 2, Box 43
Highway 182 South
Opelousas, LA 70570
U.S.A.

Magnolia Nursery
Rt. 1, Box 87
12615 Roberts Road
Chunchula, AL 36521
U.S.A.

Mallet Court Nursery
Curry Mallet
Taunton TA3 6SY
United Kingdom

Mark Jury Nursery
P.O. Box 65
North Taranaki
New Zealand

Pickards Magnolia Gardens
Stodmarsh Road
Canterbury CT3 4AG
United Kingdom

Rosemoor Gardens
Torrington
Devon EX2 5QE
United Kingdom

Savill Garden Plant Centre
Englefield Green
Egham
Surrey TW20 0UU
United Kingdom

Simpson, W. J.
Wayside
602 Nepean Highway
Frankston, Victoria 3199
Australia

Tom Dodd Nurseries
P.O. Drawer 45
U.S. Highway 98
Semmes, AL 36575
U.S.A.

Trewithen Nurseries
Trewithen Grampound Road
Nr. Truro
Cornwall TR2 4DD
United Kingdom

ZETAS Garden Center
Huddinge
Stockholm
Sweden

INDEX OF PLANT NAMES

BOTANIC NAMES

251

COMMON NAMES

CULTIVAR NAMES

This index contains cultivar references. Cultivars selected from a species or hybrid with a botanic name are followed by that botanic name. Cultivars listed without a corresponding botanic name are hybrids whose parentage can be found in the text.

GENERAL INDEX

CPSIA information can be obtained at www.ICGtesting.com

264925BV00002B/6/P